CZECHOSLOVAK ACADEMY OF SCIENCES

CZECHOSLOVAK ACADEMY OF SCIENCES

TRANSACTIONS

of the

TENTH PRAGUE CONFERENCE

on

INFORMATION THEORY, STATISTICAL DECISION FUNCTIONS, RANDOM PROCESSES

held at

Prague, from July 7 to 11, 1986

VOLUME B

SPRINGER-SCIENCE+BUSINESS MEDIA, B.V.

Library of Congress Cataloging-in-Publication Data CIP

Prague Conference on Information Theory, Statistical
Decision Functions, Random Processes (10th :
1986 : Prague, Czechoslovakia)
Transactions of the Tenth Prague Conference on
Information Theory, Statistical Decision Functions,
Random Processes, held at Prague, from July 7 to 11,
1986.

Sponsored by Czechoslovak Academy of Sciences.
1. Probabilities--Congresses. 2. Statistical
decision--Congresses. 3. Information theory--
Congresses. I. Československá akademie věd.
II. Title.
QA273.A1P73 1986 519.2 87-16658

TRANSACTIONS include contributions of authors reprinted directly in a photographic way.
For this reason the authors are fully responsible for the correctness of their text.

Originally published by Academia, Publishing House of the Czechoslovak Academy
of Sciences, Prague, 1988

ISBN 978-94-010-9915-8 ISBN 978-94-010-9913-4 (eBook)
DOI 10.1007/978-94-010-9913-4

CONTENTS

COMMUNICATION (continued)

COMMUNICATION (continued)

A PARTIAL SOLUTION OF THE MARGINAL PROBLEM

R. Jiroušek and A. Perez

Prague

Key words: marginal problem, multidimensional distribution, uncertainty management, dependence structure simplification

ABSTRACT

The marginal problem considered here arised from an intensional approach to knowledge integration and missing inference weight approximation in expert systems operating under uncertainty. This task is reduced to the approximation of the N-dimensional weight distribution of the N state variables assigned to the N nodes of the system, assumed to be of probabilistic type, given a set M of its marginal distributions (of considerably lower dimensionality than N as a rule; N of the order of 100), representing the pieces of knowledge accumulated in the Knowledge Base. The approximated missing inference weights will be then derived from it as one derives conditional probabilities from a joint probability distribution. Using approximations of the d.s.s. (dependence structure simplification) type, it was shown that this problem is easily solvable when M contains at most two-dimensional distributions. Now, we show that the same takes place when M is representable by a triangulated graph.

1. INTRODUCTION

The marginal problem considered here arised from an intensional approach to knowledge integration and approximation of the missing inference weights (truth values, certainty factors, subjective or statistical probabilities, etc.) in expert systems operating under uncertainty (Perez (1983), Perez and Jiroušek (1985)). The formal structure of multilateral relationships or interdependencies between the N state- or stochastic variables (s.v.) assigned to the N nodes of the system may be well represented by the depencence structure of N (finite-valued) random variables given by their joint N-dimensional probability distribution (PD). As a rule, the pieces of knowledge originally obtained in the Knowledge Base (KB) will represent relationships or interdependencies of considerably lower dimensionality than N (N of the order of 100).

The task is reduced to the following *marginal problem*: To approximate as well as possible the joint N-dimensional PD given a set M of its marginal distributions representing the pieces of knowledge accumulated in the KB. Further, to derive from it in a tractable way also the needed conditional probabilities, representing the missing inference weights, reducing as far as possible the integration work connected with the construction of the new implied marginals - restrictions of the approximation above.

Using approximations of the d.s.s. (dependence structure simplification) type as introduced in Perez (1977) , it was shown in Perez (1983), Perez and Jiroušek (1985) that the marginal problem is easily solvable when M contains at most two-dimensional distributions. Now, we show that the same takes place when M is representable by a triangulated graph. In this case, however, the tractability of the second part of the problem by the algorithm we propose is not absolutely guaranteed since exceptionally for some restrictions it may lead to an integration work untractable on any computer.

There are cases where one cannot without testing be sure that the probability distributions contained in M are really marginals of some joint distribution, i. e. that they are compatible. The problem of compatibility was considered by many authors, in particular by Kellerer (1964). A necessary but not always sufficient condition is the so-called weak compatibility: the restrictions on any subset of random variables common to more distributions in M must coincide. In the case of M's representable by triangulated graphs the compatibility is ensured by the weak compatibility.

2. APPROXIMATION OF JOINT PD GIVEN A SET OF ITS MARGINALS

Using the results of Perez (1977) we proceeded in Perez (1983), Perez and Jiroušek (1985) as follows: Let $A := \{X_1, \ldots, X_N\}$ be the set of finite-valued s.v.'s assigned to the N nodes of the system, let $S_1, \ldots, S_K$ be K subsets of A and let $M := \{P_{S_1}, \ldots, P_{S_K}\}$ be the given set of marginals of an (unknown) joint distribution P_A on A.

For every permutation $i_1, \ldots, i_K$ of the indices $1, \ldots, K$ it is possible to construct a *d.s.s.* (*dependence structure simplification*) approximation $\bar{P}_A^{i_1 \ldots i_K}$ of P_A by taking

$$(1) \qquad \bar{P}_A^{i_1 \ldots i_K} := P_{S_{i_1}} \ldots P_{S_{i_K}} \,/\, P_{S_{i_2} \cap S_{i_1}} \ldots P_{S_{i_K} \cap \bigcup_{k=1}^{K-1} S_{i_k}}$$

Taking as measure of approximation the *relative Shannon's entropy* or *Kullback-Leibler divergence*,

$$H(P_A, \bar{P}_A^{i_1 \ldots i_K}) := \int \log \frac{dP_A}{d\bar{P}_A^{i_1 \ldots i_K}} \, dP_A,$$

of P_A with respect to $\bar{P}_A^{i_1 \ldots i_K}$ one easily verifies that

$$(2) \quad H(P_A, \bar{P}_A^{i_1 \ldots i_K}) = I(P_A) - I_{P_A}(\bar{P}_A^{i_1 \ldots i_K}) = I(P_A) - I_M(\bar{P}_A^{i_1 \ldots i_K})$$

where $I(P_A) := H(P_A, P_{\{X_1\}} \times \ldots \times P_{\{X_N\}})$ is what we call the *dependence tightness* (d.t.) of the (unknown) P_A and

$$I_{P_A}(\bar{P}_A^{i_1 \ldots i_K}) := \int \log \frac{d\bar{P}_A^{i_1 \ldots i_K}}{dP_{\{X_1\}} \times \ldots \times dP_{\{X_N\}}} \, dP_A =$$

$$= \sum_{\ell=1}^{K} I(P_{S_{i_\ell}}) - \sum_{k=2}^{K} I(P_{S_{i_k} \cap \bigcup_{\ell=1}^{k-1} S_{i_\ell}}) := I_M(\bar{P}_A^{i_1 \ldots i_K})$$

is the d.t. of $\bar{P}_A^{i_1 \ldots i_K}$ relative to (the unknown) P_A or to (the known) M and, thus, calculable. Note that the notation used here for the latter concept is more adequate than that used in Perez (1977), Perez (1983), Perez and Jiroušek (1985). The best approximation of P_A is thus given by that d.s.s. approximation which maximizes the d.t. $I_M(\bar{P}_A^{i_1 \ldots i_K})$ with respect to all permutations $i_1, \ldots, i_K$.

This task is relatively easily solved in the case M contains at most two-dimensional marginals by the well-known greedy algorithm which is of linear complexity. Unfortunately, no equivalence to the greedy algorithm is known for the general case. The graph-aided algorithm described in Perez and Jiroušek (1985) is sufficiently quick (for N of the order of 100, typical in expert systems) but it is not proved to yield optimal solutions.

In the next section we shall see that the above task is also easily solvable when M is representable by a triangulated graph.

3. SETS M REPRESENTABLE BY TRIANGULATED GRAPHS

For M as introduced in the preceding section let us define the graph $G_M = (V,E)$ as follows:

$$V := A = \{X_1,\dots,X_N\},$$

$$E := \{\{X_i,X_j\} : \exists P_S \in M \text{ with } \{X_i,X_j\} \subset S\}$$

A set M is said to be *representable by a graph* if for every clique $\{X_{i_1},\dots,X_{i_J}\}$ of the graph G_M (clique is a subgraph where every pair of its vertices is an edge) there exists a distribution $P_S \in M$ such that $\{X_{i_1},\dots,X_{i_J}\} \subset S$.

A set M is said to be *representable by a triangulated graph* if it is representable by a graph and the graph G_M is triangulated (i. e. each cycle of G_M greater than 3 possesses a chord).

Triangulated graphs have some advantageous characteristics. One of them is that all maximal cliques of a triangulated graph can be put into a sequence $C_1,C_2,\dots,C_L$ in such a way that for every $s = 2,\dots,L$, the set

$$C_s \cap \bigcup_{k=1}^{s-1} C_k$$

is a mini-articulation set, i. e. an articulation set obtainable as an intersection of two maximal cliques (Jiroušek (1985)).

Thus, if the set $M = \{P_{S_1},\dots,P_{S_K}\}$ is representable by a triangulated graph then

(*) $\exists$ permutation $i_1,\dots,i_K$ $\forall k = 2,\dots,K$ $\exists \ell_k (1 \le \ell_k < k)$

$$\text{such that } S_{i_k} \cap \bigcup_{m=1}^{k-1} S_{i_m} \subset S_{i_{\ell_k}}$$

It is possible to deduce from Kellerer (1964) that when a sequence of indices $i_1,\dots,i_K$ meets the condition (*) then:

(i) $\overline{P}_A^{i_1 \dots i_K}$ coincides with P_{S_i} on S_i for all $i = 1,\dots,K$;

(ii) $\overline{P}_A^{i_1 \dots i_K}$ is unique, i. e. if two different permutations of indices meet the condition (*) then both of them define the same approximation.

Moreover, due to (i) one simply proves the d.t. property:

(iii) $\overline{P}_A^{i_1 \dots i_K}$ maximizes the dependence tightness $I_M(\overline{P}_A^{i_1 \dots i_K})$ with respect to all the permutations of the indices $1,\dots,K$, being thus the best d.s.s. approximation (cf. preceding section).

The combinatoric property of the set M to be representable by a triangulated graph or, equivalently, to satisfy the condition (*) permits, thus, a complete solution of the first part of the marginal problem. The following section is concerned with the second part of the marginal problem (cf. Introduction).

4. RESTRICTIONS OF D.S.S. APPROXIMATIONS

The d.s.s. approximations of multidimensional probability distributions have several advantages. Namely:

- possibility of representation in a computer;
- possibility to characterize the best approximation (with respect to the Kullback-Leibler divergence) even for unknown approximated distribution P_A;
- rather simple algorithmical solution of the problem.

Now, we would like to show another advantage of the d.s.s. approximations connected with the solution of the second part of the marginal problem. Having, namely, a d.s.s. approximation $\overline{P}_A$ of an N-dimensional probability distribution P_A let us turn our attention to the problem how to derive from it the conditional probability distribution $\overline{P}_A(X_{j_1} | X_{j_2},\dots,X_{j_n})$ of one random variable X_{j_1} given the values of a subset of some other random variables $\{X_{j_2},\dots,X_{j_n}\}$ of A. Since this conditional probability distribution is given by the quotient of $\overline{P}_{\{X_{j_1},\dots,X_{j_n}\}}$ with respect to $P_{\{X_{j_2},\dots,X_{j_n}\}}$ (corresponding restrictions of the d.s.s. approximation $\overline{P}_A$), from the theoretical point of view the problem is very simple. It suffices to integrate $\overline{P}_A$ over all random variables from the set $A - \{X_{j_1},\dots,X_{j_n}\}$. But when $n \ll N$ it might become to be an unfeasible task. Therefore we have to look for a restriction $\overline{P}_B$ of $\overline{P}_A$ such that $B \supset \{X_{j_1},\dots,X_{j_n}\}$ and, at the same time, such that the set of undesirable random variables $B - \{X_{j_1},\dots,X_{j_n}\}$ is as small as possible.

Useful and simple restrictions of a d.s.s. approximation $\overline{P}_A^{i_1\dots i_K}$ are distributions on B's of the type

$$B(i_1,\dots,i_R) = \bigcup_{k=1}^{R} S_{i_k} \qquad (1 \le R < K)$$

defined as follows:

$$\overline{P}_{B(i_1,\dots,i_R)}^{i_1\dots i_R} := P_{S_{i_1}} \dots P_{S_{i_R}} / P_{S_{i_2}\cap S_{i_1}} \dots P_{S_{i_R}\cap \bigcup_{k=1}^{R-1} S_{i_k}}$$

In this way, for a given permutation $i_1,\dots,i_K$, K-1 restrictions are defined but we want the class of restrictions to be as abundant as possible. So we have to consider not only one permutation $i_1,\dots,i_K$ but the class of all permutations defining the same approximation, or the class of all permutations for which the condition (*) is fulfilled (cf. point (ii) of preceding section). As this class contains at least

$$K + \sum_{k=2}^{K-1} 2^{k-2}\,(K-k) = 2^{K-1}$$

permutations (this lower bound is achieved e. g. when, for all $k = 2,\dots,K$, $S_k \cap \bigcup_{m=1}^{k-1} S_m \subset S_{k-1}$) we get a broad class of restrictions of the approximation $\overline{P}_A^{i_1,\dots,i_K}$.

Formally, this class of restrictions can be expressed in the following way:

$$\{\overline{P}_B : \exists i_1,\dots,i_K \ (\forall k=2,\dots,K \ \exists \ell (1 \le \ell < k) \ S_{i_k} \cap \bigcup_{m=1}^{k-1} S_{i_m} \subset S_{i_\ell})$$
$$\& \ \exists R \ (1 \le R < K) \quad \overline{P}_B = \overline{P}_{B(i_1,\dots,i_R)}^{i_1\dots i_R}\}.$$

We would like to stress that a sequence $i_1,\dots,i_R$ defining a restriction necessarily must be a prefix of some sequence $i_1,\dots,i_K$ meeting the condition (*). It does not suffice to meet the condition:

$$\forall\ k = 2,\dots,R \ \ \exists \ell (1 \le \ell < k) \quad S_{i_k} \cap \bigcup_{m=1}^{k-1} S_{i_m} \subset S_{i_\ell}.$$

This can be seen from the fact that for R = 2 any couple of indices i_1, i_2 meets the above condition but not all of them define a restriction of $\bar{P}_A^{i_1 \ldots i_K}$. On the other hand, for any $i \in \{i, \ldots, K\}$ there is a sequence $i = i_1', \ldots, i_K$ meeting the condition (*) and therefore the considered class of restrictions includes all $P_{S_i} = \bar{P}^i_{B(i)}$.

The more extensive is the class of restrictions considered the smaller is $B \supset \{X_{j_1}, \ldots, X_{j_n}\}$ we can find, but, at the same time, the greater difficulties we can meet when looking for the optimal solution. A satisfactory solution of the problem when the class of restrictions defined above is taken into account is given by the following simple algorithm. We would like to stress its rapidity enabling its use for computations in real time.

5. ALGORITHM

For a convenient description of the algorithm we introduce an auxiliary concept. We call two marginal distributions $P_{S_{k_1}}$, $P_{S_{k_2}}$ from M to be adjacent if there exists a sequence of indices $i_1, \ldots, i_K$ meeting the condition (*) such that $k_1 = i_\ell$, $k_2 = i_k$ and

$$S_{i_k} \cap \bigcup_{m=1}^{k-1} S_{i_m} \subset S_{i_\ell} \quad \text{with } \ell < k.$$

Though it cannot be seen at the first sight, it is very simple to decide whether two marginal distributions are adjacent or not when graph representation is considered.

The algorithm.

Input data: list of all random variables $A = \{X_1, \ldots, X_N\}$; system of marginal distributions (representable by a triangulated graph $M = \{P_{S_1}, \ldots, P_{S_K}\}$; list of random variables $\{X_{j_1}, \ldots, X_{j_n}\}$.

Output: sequence of indices $i_1, \ldots, i_R$ defining the restriction.

Used structures: SI stack containing sequence of indices; SV stack containing a set of variables.

Subroutine: Find the shortest sequence of distributions $P_{S_{k_1}},\dots,P_{S_{k_s}}$ such that all consecutive distributions are adjacent, $k_1 \in SI$ and $S_{k_s} \cap SV \neq \emptyset$.

Step I (test for non-triviality):

If there is S_k such that $S_k \supset \{X_{j_1},\dots,X_{j_n}\}$ choose the smallest S_k among them, set $k \to SI$ and finish.

Else go to Step II.

Step II (initialization):

Take any S_k such that $S_k \cap \{X_{j_1},\dots,X_{j_n}\} \neq \emptyset$ and set $k \to SI$, $(\{X_{j_1},\dots,X_{j_n}\} - S_k) \to SV$.

Call Subroutine.

Set $k_s \to SI$ (deleting the previous content).

Step III (creation):

Set $(\{X_{j_1},\dots,X_{j_n}\} - \bigcup_{i \in SI} S_i) \to SV$.

If $SV = \emptyset$ then finish.

Else call Subroutine.

Add $(k_2,\dots,k_s)$ to SI.

Repeat Step III.

The previous algorithm is correct in the sense that if $i_1,\dots,i_n$ is the content of SI in Step III then $i_1,\dots,i_n$ is a prefix of a permutation $i_1,\dots,i_K$ meeting the condition (*).

Indeed, since we not pretend here in the minimality of the constructed sequence of indices (this is rather a complicated problem) we can replace, at any step, the set $\{X_{j_1},\dots,X_{j_m}\}$ by A and then the algorithm leads to a permutation of all indices $(1,\dots,K)$. Thus, any content of SI can be prolonged to a permutation of all indices and therefore it suffices to show that

$$\{i_1,\dots,i_n\} = SI \Rightarrow \forall\, k=2,\dots,n \;\; \exists\, \ell_k \;(1 \le \ell_k < k)\;\; S_{i_k} \cap \bigcup_{m=1}^{k-1} S_{i_m} \subset S_{i_{\ell_k}}$$

At step III of the algorithm, indices are added to SI in such a way that if $\{i_1,\dots,i_n\} = SI$ then $\forall\, k = 2,\dots,n \;\exists\, \ell_k$ $(1 \le \ell_k < k)$ such that S_k, S_{ℓ_k} are adjacent. Moreover, the shortest possible sequences (constructed by Subroutine) are added at each repetition of the step. Therefore, if $r, k \in \{i_1,\dots,i_n\} = SI$ (r precedes k) then there exists a sequence $\{j_1,\dots,j_m\} \subset \{i_1,\dots,i_n\} \subset SI$ such that $r = j_1$, $k = j_m$ and $S_{j_1},\dots,S_{j_m}$ is the shortest sequence of consecutively adjacent maximal cliques in G_M. Denote $\ell_k = j_{m-1}$ the index for which S_{ℓ_k} and S_k are adjacent. If $S_r \cap S_k = \emptyset$ then $S_r \cap S_k \subset S_{\ell_k}$ is self-evident, otherwise it can be shown from the definition of adjacency. As r was arbitrary

$$S_k \cap \bigcup_{r=1}^{m-1} S_r \subset S_{\ell_k},$$

which proves the correctness of the algorithm.

REFERENCES

Perez A. (1977): ε-admissible Simplifications of the Dependence Structure of a Set of Random Variables. Kybernetika 13, 439-449.

Perez A. (1983) Probability approach in integrating partial knowledges for medical decision-making (in Czech). Trans. BMI 83 Conf. on biomedical engineering, Mariánské Lázně, 221-226.

Perez A. and Jiroušek R. (1985): Constructing an intensional expert system (INES). In: Medical Decision Making: diagnostic strategies and expert systems (Eds. J. H. van Bemmel, F. Grémy, J. Zvárová). North Holland, 1985, 307-315.

Kellerer H. G. (1964): Verteilungsfunktionen mit gegeben Marginalvarteilungen. Z. Wahrscheinlichkeitstheorie 3, 247-270.

Jiroušek R. (1985): Triangulated Graphs. Res. rep. No 1363, ÚTIA ČSAV Prague.

Czechoslovak Academy of Sciences
Institute of Information Theory and Automation
182 08 Prague 8
Pod vodárenskou věží 4
Czechoslovakia

there exists a sequence $\{j_1,\dots,j_m\} \subset \{1,\dots,n\}$ such that $r = j_1$, $k = j_m$ and $S_{j_1},\dots,S_{j_m}$ is the shortest sequence of consecutively adjacent maximal cliques in G_M. Denote $t_k = j_{m-1}$ the index for which S_{t_k} and S_k are adjacent. If $S_r \cap S_k = \emptyset$ then $S_r \cap S_k \subset S_{t_k}$ is self-evident, otherwise it can be shown from the definition of adjacency. As r was arbitrary

$$S_k \cap \bigcup_{r=1}^{m-1} S_r \subset S_{t_k}$$

which proves the correctness of the algorithm.

REFERENCES

Perez A. (1977): ε-admissible Simplifications of the Dependence Structure of a Set of Random Variables. Kybernetika 13, [illegible]

Perez A. (1983): [illegible] knowledge [illegible] for medical decision-making [illegible]. Proc. [illegible] medical engineering, Mariánské [illegible], 221-[illegible].

Perez A. and Jiroušek R. (1985): [illegible] expert system [illegible] In: Medical [illegible] Systems. North Holland, [illegible]

Kellerer H. G. (1964): Verteilungsfunktionen mit gegeben Marginalverteilungen. Z. Wahrscheinlichkeitstheorie 3, 247-270.

Jiroušek R. (1985): Triangulated Graphs. Res. Rep. No 1363, ÚTIA ČSAV Prague.

Czechoslovak Academy of Sciences
Institute of Information Theory and Automation
182 08 Prague 8
Pod vodárenskou věží 4
Czechoslovakia

EMPIRICAL ESTIMATES IN STOCHASTIC PROGRAMMING

Vlasta Kaňková

Prague

Key words: Stochastic optimization problem, stochastic programming problem, optimal value, optimal solution, estimate, stochastic decision problem, random sample, sample distribution function, convergence rate, stationary stochastic sequence, ergodic stochastic sequence, ϕ-mixing stochastic sequence

ABSTRACT

One of the main problems in stochastic programming is to find estimates of the optimal value and the optimal solution. Solving practical problems we often need to solve some stochastic optimization problem without knowledge of the probability laws. From the mathematical point of view we have to associate with the original optimization problem an additional stochastic decision problem.

If the mathematical solution is sought with respect to the mathematical expectation then some approximate problems can be get setting an empirical distribution function instead of the theoretical one. The mentioned access was considered in the literature many times, Dupačová (1976), Kaňková (1974, 1978), and Цыбаков (1981) for example. There have been given rather general conditions under which the corresponding estimates converge to the theoretical values with probability one. Further, they studied the convergence rate. However, these results were based on the random samples with stochastic independent members only. The only exception was in Kaňková (1974).

In this paper we shall consider these approximate methods too. Especially, it will be the aim to study the estimates behaviour based on the random samples with stochastic dependent members.

INTRODUCTION

In this paper we shall deal with stochastic optimization problems in which the optimum is sought with respect to the mathematical expectation. We shall assume that the probability laws of the random elements are unknown. First, we shall summarize till now achieved results for independent random samples. Further, we shall try to get the similar results for the case, when the members of random samples are dependent. More precisely, we shall admit some type of stochastic dependence.

Let (Ω, S, P) be a probability space, where P is a complete probability measure,

$T=T(\omega) = [T_1(\omega),\dots,T_s(\omega)]$ be an s-dimensional random vector defined on (Ω, S, P). We denote by $F(t)$ the distribution function of $T(\omega)$.

Let, further, $g(t,x)$ be a real-valued continuous function defined on $\tau \times K$ where $K \subset E_r$ is a non-empty, compact convex set and $\tau \subset E_s$ satisfies the condition $P\{\omega: T(\omega) \in \tau\} = 1$.

(E_r denotes r-dimensional Euclidean space).

It is easy to see that $g(T,x)$ is for every $x \in K$ a random variable defined on (Ω, S, P). We shall introduce the stochastic optimization problem as the problem to find

$$\max_{x \in K} Eg(T,x). \tag{1}$$

(E denotes the operator of the mathematical expectation.)

If $T_i(\omega) = [T_{i1}(\omega),\dots,T_{is}(\omega)] = T_i$, $i=\dots-1,0,1,2,\dots$ are random vectors we shall define the functions $U_i(t,\omega)$, $F_n(t,\omega)$, $t = [t_1,\dots,t_s] \in E_s$, $\omega \in \Omega$, $i = 1,2,\dots$, $n = 1,2,\dots$, by

$$U_i(t,\omega) = 1 \iff T_{ij}(\omega) < t_j \text{ for all } j = 1,2,\dots,s,$$

$$U_i(t,\omega) = 0 \iff T_{ij}(\omega) \ge t_j \text{ for at least one } j \in \{1,2,\dots,s\}$$

$$F_n(t,\omega) = \frac{1}{n} \sum_{i=1}^{n} U_i(t,\omega) \tag{2}$$

Further we shall define $I(x)$, $I_n(x,\omega)$, $n = 1,2,\dots$, $x \in K$, $\omega \in \tau$ by

$$(3)\qquad I_n(x,\omega) = \int g(t,x)\,dF_n(t,\omega)$$
$$I(x) = \int g(t,x)\,dF(t)$$

(The integrals are taken in the sense of Rieman-Stieltjes.)

It is easy to see that if

(i) $\{T_i(\omega)\}_{i=-\infty}^{\infty}$ is a random sequence of s-dimensional random vectors having the same distribution function $F(t)$ as the random vector $T(\omega)$,

then $\max_{x\in K} I_n(x,\omega)$ estimates the theoretical value $\max_{x\in K} I(x)$. It holds, if $g(t,x)$ is a continuous, bounded function and $\{T_i\}_{i=1}^{\infty}$ is an ergodic sequence then

$$(4)\qquad P\{\omega:\ \max_{x\in K} I_n(x,\omega) \underset{(n\to\infty)}{\to} \max_{x\in K} I(x)\} = 1.$$

However the convergence rate was investigated only in the independent random sample case.

In this paper we shall try to discuss these empirical estimates for dependent random sample.

STATIONARY RANDOM SAMPLE CASE

First, we shall present an elementary assertion.

Theorem 1. Let condition (i) be fulfilled. If the random sequence $\{T_i(\omega)\}_{i=0}^{\infty}$ is ergodic and if $g(t,x)$ is a continuous and bounded function then

$$(5)\qquad E\max_{x\in K} I_n(x,\omega) \underset{(n\to\infty)}{\to} \max_{x\in K} I(x).$$

Proof: Under the assumptions of Theorem 1 it was proved in Kaňková (1974) that

$$(6)\qquad P\{\omega:\ \max_{x\in K} I_n(x,\omega) \underset{(n\to\infty)}{\to} \max_{x\in K} I(x)\} = 1.$$

Let $\varepsilon > 0$ be arbitrary. According to the assumptions there exists a constant $M\in E$, such that

$$|g(t,x)| < M \quad \text{for all } t\in\tau,\ x\in K.$$

But from (3) it is easy to see that also

$$|I_n(x,\omega)| \le M,$$

$$|I(x)| \le M \text{ for all } x \in K,\ \omega \in \Omega,\ n = 1,2,\ldots$$

Further, using (6) we get successively

$$|E \max_{x\in K} I_n(x,\omega) - \max_{x\in K} I(x)| \le \int_{\Omega_n(x)} |\max_{x\in K} I_n(x,\omega) - \max_{x\in K} I(x)|dP +$$

(7) $$+ \int_{\Omega-\Omega_n(x)} |\max_{x\in K} I_n(x,\omega) - \max_{x\in K} I(x)|dP \le$$

$$\le \varepsilon/2 + 2M \,.\, P\{\omega: |\max_{x\in K} I_n(x,\omega) - \max_{x\in K} I(x)| > \varepsilon/2\}$$

where $\Omega_n(x) = \{\omega \in \Omega: |\max_{x\in K} I_n(x,\omega) - \max_{x\in K} I(x)| \le \varepsilon/2\}$.

As $\varepsilon > 0$ was arbitrary, it is easy to see that from (7) and (6) follows the assertion of the Theorem.

Now we can easily see that, under general conditions, $\max_{x\in K} I_n(x,\omega)$ estimates the theoretical value $\max_{x\in K} I(x)$ even in the case, when the random sample fulfils the ergodic condition only. More precisely, it means without the assumptions of independence. Moreover,this estimate is consistent and asymptotically unbiased.

Now, we shall try to investigate the rate convergence of these empirical estimates in the dependence case. However, we shall restrict our consideration to random sequences $\{T_i(\omega)\}_{i=-\infty}^{+\infty}$ which fulfil the ϕ-mixing condition.

Let $\{\xi_i\}_{i=-\infty}^{+\infty}$ be a s-dimensional strong stationary random sequence defined on (Ω,S,P).

Let, further, $B(-\infty,a)$ be the σ-algebra given by $\ldots,\xi_{a-1},\xi_a$
$B(b,+\infty)$ be the σ-algebra given by $\xi_b,\xi_{b+1},\ldots$
(a, b are an integrel).

If N denotes the set of the natural numbers, $\phi(.)$ a non-negative real-valued function defined on N then we can define the ϕ-mixing random sequence by the following definition.

Definition: We say that a strong stationary random sequence $\{\xi_i\}_{i=-\infty}^{\infty}$ fulfils the condition of ϕ-mixing if

$$|P(A_1 \cap A_2) - P(A_1)P(A_2)| \le \phi(n).P(A_1)$$

for $A_1 \in B(-\infty,k)$, $A_2 \in B(k+n,+\infty)$, $-\infty < k < +\infty$, $n \ge 1$.

Remark: Of course, it is assumed that

$$\lim_{n\to\infty} \phi(n) = 0.$$

Let the function $g(t,x)$ satisfy the conditions

(ii) $g(t,x)$ is a continuous, bounded function,

(8) $$|g(t,x)| \le M \text{ for all } t \in \tau,\ x \in K,$$

(iii) $g(t,x)$ is for every $t \in \tau$ a Lipschitz function of x with the Lipschitz constant L not depending on t.

Let, further the constant R satisfies the condition

$$\sup_{x,x' \in K} \|x-x'\| \le R$$

with $\|.\|$ denoting the Euclidean norm in E_r.

If we define a number $N(z)$ by

(9) $$N(z) = \min\{n \text{ integral: } n \ge \frac{3LR}{z}\} \quad \text{for } z \in E_1,\ z > 0$$

then the following statement is valid.

Theorem 2: Let $\{T_i\}_{i=-\infty}^{\infty}$ be a ϕ-mixing random sequence. If the conditions (i), (ii), (iii) are satisfied and if $I(x)$, $I_n(x,\omega)$, $n = 1,2,\ldots$ are given by (3), $N(z)$ by (9) then

$$P\{\omega: |\max_{x \in K} I_n(x,\omega) - \max_{x \in K} I(x)| \ge z\} \le [N(z)]^r \frac{36M^2}{z^2n^2}(1+\sum_{k=1}^{n}(n-k)\phi(k))$$

for $z \in E_1$, $z \ge 0$, $n = 1,2$.

Proof: Let $z_0 > 0$, $z_0 \in E_1$ be arbitrary. From (9) follows the existence of points $x_1, x_2, \ldots, x_{N_0}$, $N_0 = [N(z_0)]^r$ such that

$$\sup_{x \in K} \inf_{x_i \in [x_1,\ldots,x_{N_0}]} |g(t,x)-g(t,x_i)| \le \frac{z_0}{3} \text{ for all } t \in \tau$$

so that

$$\sup_{x \in K} \inf_{x_i \in [x_1,\ldots,x_{N_0}]} |I(x) - I(x_i)| \le \frac{z_0}{3}$$

and

$$\sup_{x \in K} \inf_{x_i \in [x_1,\ldots,x_{N_0}]} |I_n(x,\omega)-I_n(x_i,\omega)| \le \frac{z_0}{3},\ n=1,2,\ldots,\ \omega \in \Omega$$

Since it, further, follows from the triangular inequality

$$|I_n(x,\omega)-I(x)| \le |I_n(x,\omega)-I_n(x_i,\omega)|+|I_n(x_i,\omega)-I(x_i)|+|I(x_i)-I(x)|$$

for $x \in K$, $x_i \in [x_1,\ldots,x_{N_0}]$, $n = 1,2,\ldots$

we get

(10) $$P\{\omega: |\max_{x\in K} I_n(x,\omega) - \max_{x\in K} I(x)| \geq z_o\} \leq$$

$$\leq P\{\omega: |I_n(x,\omega) - I(x)| \geq z_o \text{ for at least one } x \in K\} \leq$$

$$\leq P\{\omega: |I_n(x_i,\omega) - I(x_i)| \geq \frac{z}{3} \text{ for at least one } x_i \in [x\ ,\dots,x_{N_o}]\} \leq$$

$$\leq \sum_{i=1}^{N_o} P\{\omega: |I_n(x_i,\omega) - I(x_i)| \geq \frac{z_o}{3}\}$$

As $E[I_n(x_i,\omega) - I(x_i)] = 0$ for $i = 1,2,\dots,N_o$, $x_i \in K$, $n = 1,2,\dots$ we get further using Chebyshev's inequality

(11) $$P\{\omega: |I_n(x_i,\omega) - I(x_i)| \geq \frac{z_o}{3}\} \leq \frac{9}{z^2} E[I_n(x_i,\omega) - I(x_i)]^2, \quad i=1,2,\dots,N_o$$

As in Billingsley (1977) is proved

$$|E[g(T_i,x) - I(x))(g(T_j,x) - I(x))] \leq 4M^2 \phi(|j-i|)$$

for $j, i = 1,2,\dots,n$, $i \neq j$ and every $x \in K$ we can obtain

$$P\{\omega: |I_n(x_i,\omega) - I(x_i)| > \frac{z_o}{3}\} \leq \frac{4M^2}{n^2}[1 + \sum_{k=1}^{n} (n-k)\phi(k)]$$

From this inequality and from (10), (11) follows immediately the assertion of the Theorem.

NONSTATIONARY RANDOM SAMPLE CASE

In the former part we treated only the stationary an ergodic random sequence $\{T_i\}_{i=-\infty}^{\infty}$. Now we shall try to drop these assumptions. Of course, we shall restrict our consideration to the special case. However, we shall find assumptions under which $\max_{x\in K} I_n(x,\omega)$ converges in probability to the theoretical value $\max_{x\in K} I(x)$. More precisely we shall present the following statement.

Theorem 3. Let $z > 0$, $z \in E_1$ be arbitrary. If

1. $g(t,x) = \sum_{i=1}^{s} t_i h_i(x)$, where $h_i(x)$ is a continuous, real-valued function defined on E_r such that $|h_i(x)| \leq M_1$ for all $x \in K$, $i=1,2,\dots,s$ and a constant $M_1 \in E_1$,

2. $E\, T_{nj}(\omega) = \mu_j = E\, T_j(\omega)$ for all $j = 1,2,\ldots,s$, $n = 1,2,\ldots$

3. $E|(T_{nj}-\mu_j)(T_{n+m,j}-\mu_j)| \le M(m)$ for all $n = 1,2,\ldots$, $m = 1,2,\ldots$, $j = 1,2,\ldots,s$,

where $M(m)$ fulfil the condition $M(m) \to 0$ $(m\to\infty)$ then

$$P\{\omega: |\max_{x\in K} I_n(x,\omega) - \max_{x\in K} I(x)| > z\} \to 0, \quad (n\to\infty)$$

Proof: Let $\varepsilon > 0$, $z > 0$, $z \in E_1$ be arbitrary. It is easy to see that $g(T,x)$ for every $x \in K$ is a random variable defined on (Ω, S, P). So we can consider $Eg(T,x)$ for every $x \in K$. Further, it follows from the assumptions that $Eg(T,x)$ is a continuous function on K. So we have

$$P\{\omega: |\max_{x\in K} I_n(x,\omega) - \max_{x\in K} I(x)| > z\} =$$

$$(12) \quad = P\{\omega: |\max_{x\in K} \frac{1}{n}\sum_{i=1}^{n} g(T_i,x) - \max_{x\in K} Eg(T,x)| > z\} \le$$

$$\le P\{\omega: |\frac{1}{n}\sum_{i=1}^{n} g(T_i,x) - Eg(T,x)| > \frac{z}{3} \text{ for at least one } x \in K\}.$$

Using the assumption 1 we get further

$$(13) \quad P\{\omega: |\frac{1}{n}\sum_{i=1}^{n} g(T_i,x) - Eg(T,x)| > \frac{z}{3} \text{ for at least one } x \in K\}$$

$$\le \sum_{j=1}^{s} D(\sum_{i=1}^{n} T_{ij}) \cdot \frac{9M^2s^2}{n^2 . z^2}$$

As

$$D(\sum_{i=1}^{n} T_{ij}) \le n.M(0) + (n-1)M(1) + \ldots + (N-m)M(m) + \ldots + M(n-1)$$

$$\le n.M(0) + (n-1)M(1) + \ldots + (n-m)M(m) + \frac{(n-m)(n-m-1)}{2} \max[M(m+1),\ldots \ldots, M(n-1)]$$

for $m \in \{1,\ldots,(n-1)\}$, $j = 1,2,\ldots,s$

we get from (13)

$$P\{\omega: |\frac{1}{n}\sum_{i=1}^{n} g(T_i,x) - Eg(T,x)| > \frac{z}{3} \text{ for at least one } x \in K\}$$

$$\le \frac{9M^2s^3}{n^2.z^2}[nM(0) + (n-1)M(1) + \ldots + (n-m)M(m)] + \frac{(n-m)(n-m-1)}{2}\max[M(m+1),\ldots \ldots, M(n-1)]$$

Now from (12) it is easy to see that there exists n_0 such that for $n > n_0$

$$P\{\omega : |\max_{x \in K} I_n(x, \ ) - \max_{x \in K} I(x)| > z\} \leq \varepsilon$$

Since $\varepsilon > 0$, $z \in E_1$, $z > 0$ were arbitrary we have proved the Theorem.

In this paper we have tried to generalize the former results obtained for independent random sample to some type of dependent samples. We believe that the reported results may enable to use the empirical estimates in more practical problems.

REFERENCES

Billingsley P. (1965): Ergodic Theory and Information.Wiley, New York.

Billingsley P. (1977): Convergence of probability Measures, Wiley & Sons, New York.

Цыбаков А. В. (1981): Оценки точности метода минимизации эмпирического риска, Проблемы передачи информации, 17, № 1, 50-61.

Dupačová J. (1976): Experience in Stochastic Programming Models. IX International symposium on Mathematical Programming, Budapest 1976.

Hoeffding W. (1963): Probability Inequalities for Sums of Bounded Random Variables. Journal of the Americ. Statist. Ass. 58, N 301, 13-30.

Kaňková V. (1974): Optimum Solution of a Stochastic Optimization. Problem with Unknown Parameters. In: Trans. of the Sevent Prague Conference 1974, Academia, Prague 1977, 239-244.

Kaňková V. (1978): An Approximative Solution of Stochastic Optimization Problem. In: Trans. of the Eighth Prague Conference, Academia, Prague, 327-332.

Loév M. (1959): Probability Theory. Second edition D. van Nostrand company, New York.

Тарасенко Р. С. (1980): Об оценке скорости сходимости адаптивного метода случайного поиска, Проблемы случайного поиска,8,182-185.

Czechoslovak Academy of Sciences
Institute of Information Theory and Automation
182 08 Prague 8
Pod vodárenskou věží 4
Czechoslovakia

AN AXIOMATIC CHARACTERIZATION OF J-DIVERGENCE

Pl. Kannappan, P.N. Rathie

Waterloo, Campinas

Key words: directed divergence, J-divergence, characterization, functional equations

ABSTRACT

The J-divergence or symmetric divergence is axiomatically characterized through a set of five postulates with the help of functional equations.

1. INTRODUCTION

Let $\Delta_n = \{P = (p_1,\ldots,p_n): p_i \geq 0, \sum_{i=1}^{n} p_i = 1\}$ denote the set of discrete complete probability distributions. Then for two distributions $P \in \Delta_n$ and $Q \in \Delta_n$, the measure the directed divergence (Kullback (1959) or the information gain (Renyi (1961)) is defined as

$$(1.1) \qquad I_n(P\|Q) = I_n\begin{pmatrix} p_1,\ldots,p_n \\ q_1,\ldots,q_n \end{pmatrix} = \sum_{i=1}^{n} p_i \log(p_i/q_i),$$

where whenever a q_i is zero the corresponding p_i is also zero and the base of the logarithm is 2.

For $n = 2$, (1.1) yields

$$(1.2) \qquad I_2\begin{pmatrix} p,1-p \\ q,1-q \end{pmatrix} = p \log(p/q) + (1-p)\log\{(1-p)/(1-q)\},$$

for $(p,q) \in]0,1[\times]0,1[\cup\{(0,y)\}\cup\{(1,z)\}$ with $y \in [0,1[$ and $z \in]0,1]$.

A characterization theorem for (1.1) was given in Kannappan, Rathie (1973).

The J-divergence or symmetric divergence between P and Q is defined as (Kullback (1959))

(1.3) $$J_n(P\|Q) = I_n(P\|Q) + I_n(Q\|P) = \sum_{i=1}^{n}(p_i-q_i)\log(p_i/q_i)$$

where whenever a p_i is zero the corresponding q_i is zero and vice versa.

For $n = 2$, (1.3) takes the following form,

(1.4) $$J_2\begin{pmatrix}p,1-p\\q,1-q\end{pmatrix} = I_2\begin{pmatrix}p,1-p\\q,1-q\end{pmatrix} + I_2\begin{pmatrix}q,1-q\\p,1-p\end{pmatrix} = (p-q)\log[p(1-q)/\{q(1-p)\}]$$

for $(p,q) \in K =]0,1[\times]0,1[\cup \{(0,0)\} \cup \{(1,1)\}$.

For the two hypotheses H_1:P and H_2:Q, the directed divergence (1.1) is the expected information in a single observation for discriminating in favour of hypothesis H_1 against H_2 (Kullback (1959)) while the symmetric divergence (1.3) is the expected information for discriminating between H_1 and H_2.

For application of symmetric divergence (1.3) to "pattern recognition" and "questionnaire analysis" the reader is referred to Ton, Gonzales (1974) and Haaland, Brockett, Levine (1978), respectively.

The object of this paper is to give a characterization theorem for (1.3) by assuming the following set of five postulates.

Let $J_n, I_n : \Delta_n^2 \to R$ (reals) for $n \geq 2$.

Postulate 1: (*Recursivity*). For p_1+p_2, $q_1+q_2 > 0$ and for all $n = 3,4,\ldots$,

$$J_n\begin{pmatrix}p_1,\ldots,p_n\\q_1,\ldots,q_n\end{pmatrix} = J_{n-1}\begin{pmatrix}p_1+p_2,p_3,\ldots,p_n\\q_1+q_2,q_3,\ldots,q_n\end{pmatrix} + (p_1+p_2)I_2\begin{pmatrix}p_1/(p_1+p_2),p_2/(p_1+p_2)\\q_1/(q_1+q_2),q_2/(q_1+q_2)\end{pmatrix}$$
$$+ (q_1+q_2)I_2\begin{pmatrix}q_1/(q_1+q_2),q_2/(q_1+q_2)\\p_1/(p_1+p_2),p_2/(p_1+p_2)\end{pmatrix}.$$

Postulate 2: (*Symmetry*). For any arbitrary permutation (a_1,a_2,a_3) of (1,2,3),

$$J_3\begin{pmatrix} p_1,p_2,p_3 \\ q_1,q_2,q_3 \end{pmatrix} = J_3\begin{pmatrix} p_{a_1},p_{a_2},p_{a_3} \\ q_{a_1},q_{a_2},q_{a_3} \end{pmatrix} .$$

Postulate 3: (*Derivative*). Let $f(x,y) = I_2\begin{pmatrix} x,1-x \\ y,1-y \end{pmatrix}$, for $(x,y) \in K$ have partial derivatives with respect to $x,y \in]0,1[$ up to the second order.

Postulate 4: (*Nullity*). For $x \in [0,1]$

$$J_2\begin{pmatrix} x,1-x \\ x,1-x \end{pmatrix} = 0.$$

Postulate 5: (*Normalization*).

$$J_2\begin{pmatrix} 2/3, & 1/3 \\ 1/3, & 2/3 \end{pmatrix} = 2/3 .$$

The postulate 1 suggests the way in which the J-divergence and the directed divergences are added up when the union of two mutually exclusive events are considered. The postulate 2 says that the value of J-divergence does not depend on the order in which the possible events are labelled. The postulate 3 is a regularity condition. The postulate 4 says that the J-divergence between the two distributions P and Q vanishes when the distributions coincide. The postulate 5 is a normalization postulate.

The characterization theorem for the J-divergence (1.3) is proved in the next section. This is the first attempt to give an axiomatic foundation for the J-divergence between two discrete distributions.

2. CHARACTERIZATION THEOREM

In this section we prove the following characterization theorem for the J-divergence.

THEOREM. The postulates 1 to 5 determine the function $J_n(P\|Q)$ uniquely as given by (1.3).

Proof: Denoting

(2.1) $$g(x,y) = J_2\begin{pmatrix}x,1-x\\y,1-y\end{pmatrix} \qquad (x,y) \in K$$

we get from the postulates 1 and 2,

(2.2) $$g(x,y)+(1-x)f\{u/(1-x),v/(1-y)\}+(1-y)f\{v/(1-y),u/(1-x)\}$$
$$= g(u,v)+(1-u)f\{x/(1-u),y/(1-v)\}+(1-v)f\{y/(1-v),x/(1-u)\},$$

for $x,y,u,v,x+u,y+v \in]0,1[$, $x = 0 = y$, $u = 0 = v$, $x+u = 1 = y+v$ and

(2.3) $$\text{L.H.S. of (2.2)} = g(x+u,y+v)+(x+u)f\{x/(x+u),y/(y+v)\}$$
$$+(y+v)f\{y/(y+v),x/(x+u)\},$$

for $x,y,u,v,x+u,y+v \in]0,1[$, $x = 0 = y$, $u = 0 = v$, $x+u = 1 = y+v$, $x+u,y+v > 0$.

For $u = 0 = v$, (2.2.) yields

(2.4) $$g(x,y) = f(x,y)+f(y,x)+(x+y-2)D+E, \qquad (x,y) \in K,$$

where $D = f(0,0), E = g(0,0)$.

Since f has partial derivaties up to the second order so has g.

The proof of the theorem is based on the solution of the functional equation (2.2) which is obtained in the following lemma:

LEMMA. The solutions of the functional equation (2.2) where $f(x,y)$ for $(x,y) \in K$ has partial derivatives with respect to $x,y \in]0,1[$ up to the second order are given by

(2.5f) $$f(t,s) = A[t \log s + (1-t)\log(1-s)]$$
$$+B[t \log t + (1-t)\log(1-t)]+Ct+D,$$

and

(2.5g) $$g(t,s) = A[t \log s + (1-t)\log(1-s) + s \log t + (1-s)\log(1-t)]$$
$$+B[t \log t + (1-t)\log(1-t) + s \log s + (1-s)\log(1-s)]$$
$$+(C+D)(t+s)+E,$$

for $t,s \in]0,1[$, where A,B,C,D and E are arbitrary constants.

Proof of the lemma: For fixed y and v in $]0,1[$, the functional equation (2.2) yields

$$(2.6)\quad G(x)+(1-x)F\{u/(1-x)\}+H\{u/(1-x)\} = N(u)+(1-u)L\{x/(1-u)\}+M\{x/(1-u)\},\quad x,u,x+u \in]0,1[,$$

where

$$(2.7)\quad \begin{cases} G(t) = g(t,y) & N(t) = g(t,v) \\ F(t) = f(t,v/(1-y)) & L(t) = f(t,y/(1-v)) \\ H(t) = (1-y)f(v/(1-y),t) & M(t) = (1-v)f(y/(1-v),t), \quad t \in]0,1[. \end{cases}$$

Differentiating (2.6) first with respect to x and then the resulting expression with respect to u and making the substitutions $u/(1-x) = t$, $x/(1-u) = s$, we have

$$(2.8)\quad t(1-t)F''(t)-s(1-s)L''(s) = [(1-st)/\{(1-s)(1-t)\}][(1-s)^2M'(s) +s(1-s)^2M''(s)-(1-t)^2H'(t)-t(1-t)^2H''(t)],\quad \text{for } t,s \in]0,1[.$$

This implies that

$$(2.9)\quad t(1-t)F''(t) = s(1-s)L''(s) = c_1 \quad \text{say}$$

and

$$(2.10)\quad (1-s)^2M'(s)+s(1-s)^2M''(s) = (1-t)^2H'(t)+t(1-t)^2H''(t) = a_1, \quad \text{say.}$$

Solving the differential equation (2.9), we have

$$f\{t,v/(1-y)\} = F(t) = c_1(v/(1-y))[t\log t + (1-t)\log(1-t)] +c_2(v/(1-y))\, t+c_3(v/(1-y)),\quad t,v,y \in]0,1[,$$

implying

$$(2.11)\quad f(t,s) = c_1(s)[t\log t + (1-t)\log(1-t)]+c_2(s)t+c_3(s),\quad t,s \in]0,1[,$$

where c_1, c_2 and c_3 are arbitrary functions of s.

Similarly, the differential equation (2.10) yields

$$(1-v)f\{y/(1-v),s\} = M(s) = -a_1(y,v)\log(1-s)+a_2(y,v)\log s+a_3(y,v)$$

implying

$$f(t,s) = -(1-v)^{-1}a_1(t(1-v),v)\log(1-s)+a_2(t(1-v),v)\log s+a_3(t(1-v),v),$$

$$t,s,v \in]0,1[.$$

Since the left hand side of the above equation does not contain v, we have

(2.12) $$f(t,s) = \alpha_1(t)\log(1-s)+\alpha_2(t)\log s+\alpha_3(t), \qquad t,s \in]0,1[$$

where α_1, α_2 and α_3 are arbitrary functions of t.

Hence from (2.11) and (2.12), we get

(2.13) $$c_1(s)[t \log t + (1-t)\log(1-t)]+c_2(s)t+c_3(s)$$
$$= \alpha_1(t)\log(1-s)+\alpha_2(t)\log s+\alpha_3(t), \qquad t,s \in]0,1[.$$

Since f has partial derivatives of order two, c_i's and α_i's have derivatives of order 2. Differentiating (2.13) twice with respect to t, we have

(2.14) $$c_1(s)[t(1-t)]^{-1} = \alpha_1''(t)\log(1-s)+\alpha_2''(t)\log s+\alpha_3''(t), \qquad t,s \in]0-1[.$$

Hence

(2.15) $$c_1(s) = d_1\log(1-s)+d_2\log s + d_3, \qquad s \in]0,1[$$

where d_1, d_2 and d_3 are arbitrary constants.

Substituting the expression for $c_1(s)$ from (2.15) into (2.14) and comparing the corresponding coefficients, we have $\alpha_i''(t) = d_i/\{t(1-t)\}$, $i = 1,2,3$ giving

(2.16) $$\alpha_i(t) = d_i[t \log t+(1-t)\log(1-t)]+e_it+b_i , \qquad t \in]0,1[$$

for $i = 1,2,3\ldots$.

Hence (2.12) and (2.16) yield

$$(2.17)\quad f(t,s) = [d_1\{t \log t +(1-t)\log(1-t)\}+e_1t+b_1]\log(1-s) +[d_2\{t \log t +(1-t)\log(1-t)\}+e_2t+b_2]\log s + d_3\{t \log t +(1-t)\log(1-t)\}+e_3t+b_3 , \qquad t,s \in]0,1[.$$

Substituting the expression for f from (2.17) and of g from (2.4) into (2.2) and comparing the coefficients of $\log v$, $\log(1-y-v)$ and $\log(1-y)$ respectively, we have $d_1 = d_2 = 0$, $e_1 = -b_1$, $e_1+e_2 = -b_2 = 0$. Hence (2.17) is of the form (2.5f) and (2.4) is of the form (2.5g). This proves the lemma.

Continuation of the proof of the theorem. The substitution of $u = v = 0$ in (2.3) yields

$$(2.18)\qquad f(1,1) = 0 = f(0,0) = D.$$

The postulate 4 and (2.5g) yield $E = g(0,0) = 0$, $B = -A$, $C = -D = 0$, and the postulate 5 and (2.5g) give $A = -1$, so that

$$(2.19)\quad \begin{cases} f(t,s) = t \log(t/s)+(1-t)\log\{(1-t)/(1-s)\}], \\ g(t,s) = t \log(t/s)+(1-t)\log\{(1-t)/(1-s)\}+ s \log(s/t) \\ \qquad +(1-s)\log\{(1-s)/(1-t)\}, \qquad \text{for } (t,s) \in K. \end{cases}$$

Successive use of the postulate 1 gives

$$(2.20)\quad J_n(P\|Q) = J_2\begin{pmatrix} p_1+\ldots+p_{n-1}, p_n \\ q_1+\ldots+q_{n-1}, q_n \end{pmatrix} + \sum_{i=2}^{n-1} r_i f(p_i/r_i, q_i/s_i) + \sum_{i=2}^{n-1} s_i f(q_i/s_i, p_i/r_i),$$

where $r_i = p_1+\ldots+p_i$ and $s_i = q_1+\ldots+q_i$.

Hence (2.20), (2.19) and (2.1) give

$$(2.21)\qquad J_n(P\|Q) = \sum_{i=1}^{n} (p_i-q_i)\log(p_i/q_i),$$

proving the theorem.

Acknowledgement: This work is supported by grants from NSERC of Canada and Instituto de Matematica Pura e Aplicada, Brazil

REFERENCES

Haaland, P., Brockett, P.L. and Levine, A. (1978): A characterization of divergence with applications to questionnaire information.

Kannappan, Pl. and Rathie, P.N. (1973): On a characterization of directed divergence, Information and Control, 22, 163-171.

Kullback, S. (1959): Information Theory and Statistics. Wiley, New York.

Rényi, A., (1961): On measures of entropy and information, Proc. Fourth Berkeley Symp. Math. Statist. Probabil., 1, 547-561.

Ton, J.T. and Gonzales, R.C. (1974): Pattern Recognition Principles, Addison-Wesley, Reading.

Department of Pure Mathematics
University of Waterloo
Waterloo, Ontario
Canada N2L 3G1

Institute of Mathematics
Statistics and Computer Science
State University of Campinas
Campinas, SP, Brazil

THE METHOD OF WORD SYNCHRONIZATION IN ALPHABETIC

CODES FOR THE DIGITAL TRANSMISSION SYSTEMS

Gurgen G.Khachatrian, Levon G.Khachatrian, Garegin S.Markarian

YEREVAN

Key words: alphabetic codes, word synchronization, ability of synchronization, 6B-4T ternary code,probability of achieving synchronization, digital transmission system.

ABSTRACT

The new algorithm pf word synchronization for the alphabetic codes is presented. The algorithm allows to use the ability of synchronization completely. Finally, the probability of achieving synchronization after N checks, for the 6B-4T code is calculated.

I. INTRODUCTION

As was shown in an KB-MT alphabetic line code maps block of K binary information digits into block of M ternary channel symbols where M is the least integer such that $M > K/\log 3$. This codes are used in digital transmission systems (DTS) for the increase of the information rate, spectrum shaping and error control.

However the utilization of alphabetic codes is impossible without the highefficient word synchronization system (WSS).

The WSS for the multilevel codes was described in [2] and WSS for the ternary 6B-4T code were described in [3]. The efficiency of these WSS is far from optimal. In this paper we consider the algorithm of word synchronization, which allows to minimize the time of entrance in synchronization and provides the defence from the errors in digital repeater line (DRL).

II. THE ALGORITHM OF SYNCHRONIZATION

The algorithm of synchronization for the highredundance block codes have been described in [4]. However, it is impossible to use this algorithm in alphabetic codes, because the redundancy of these codes is very low.

The algorithm of the synchronization is the following. At the beginning, M accidental neighbouring phases of the word synchronization are selcted and at the same time M -ary code words in all M phases are checked. In each i-th phase, check continues before the number of forbidden code words N_1^i is less then N_0, which is determined by the compromise between the time of the entrance in synchronization and defence from the errors in communication channel. If $N_1^i > N_0$, the phase is rejected and the checking procedure continues in the remained phases.

So, the considered algorithm of word synchronization is an optimal in the sense, that it uses the ability of synchronization in the alphabetic codes completely.

III. CALCULATION OF THE ABILITY OF SYNCHRONIZATION

Let's denote by P_N the probability of achieving synchronization after N checks for the noiseless channel. The P_N is called to be the ability of synchronization for the given alphabetic code.

We examine the check of N code words with the shifting for i positions, $i=\overline{1,M}$. Let, $\varphi(\beta_j;\beta_{j+1})$ be the number of combinations for which, segments of the lengths $M,(a_j;b_j)$ and $(a_j^i;b_j^i)$ are the code words and the segments $(a_j;a_j^i)$ and $(b_j;b_j^i)$ of the lengths i are equal to β_j and β_j^i accordingly. Let, $C(\alpha)$ be the number of combinations, which begin with α. Then, for the i-th phase:

(1) $$f_N^i(\alpha) = \sum_{\beta_1 \ldots \beta_{N-1}} \varphi(\beta_1;\beta_2)\cdot\varphi(\beta_2;\beta_3)\cdot\ldots\cdot\varphi(\beta_{N-1};\alpha)$$

and from the (1) follows:

(2) $$f_N^i = \sum_{\alpha} f_N^i(\alpha); \quad F_N^i(\alpha) = f_N^i(\alpha)\cdot C(\alpha); \quad F_N^i = \sum_{\alpha} F_N^i(\alpha).$$

The value of F_N^i is the number of combination for which all the N segments shifted by the i positions are the code words. Probability P_0^i of this event is equal to

(3) $$P_0^i = F_N^i / M_0^N$$

there M_0 is the cardinality of the code. Thus, the probability P_N is defined by

(4) $$P_N = \min_i \{1 - P_0^i\}$$

As follows from (1) - (4), P_N depends only from the code table. Therefore, such a parameter can be used to characterize the ability of synchronization of the alphabetic code.

The ternary alphabetic 6B-4T code have been described in [5], (1984) and now greater attention is paid to him. In this paper we do the estimation of the ability of synchronization for the 6B-4T code. The cardinality of this code $M_0 = 64$, and it follows from the code symmetry that $P_0^1 = P_0^3$ and $P_0^2 \ll P_0^1$. Therefore, we will calculate only P_0^1. We note, that in 6B-4T code

(5) $$\varphi(0;0)=22 \qquad \varphi(0;1)=19 \qquad \varphi(0;2)=19$$

$$\varphi(1;0)=19 \qquad \varphi(1;1)=18 \qquad \varphi(1;2)=16$$

$$\varphi(2;0)=19 \qquad \varphi(2;1)=16 \qquad \varphi(2;2)=20$$

From the (1) and (2) we obtain:

(6) $$\begin{cases} f^1_{N+1}(0) = 22\cdot f^1_N(0) + 19\cdot f^1_N(1) + 19\cdot f^1_N(2) \\ f^1_{N+1}(1) = 19\cdot f^1_N(0) + 18\cdot f^1_N(1) + 16\cdot f^1_N(2) \\ f^1_{N+1}(2) = 19\cdot f^1_N(0) + 16\cdot f^1_N(1) + 20\cdot f^1_N(2) \end{cases}$$

and after the transformations in (6) we obtain the following equation:

(7) $$f^1_{N+3}(\alpha) = 60\cdot f^1_{N+2}(\alpha) - 218\cdot f^1_{N+1}(\alpha) + 122\cdot f^1_N(\alpha), \quad \alpha = 0;1;2.$$

The f^1_N defines from the (7):

(8) $$f^1_N = C_1\cdot X_1^N + C_2\cdot X_2^N + C_3\cdot X_3^N$$

were $X_1;X_2;X_3$-are roots of following equation :

(9) $$X^3 - 60\cdot X^2 + 218\cdot X - 122 = 0$$

$C_1;C_2;C_3$ - are defined by $f_1;f_2;f_3$ and $C_1 \approx 1$.

The $X_1 = 56{,}1567; X_2 = 3{,}15$ and $X_3 = 0{,}7$ are the approximate solutions of (9). Therefore: $F^1_N \approx (56{,}1567)^N$, and $P^1_0 = (56{,}1567/64)^N$.

The ability of synchronization for the $6B\text{-}4T$: $P_N = 1 - (56{,}1567/64)^N$.

As follows from the received estimations, the $6B\text{-}4T$ has good synchronization properties so in particular the probability of achieving synchronization after $N=250$ checks, for the noiseless channel, is $P_{250} = 1 - 10^{-17}$.

IV. CONCLUSIONS

So, the considered algorithm allows to minimize the time of entrance in synchronization, provides the defence from the errors in communication channel and discovers the exit from the synchronization with the great efficiency. The new criterion, called the ability of synchronization, allows to compare the properties of synchronization of the alphabetic codes.

REFERENCES

1. Bellamy J. (1982): Digital Telephony. John Wiley and Sons, New York

2. Miata M. (1975): "Transactions on Communications", Vol.COM-23, No.4, p.p448-452.

3. Patent USA, No.4244052.

4. Peterson W.W., Weldon E.J. (1972): Error-Correcting Codes. The Mit Press, London.

5. Markarian G.S. et al.: (1984): (The block balanced code for the ΔTS). Radiotechnika, No.4, p.48-50.

Armenian Academy of Sciences,
Computing Centre,
P.Sévak Str. 1.,
375044, YEREVAN,
U;S;S;R.

G.C.[illegible], [illegible], A.E.[illegible]

2. [illegible] M. (19[illegible]): "[illegible]", Vol.[illegible], No.[illegible], p.[illegible]

3. Patent USA, No.[illegible].

4. Peterson W.W., Weldon E.J. (1972): Error-Correcting Codes. The MIT Press, London.

5. [illegible] G.S. et al. (199[illegible]): (The block-balanced code [illegible] [illegible], No.[illegible], p.[illegible]

ON THE ASYMPTOTIC APPROACH TO TIME SERIES RECOGNITION

Nerutė Kligienė

Vilnius

Key words: asymptotic eigenvalue distribution, autoregression

ABSTRACT

A stationary autoregressive sequence $\{X_t, t=0,\pm 1,\ldots\}$ denoted by AR(p) is considered in order to investigate the asymptotic distribution of the eigenvalues of some matrix R_N. This matrix is essential in evaluating the probability of misrecognition of the observations $x_1,\ldots,x_N$. The aim is to construct to R_N an asymptotically equivalent Toeplitz matrix T_N, $N \gg p$, what enables us to evaluate the eigenvalues $\lambda_{0N}^{(R)}, \lambda_{1N}^{(R)}, \ldots, \lambda_{N-1,N}^{(R)}$ of matrix R_N asymptotically as $N \to \infty$. The main result relates a frequency domain analysis and a time domain analysis of a time series.

INTRODUCTION

Let $\{X_t, t=0,\pm 1,\ldots\}$ be a stationary sequence with zero mean and be described by a finite order model depending on parameters θ, $\theta \in R^{p+1}$. We are going to test the hypothesis H_1: $\theta = \theta^{(1)}$ against the hypothesis H_2: $\theta = \theta^{(2)}$ by means of likelihood ratio statistics on the base of observations $x_{1N} = (x_1, x_2,\ldots,x_N)$ of a sequence $\{X_t, t=1,2,\ldots N\}$. In order to evaluate the probability of the error we need to compute the eigenvalues $\lambda_{0N}^{(R)},\ldots,\lambda_{N-1,N}^{(R)}$ of the matrix $R_N = [\Sigma_N^{(2)}]^{-1} \cdot \Sigma_N^{(1)}$, where $\Sigma_N^{(i)} = E\{X_{1N} \cdot X_{1N}^T\}, i=1,2$ is a covariance matrix of vector X_{1N}, corresponding to parameters $\theta = \theta^{(i)}$, $i=1,2$ and the matrix $[\Sigma_N^{(i)}]^{-1}$ is an inverse of $\Sigma_N^{(i)}$. It is an untractable problem to compute the values $\lambda_{0N}^{(R)}, \lambda_{1N}^{(R)}, \ldots, \lambda_{N-1,N}^{(R)}$ if N is large.

THE ASYMPTOTIC BEHAVIOUR OF MATRICES

Consider a case when $\{ X_t, t=0, \pm 1, \pm 2, \dots \}$ is an autoregressive sequence of the finite order p, described by the equation

$$\sum_{j=0}^{p} a_j^{(i)} X_{t-j} = \varepsilon_t , \tag{I}$$

where the sequence $\{ \varepsilon_t, t=0, \pm 1, \pm 2, \dots \}$ consists of independent identically distributed random variables ε_t. As usual, the model (I) will be denoted in brief by AR(p). Let AR(p) parameters $\theta^{(i)} = (a_0^{(i)}, a_1^{(i)}, \dots, a_p^{(i)})$ be such that all the roots of the equation $\sum_{j=0}^{p} a_j^{(i)} z^{p-j} = 0$ are less than I in absolute value. The inverse of the covariance matrix, $[\Sigma_N^{(i)}]^{-1}$ of process AR(p) has a specific structure, i.e. it is symmetric about both diagonals (persymmetric), asymptotically Toeplitz and its entries ${}^{-1}\sigma_{jk}^{(i)}$ $j,k = 1,2,\dots,N$ have the following property

$${}^{-1}\sigma_{jk}^{(i)} = 0 \quad \text{if} \quad |j-k| > p, \quad \forall\ j,k = 1,2,\dots,N. \tag{2}$$

Recall that $N \times N$ matrix $T_N = (t_{jk})$ is Toeplitz if its entries t_{jk} satisfy the invariance property $t_{jk} = t_{j-k}$ for all j and k. In spite of all nice properties enumerated above it is impossible to derive any expressions of the eigenvalues in case of finite N.

Let us consider the asymptotic behaviour of eigenvalues of finite order Toeplitz matrices by constructing an asymptotically equivalent circulant matrix.

It is well known by Anderson (I97I) that a matrix $[\Sigma_N^{(i)}]^{-1}$ of any stationary sequence $\{ X_t, t=0, \pm 1, \pm 2, \dots \}$ may be factorized in the following way

$$[\Sigma_N^{(i)}]^{-1} = A_N^{(i)T} \cdot A_N^{(i)} , \tag{3}$$

where $A_N^{(i)}$ is a triangular matrix, $A_N^{(i)T}$ is its transpose. In the AR(p) case $A_N^{(i)}$ entries may be expressed simply in terms of coefficients $a_0^{(i)}, a_1^{(i)}, \dots a_p^{(i)}$. The entries $\Delta_{jk}^{(i)}$ at the left upper corner of $A_N^{(i)}$ are also expressable in terms of $a_0^{(i)}, a_1^{(i)}, \dots, a_p^{(i)}$ but due to the more complicated expression they are omitted here. It is not difficult to derive from the model (I) the expression of the matrix $A_N^{(i)}$ as in (4), denoting by **O** the part of any matrix consisting of zero entries

$$(4)\qquad A_N^{(i)} = \begin{bmatrix} \Delta_{11}^{(i)} & & & & & & \\ \Delta_{21}^{(i)} & \Delta_{22}^{(i)} & & & & 0 & \\ \cdots & \cdots & & & & & \\ \Delta_{p1}^{(i)} & \Delta_{p2}^{(i)} & \cdots & \Delta_{pp}^{(i)} & & & \\ a_p^{(i)} & a_{p-1}^{(i)} & \cdots & a_1^{(i)} & a_0^{(i)} & & \\ & \ddots & & & & \ddots & \\ 0 & & a_p^{(i)} & a_{p-1}^{(i)} & \cdots & a_1^{(i)} & a_0^{(i)} \end{bmatrix}.$$

The matrix $R_N = [\Sigma_N^{(2)}]^{-1} \cdot \Sigma_N^{(1)}$, therefore, may be rewritten like this

$$(5)\qquad R_N = [A_N^{(1)T}]^{-1} \cdot [\Sigma_N^{(2)}]^{-1} \cdot [A_N^{(1)}]^{-1}$$

For a while, when considering $\Sigma_N^{(2)}$ and $A_N^{(1)}$ separately, let us omit the upper indices $i = 1, 2$. It is not difficult to write out the entries ${}^{-1}\sigma_{st}$ of any matrix Σ_N^{-1} on the base of expressions (3), (4):

$$(6)\qquad {}^{-1}\sigma_{st} = \sum_{k=0}^{\min(s,t)-1} a_k a_{k+|s-t|}, \quad \text{if } s,t \leq p \text{ or } s,t \geq N-p+1;$$

$$(7)\qquad {}^{-1}\sigma_{st} = {}^{-1}\sigma_{s-t} = \sum_{k=0}^{p-s+t} a_k a_{k+s-t} \quad \text{otherwise}.$$

This leads us to the matrix Σ_N^{-1} of the following structure

$$(8)\qquad \Sigma_N^{-1} = \begin{bmatrix} \sigma_{st},\ s,t=1\div p & {}^{-1}\sigma_p & & & & 0 \\ & {}^{-1}\sigma_1 & \ddots & & & \\ {}^{-1}\sigma_p \cdots {}^{-1}\sigma_1 & {}^{-1}\sigma_0 & & \ddots & & \\ & \ddots & {}^{-1}\sigma_0 & {}^{-1}\sigma_1 & \cdots & {}^{-1}\sigma_p \\ 0 & & \ddots & {}^{-1}\sigma_1 & \sigma_{st},\ s,t=N-p+1\div N & \\ & & & {}^{-1}\sigma_p & & \end{bmatrix}.$$

After such preliminaries we are able to construct to R_N an asymptotically equivalent Toeplitz matrix T_N, $N \gg p$ in order to evaluate the asymptotic distribution of eigenvalues $\lambda_{0N}^{(R)}, \ldots, \lambda_{N-1,N}^{(R)}$ as $N \to \infty$.

Lemma I. Matrix A_N^{-1}, inverse to A_N, is asymptotically Toeplitz.

Proof. The model (I) may be rewritten in such a way

$$(9)\qquad X_t = \sum_{j=0}^{\infty} c_j \varepsilon_{t-j}, \quad t = 0, \pm 1, \pm 2, \ldots$$

It is easy to check up that matrix A_N^{-1}, inverse to A_N, may be written out as

$$\text{(10)} \qquad A_N^{-1} = \begin{pmatrix} {}^{-1}\Delta_{11} & & & & & 0 \\ {}^{-1}\Delta_{21} & {}^{-1}\Delta_{22} & & & & \\ \cdots & \cdots & & & & \\ {}^{-1}\Delta_{p1} & {}^{-1}\Delta_{p2} & \cdots & {}^{-1}\Delta_{pp} & & \\ c_p & c_{p-1} & \cdots & c_1 & c_0 & \\ & & \ddots & & \ddots & \ddots \\ c_{N-1} & c_{N-2} & \cdots & c_{p-1} & \cdots & c_1 \; c_0 \end{pmatrix},$$

where the entries, except those in the upper left corner, coincide with the coefficients c_j of the decomposition (9). Let us construct a triangular matrix C_N with entries

$$\text{(11)} \qquad c_{st} = \begin{cases} c_{s-t}, & s, t = 1, 2, \ldots, N, \; s \geq t; \\ 0, & s < t \end{cases}$$

and prove that sequences of matrices $\{ A_N^{-1}, \; N = p+1, p+2, \ldots \}$ and $\{ C_N, \; N = p+1, p+2, \ldots \}$ are asymptotically equivalent.

Recall that two sequences of $N \times N$ matrices $\{ A_N, N = 1, 2, \ldots \}$ and $\{ C_N, \; N = 1, 2, \ldots \}$ are said to be asymptotically equivalent if

i) A_N and C_N are uniformly bounded in strong norm:

$$\text{(12)} \qquad \exists M, \quad \| A_N \| \leq M, \quad \| C_N \| \leq M, \quad M < \infty;$$

ii) $D_N = A_N - C_N$ tends to zero in weak norm as $N \to \infty$:

$$\text{(13)} \qquad \lim_{N \to \infty} | D_N | = 0.$$

If one of the two matrices is Toeplitz then the other is said to be asymptotically Toeplitz.

Stationarity of the sequence $\{ X_t, \; t = 0, \pm 1, \ldots \}$ involves the validity of spectral representation of the covariance function $\sigma_u = E\{ X_t X_{t+u} \}, \; u = 0, \pm 1, \pm 2, \ldots, \; \forall t$

$$\text{(14)} \qquad \sigma_u = \frac{1}{2\pi} \int_0^{2\pi} e^{-iu\lambda} f(\lambda) \, d\lambda,$$

$$\text{(15)} \qquad f(\lambda) = \sum_{k=-\infty}^{\infty} \sigma_k e^{ik\lambda},$$

where $f(\lambda)$ is a spectral density of the sequence $\{ X_t, t = 0, \pm 1, \ldots \}$. Consequently

$$\| A_N^{-1} \|^2 = \max_{\|x\|=1} x^* A_N^{-1\,T} \cdot A_N^{-1} x = \max_{\|x\|=1} x^* \Sigma_N x =$$

$$= \max_{\|x\|=1} \sum_{s,t=1}^{N} x_s^* \sigma_{s-t} x_t = \max_{\|x\|=1} \left\{ \frac{1}{2\pi} \int_0^{2\pi} \left| \sum_{t=1}^{N} x_t e^{it\lambda} \right|^2 \cdot \right.$$

(I6) $$\left. \cdot f(\lambda) d\lambda \right\} \le \max_{\lambda} | f(\lambda) | \le \sum_{u=-\infty}^{\infty} |\sigma_u| \triangleq M .$$

Denote by $D_N = A_N^{-1} - C_N = \{ d_{st},\ s,t = 1,2,\dots,N\}$ a matrix having only the p first entries non-zero, what follows from (I0) and (II). Thus inequality (I6) also involves the boundedness of the norm $\| C_N \|^2$. Let us prove the validity of (I3) in our case. For any matrix D_N it is true

(I7) $$| \tfrac{1}{N} \text{trace } D_N | \le | D_N |,$$

so it is necessary to prove $\lim_{N\to\infty} \frac{1}{N} \text{trace } D_N = 0$,

(I8) $$\lim_{N\to\infty} \frac{1}{N} \text{trace } D_N = \lim_{N\to\infty} \frac{1}{N} \sum_{s=1}^{p} d_{ss} = 0,$$

what completes the proof of Lemma I.

<u>Lemma 2</u>. The matrix $R_N = \{ r_{st},\ s,t = 1,2,\dots,N\}$ is asymptotically equivalent to the Toeplitz matrix $T_N = \{ r_{s-t},\ s,t=1,\dots,N\}$ with

(I9) $$r_u = \sum_{j=0}^{N-1} \sum_{k=0}^{N-u-1} c_j^{(1)} c_k^{(1)} \overset{-1}{\sigma}_{|j-k+u|},\quad u = 0,1,\dots,N-1$$

where $c_j^{(1)}$ are the coefficients of the decomposition (9) if true parameters are $\theta^{(1)}$ while $a_j^{(2)}, j=0,\dots,p$ are AR(p) parameters under H_2, $\overset{-1}{\sigma}{}_s^{(2)}$ given by (7).

The proof is analogous to that of Lemma I, and it is essentially based on the expressions (5), (6)-(I0) and their transformations.

<u>Theorem I</u>. The eigenvalues $\{ \lambda_{kN}^{(R)},\ k = 0,1,\dots,N-1\}$ of the matrix $R_N = [A_N^{(1)T}]^{-1} \cdot [\Sigma_N^{(2)}]^{-1} \cdot [A_N^{(1)}]^{-1}$ are asymptotically distributed as the values

(20) $$\left\{ \mathcal{A}_2\left(\frac{2\pi k}{N}\right) / \mathcal{A}_1\left(\frac{2\pi k}{N}\right),\ k=0,1,\dots,N-1 \right\},$$

where

$$\mathcal{A}_j(\lambda) \triangleq \left| \sum_{\nu=0}^{p} a_\nu^{(j)} e^{i\nu\lambda} \right|^2, \quad j=1,2. \tag{21}$$

Proof. It is easy to prove that the sequence of matrices $\{ [\Sigma_N^{(2)}]^{-1},\ N=2p+1, 2p+2, \dots \}$ is asymptotically equivalent to the sequence of Toeplitz matrices $\{ T_N = (t_{s-j}),\ N=2p+1, 2p+2, \dots \}$ with

$$t_u = \begin{cases} \sum_{k=0}^{p-|u|} a_k^{(2)} a_{k+|u|}^{(2)} = \frac{1}{2\pi} \int_{-\pi}^{\pi} e^{iu\lambda} f(\lambda)\, d\lambda; \\ 0, \qquad u > p; \end{cases} \tag{22}$$

where

$$f(\lambda) = \sum_{k=-\infty}^{\infty} t_k e^{ik\lambda} = \sum_{k=-p}^{p} t_k e^{ik\lambda} = \sum_{k=-p}^{p} \sum_{u=0}^{p-k} a_u^{(2)} \cdot a_{u+|k|}^{(2)} \cdot e^{ik\lambda} = \left| \sum_{k=0}^{p} a_k^{(2)} e^{ik\lambda} \right|^2 \triangleq \mathcal{A}_2(\lambda). \tag{23}$$

Using the results of Grenader, Szego (I958) on the distribution of eigenvalues of the Toeplitz matrix and a matrix being a product of other matrices as well as the results in (22), (23), we are able to prove our statement.

Note that, evidently

$$\sigma_u^{(1)} = E\{ X_t X_{t+u} \} = \sum_{k=0}^{\infty} c_k^{(1)} c_{k+u}^{(1)} \tag{24}$$

and the eigenvalues of the matrix $[A_N^{(1)}]^{-1}$, that is asymptotically Toeplitz, according to Lemma I, are distributed as the corresponding points of the function

$$f_1(\lambda) = \sum_{k=-\infty}^{\infty} c_k^{(1)} e^{ik\lambda} = \sum_{k=0}^{\infty} c_k^{(1)} e^{ik\lambda}. \tag{25}$$

The Fourier transformation corresponding to $[A_N^{(1)T}]^{-1}$ is

$$\bar{f}_1(\lambda) = \sum_{k=-\infty}^{\infty} c_k^{(1)} e^{-ik\lambda} = \sum_{k=0}^{\infty} c_k^{(1)} e^{-ik\lambda}. \tag{26}$$

This way the eigenvalues of the matrix R_N are distributed as the corresponding points of the function

$$f_1(\lambda)\mathcal{A}_2(\lambda)\bar{f}_1(\lambda) = \mathcal{A}_2(\lambda)\sum_{k=0}^{\infty} c_k^{(1)} e^{ik\lambda} \sum_{r=0}^{\infty} c_r^{(1)} e^{-ik\lambda} =$$

$$= \mathcal{A}_2(\lambda)\sum_{k=0}^{\infty}\sum_{u=-\infty}^{\infty} c_k^{(1)} c_{k+u}^{(1)} e^{iu\lambda} = \mathcal{A}_2(\lambda)\cdot$$

(27) $$\sum_{u=-\infty}^{\infty} \sigma_u^{(1)} e^{iu\lambda} = \mathcal{A}_2(\lambda)\cdot \left|\sum_{k=0}^{p} a_k^{(1)} e^{ik\lambda}\right|^{-2} = \frac{\mathcal{A}_2(\lambda)}{\mathcal{A}_1(\lambda)}.$$

Theorem I is proved.

DISCUSSION

The study of Toeplitz eigenvalue problems is actual for many areas of investigation and there is much to be learned. In time series analysis, for example, Box, Jenkins (I970), a common practice is to neglect a part of the considered statistics involving the first p observations x_{1p} , what is equivalent to the approximation of $[\Sigma_N^{(i)}]^{-1}$ by the other matrix B_N - non-Toeplitz, non-persymmetric. So very essential features of the covariance matrix are violated. Our aim was to profit nice properties of Toeplitz matrices and those of their simpler and more structured cousins - circulant matrices, approximating the matrices analogous to (8).

Notice that the result (20) relates the asymptotic analysis of time series in time domain to the spectral analysis because $2\pi\cdot\mathcal{A}_i^{-1}(\lambda)$ really is a spectral density of AR(p) with parameters $\theta^{(i)}$. The relation between a frequency and time domain analysis of time series has been discussed by Durbin (I984) and it was pointed out that it was useful to switch from one to the other according to convenience.

REFERENCES

Anderson T.W. (I97I): The statistical analysis of time series. John Wiley and Sons, Inc. New York London Sydney Toronto.

Box G.E.P. and Jenkins (I970): Time series analysis: Forecasting and Control. Holden Day, San Francisco.

Durbin J. (1984): Present position and potential developments: some personal views - time series analysis, Journal Royal Statist. Soc., A., 147, Part 2, 161-173.

Grenander U. and Szego G. (1958): Toeplitz forms and their applications. Univ. of Calif. Press, Berkeley.

Lithuanian Academy of Sciences
Institute of Mathemàtics and Cybernetics
4, Akademijos str.
232021 Vilnius, Lithuanian SSR

RANDOM COEFFICIENT AR(1) PROCESS

Alena Koubková

Prague

Key words: autoregressive series, stationarity, covariance function, spectral density

ABSTRACT

One special model of AR(1) series with time-dependent random coefficients is invetigated. Conditions for stationarity and explicit forms of its covariance function and spectral density are derived.

I. INTRODUCTION

Autoregressive series with random coefficients were first defined and investigated in Anděl (1976) and then in Nicholls and Quinn (1982). All these authors assume the random coefficients to be independent in time. The simplest model of AR(1) process with time-dependent random coefficients was described in Koubková (1982). Here we shall generalize that model for the case in which the random parameters have nonvanishing means.

We suppose that a process $\{X_t\}$ is generated by

$$X_t = b(t) X_{t-1} + a^{-1} Y_t , \qquad t = 2,3,\dots,N \tag{1}$$

where X_1 is a random variable with zero mean and a variance $\sigma^2 > 0$, $Y_2,\dots,Y_N$ is a sequence of independent random variables with zero means and unit variances which are independent of X_1, $a > 0$ is a real number, and $b(2),\dots,b(N)$ is a sequence of random coefficients. In our model we assume that the coefficients are of the form

(2) $b(t) = \gamma + \beta_o Z_t + \beta_1 Z_{t-1}$, $t = 2,3,\dots,N$

where $Z_1,\dots,Z_N$ are independent random variables with zero means and the same variances $\sigma^2 > 0$, which are independent of X_1, $Y_2,\dots,Y_N$, and γ, β_o, β_1 are nonzero constants. Obviously, $\{b(t)\}$ generate MA(1) process with the mean γ and the covariance function $B(t)$ which satisfies

$$
\begin{aligned}
B(0) &= Eb^2(s) - \gamma^2 = (\beta_o^2 + \beta_1^2)\sigma^2 , \\
B(1) &= Eb(s+1)b(s) - \gamma^2 = \beta_o\beta_1\sigma^2 , \qquad (3)\\
B(t) &= Eb(s+t)b(s) - \gamma^2 = 0 \quad \text{for} \quad t = 2,\dots,N-2.
\end{aligned}
$$

II CONDITIONS FOR STATIONARITY

It is convenient to write X_t in the equivalent form

(4) $X_t = b(t)b(t-1)\dots b(2)X_1 + a^{-1}b(t)\dots b(3)Y_2 + \dots + a^{-1}b(t)Y_{t-1} + a^{-1}Y_t$.

From the assumption of independence $Z_1,\dots,Z_N$ on $X_1,Y_2,\dots,Y_N$ it is evident that

(5) $EX_t = 0$ for all t.

The covariance function $R(s,t)$ is then

(6a) $R(s,t) = EX_sX_t = Eb(s)\dots b(t+1)b^2(t)\dots b^2(2)\sigma^2 +$

$$a^{-1}E\left[b(s)\dots b(t+1)b^2(t)\dots b^2(3) + \dots + b(s)\dots b(t+1)\right]$$

for $s,t = 2,\dots,N$, $s \geq t$, and

(6b) $R(s,1) = Eb(s)b(s-1)\dots b(2)\sigma^2$

Now we shall derive conditions under which $R(s,t)$ depends only on the difference $s-t$.

Lemma 1. Let the variables $Z_1,\dots,Z_N$ have the same moments EZ_t^3 and EZ_t^4 for all t. If the series $X_1,\dots,X_N$ is stationary then

$$(7)\qquad \sigma^2 = \frac{a^{-2}}{1-[\gamma^2+(\beta_0^2+\beta_1^2)\sigma^2]} \quad \text{where} \quad \gamma^2 + (\beta_0^2+\beta_1^2)\sigma^2 < 1$$

and simultaneously one of the next two conditions is fulfilled:

$$(8)\qquad EZ_t^3 = -\frac{2\gamma\sigma^2}{\beta_0}, \qquad EZ_t^4 = \sigma^4 + \frac{4\gamma^2\sigma^2}{\beta_0^2}, \quad t = 1,\dots,N,$$

$$(9)\qquad EZ_t^3 = -\frac{2\gamma\sigma^2}{\beta_1}, \qquad EZ_t^4 = {}^4 + \frac{4\gamma^2\sigma^2}{\beta_1^2}, \quad t = 1,\dots,N,$$

Proof. We can get conditions (7),(8) and (7),(9) by solving equation $\operatorname{Var} X_2 = \operatorname{Var} X_3 = \operatorname{Var} X_4 = \sigma^2$ which is a necessary condition for stationarity of $\{X_t\}$. □

To prove sufficiency of (7) and (8) or (7) and (9) for the stationarity of $\{X_t\}$ we first prove some preliminaries.

Lemma 2. Let $Z_1,\dots,Z_N$ be independent random variables with $EZ_t = 0$, $EZ_t^2 = \sigma^2$ for all t and let $b(2),\dots,b(N)$ be generated by (2). If (8) is fulfilled then

$$(10)\qquad EZ_s\, b^2(s) \dots b^2(k) = 0,$$

$$(11)\qquad EZ_s^2\, b^2(s) \dots b^2(k) = \sigma^2(\beta_0^2\sigma^2+\gamma^2)\Big[\sum_{j=0}^{s-k-1} (\beta_1^2\sigma^2)^j\, Eb^2(s-j-1)\dots$$

$$\dots b^2(k) + (\beta_1^2\sigma^2)^{s-k}\Big] + \sigma^2(\beta_1^2\sigma^2)^{s-k+1}$$

for $s = 2,\dots,N$ and $2 \le k \le s$. If (9) fulfilled then

$$(12)\qquad EZ_s b^2(s) \dots b^2(k) = 2\gamma\beta_0\sigma^2\Big(1-\frac{\beta_0}{\beta_1}\Big)\Big[\sum_{j=0}^{s-k-1} (2\beta_0\beta_1\sigma^2)^j \cdot$$

$$\cdot Eb^2(s-j-1)\dots b^2(k) + (2\beta_0\beta_1\sigma^2)^{s-k}\Big],$$

$$(13)\quad EZ_s^2 b^2(s) \ldots b^2(k) = \left[\sigma^2(\beta_o^2\sigma^2 + \gamma^2) - 4\gamma^2\sigma^2 \frac{\beta_o}{\beta_1}(1 - \frac{\beta_o}{\beta_1})\right] \cdot$$

$$\cdot\left[\sum_{j=o}^{s-k-1} (\beta_1^2\sigma^2)^j \, Eb^2(s-j-1) \ldots b^2(k) + (\beta_1^2\sigma^2)^{s-k}\right] +$$

$$+ 4\gamma^2\beta_o\sigma^4(\beta_1 - 2\beta_o)(1 - \frac{\beta_o}{\beta_1})\left\{\sum_{j=o}^{s-k-1} (\beta_1^2\sigma^2)^j \,.\right.$$

$$\cdot\left[\sum_{i=o}^{s-j-k-2} (2\beta_o\beta_1\sigma^2)^i \; Eb^2(s-j-i-2) \ldots b^2(k) \quad + \right.$$

$$\left.\left. + (2\beta_o\beta_1\sigma^2)^{s-j-k-1}\right]\right\} + \sigma^2(\beta_1^2\sigma^2)^{s-k+1}$$

for $s = 2,\ldots,N$ and $2 \leq k \leq s$. If (8) or (9) is fulfilled then

$$(14)\qquad Eb^2(s) \ldots b^2(k) = \left[\gamma^2 + (\beta_o^2 + \beta_1^2)\sigma^2\right]^{s-k+1}$$

for $s = 2,\ldots,N$ and $2 \leq k \leq s$.

Proof. All the formulas can be proved by induction. □

Corollary 3. If (7) and (8) or (7) and (9) are satisfied then

$$(15)\qquad \mathrm{Var}\; X_s = \sigma^2 \quad \text{for all} \quad s = 1,\ldots,N\,.$$

Lemma 4. The sequence $b(2),\ldots,b(N)$ defined by (2) fulfils

$$(16)\quad Eb(s) \ldots b(k) = \gamma Eb(s-1)\ldots b(k) + \beta_o\beta_1\sigma^2\, Eb(s-2)\ldots b(k)$$

for $k = 2,\ldots,N-2$ and $k+2 \leq s \leq N$,

$$(17)\quad Eb(s)\ldots b(t+1)b^2(t)\ldots b^2(k) = \gamma Eb(s-1)\ldots b(t+1)b^2(t)\ldots b^2(k) +$$

$$+ \beta_o\beta_1\sigma^2\; Eb(s-2)\ldots b(t+1)b^2(t) \ldots b^2(k)$$

for $t = 2,\dots,N-2$, $s = t+2,\dots,N$ and $k = 2,\dots,t$.
Proof is easy. □

Corollary 5. For the covariance function $R(s,t)$ of the series $X_1,\dots,X_N$ it holds

$$(18) \qquad R(s,t) = \gamma R(s-1,t) + \beta_0\beta_1\delta^2 R(s-2,t)$$

for $t = 1,\dots,N-2$ and $s = t+2,\dots,N$.

Theorem 6. The sequence $X_1,\dots,X_N$ is stationary iff the conditions (7) and (8) are satisfied. For covariance function $R(t)$ it holds

$$(19) \qquad \begin{aligned} R(0) &= \sigma^2, \\ R(1) &= \gamma\sigma^2, \\ R(t) &= \gamma R(t-1)+\beta_0\beta_1\delta^2 R(t-2) \quad \text{for } t = 2,\dots,N-1. \end{aligned}$$

Proof: We shall prove sufficiency of (7) and (8). In view of (18) and (15) it is enough to prove that $R(s+1,s)$ does not depend on s where $R(s+1,s)$ is given by (6a) and (6b). If (7) and (8) hold we get (using (10) and (14)) $R(s+1,s) = \gamma\sigma^2$ for $s = 2,\dots N$. Evidently, $R(2,1) = \gamma\sigma^2$ and so $R(1) = \gamma\sigma^2$ for all s. If (7) and (9) hold then (using (12) and (14) we obtain that $R(s+1,s)$ depends on s and thus $\{X_t\}$ is not stationary in this case. □

III. COVARIANCE FUNCTION

Koubková (1982) proved that under the assumption $\gamma = 0$ the explicit form of the covariance function is

$$R(t) = \begin{cases} \sigma^2(\beta_0\beta_1\delta^2)^{t/2} & \text{for } t \text{ even}, \\ 0 & \text{for } t \text{ odd}. \end{cases}$$

It is the same as the covariance function of the classical AR(2) series $\{V_t\}$ generated by

$$V_t = \beta_o\beta_1\sigma^2 V_{t-2} + a^{-1}\sqrt{\frac{1-(\beta_o\beta_1\sigma^2)}{1-(\beta_o^2+\beta_1^2)\sigma^2}}\; Y_t, \qquad t = 3,\ldots,N$$

where V_1, V_2 are uncorrelated random variables with zero means and variances σ^2.

Let now $\gamma \neq 0$. Let $\{V_t\}$ be a classical autoregression generated by

$$V_t = \gamma V_{t-1} + \beta_o\beta_1\sigma^2 V_{t-2} + c^{-1}Y_t \quad \text{for} \quad t = 3,\ldots N .$$

Evidently, for its covariance function we have

$$R(t) = \gamma R(t-1) + \beta_o\beta_1\sigma^2 R(t-2), \qquad t = 1,\ldots,N-1 .$$

In particular for $t = 1$ we obtain $R(1) = \dfrac{\gamma R(0)}{1-\beta_o\beta_1\sigma^2}$ and thus $R(0) = \sigma^2$, $R(1) = \gamma\sigma^2$ cannot be simultaneously fulfilled. Therefore the covariance function of $\{V_t\}$ cannot be the same as the covariance function of $\{X_t\}$.

Theorem 7. The explicit form of the covariance function of the series $X_1,\ldots,X_N$ is

$$(20) \qquad R(t) = \frac{\sigma^2}{v-\gamma}\left[\left(\frac{v}{2}\right)^{t+1} - \left(\frac{w}{2}\right)^{t+1}\right]$$

when $\gamma^2 + 4\beta_o\beta_1\sigma^2 > 0$ where $v = \gamma + \sqrt{\gamma^2 + 4\beta_o\beta_1\sigma^2}$, $w = \gamma - \sqrt{\gamma^2 + 4\beta_o\beta_1\sigma^2}$ or

$$(21) \qquad R(t) = \sigma^2(1 + t)\left(\frac{\gamma}{2}\right)^2$$

when $\gamma^2 + 4\beta_o\beta_1\sigma^2 = 0$ or

$$(22) \qquad R(t) = \sigma^2(-\beta_o\beta_1\sigma^2)^{t/2}\left(\frac{\sin\omega t}{\operatorname{tg}\omega} + \cos\omega t\right)$$

when $\gamma^2 + 4\beta_0\beta_1\delta^2 < 0$ where $\cos\omega = \dfrac{\gamma}{2\sqrt{-\beta_0\beta_1\delta^2}}$,

$$\sin\omega = 1 + \sqrt{\frac{\gamma^2}{4\beta_0\beta_1\delta^2}}$$

Proof. The explicit form of $R(t)$ can be obtained as a solution of the system (19). □

IV. SPECTRAL DENSITY

Theorem 8. a) Let the covariance function $R(t)$ of $\{X_t\}$ be of the form (20) and let $|v| < 2$ and $|w| < 2$. Then the spectral density exists and it is equal to

$$f(\lambda) = \frac{\sigma^2}{4\pi(v-\gamma)}\left[\frac{v(4-v^2)}{4-4v\cos\lambda + v^2} - \frac{w(4-w^2)}{4-4w\cos\lambda + w^2}\right] \tag{23}$$

for $\lambda \in \langle -\pi, \pi \rangle$.

b) Let the covariance function $R(t)$ be of the form (21). Then the spectral density exists and it is equal to

$$f(\lambda) = \frac{\sigma^2}{2\pi}\left[\frac{2 - 2\gamma\cos\lambda + \frac{\gamma^2}{2}\cos 2\lambda}{(1-\gamma\cos\lambda + \frac{\gamma^2}{4})^2} - 1\right] \tag{24}$$

for $\lambda \in \langle -\pi, \pi \rangle$.

c) Let the covariance function $R(t)$ be of the form (22). Then the spectral density exists and it is equal to

$$f(\lambda) = \frac{\sigma^2}{\pi}(1 - \gamma\cos\lambda - \beta_0\beta_1\delta^2\cos 2\lambda)\left[1 - 2\gamma(1-\beta_0\beta_1\delta^2)\cos\lambda - 2\beta_0\beta_1\delta^2\cos 2\lambda + \gamma^2 + (\beta_0\beta_1\delta^2)^2\right]^{-1} - \frac{\sigma^2}{2\pi} \tag{25}$$

for $\lambda \in \langle -\pi, \pi$

Proof. A sufficient condition for existence of the spectral density is

$$\sum_{t=-\infty}^{\infty} |R(t)| < \infty$$

Then the spectral density can be computed by

$$f(\lambda) = \frac{1}{2\pi} \sum_{t=-\infty}^{\infty} e^{-it\lambda} R(t) \quad \text{for} \quad \lambda \in \langle -\pi, \pi \rangle . \qquad \square$$

REFERENCES

Anděl J. (1976): Autoregressive series with random parameters. Math. Operationsforsch. Statist. 7, 735 - 761.

Koubková A. (1982): First-order autoregressive processes with time-dependent random parameters. Kybernetika 18, 408 - 414.

Nicholls D.E. and Quinn B.G. (1982): Random coefficient autoregressive models. An introduction. Lecture Notes in Statistics 11, Springer, New York - Berlin.

Charles University
Faculty of Mathematics and Physics
Sokolovská 83
186 00 Praha 8
Czechoslovakia

EXACT SLOPES FOR KOLMOGOROV-LÉVY TYPE STATISTICS*

Andrzej Kozek

Wrocław

Key words: exact slopes, large deviations, Kolmogorov-Lévy type statistics

ABSTRACT

We derive exact slopes for Kolmogorov-Lévy type distance between empirical (edf) and true distribution functions (df's).

1. INTRODUCTION AND RESULTS

If G and F are two df's then a Kolmogorov-Lévy type metric ρ_α is given by

$$\rho_\alpha(F,G) = \inf\{\varepsilon>0:\ G(x-\alpha\varepsilon)-\varepsilon \le F(x) \le G(x+\alpha\varepsilon)+\varepsilon,\ \forall x\in R\},$$

where $\alpha \ge 0$. The case $\alpha = 0$ is usually called Kolmogorov-Smirnov metric and the case $\alpha = 1$ Prokhorov-Lévy metric. Let $X_1,\ldots, X_n$ be iid F with edf F_n. The Kolmogorov-Lévy statistic is defined by

(1.1) $$L_n = \rho_\alpha(F_n,F).$$

Throughout the paper we shall assume that F satisfies the following

CONDITION A. F has a differentiable positive density f such that $|\alpha^2 f'(x)| \le B$ and f' changes sign at most K-1 times, $B > 2$, $K \ge 1$.

*The paper contains a proof of Theorem 1 announced in Kozek (1982) and continues the research carried out while the author was at the University of Cologne supported by a fellowship of the A. von Humboldt Foundation.

The evaluation of Bahadur efficiency of tests based on statistics (1.1) requires a large deviation result given in Theorem 1 below. The proof given in Section 2 is based on a theorem of Plachky and Steinbach, (Plachky-Steinbach (1975)), which extends the celebrated Chernoff Theorem (Chernoff (1952)).

Let

$$(1.2)\qquad g(p,t) = \begin{cases} \ln \int e^{tx} P(dx) & \text{if } t \ge 0 \\ +\infty & \text{if } t < 0, \end{cases}$$

where P is the zero-one probability distribution with $P(1) = p$ and $P(0) = 1-p$, $p \in [0,1]$. For $t \ge 0$

$$(1.3)\qquad g(p,t) = \ln(pe^{t}+(1-p))$$

and for every $p\in[0,1]$ $g(p,\cdot)$ is lower semicontinuous (lsc) convex function on $[0,\infty)$. Its conjugate function given by

$$h(p,s) = \sup\{st-g(p,t),\ t\in[0,\infty)\}$$

has the following form

$$(1.4)\qquad h(p,s) = \begin{cases} +\infty & \text{if } s > 1 \text{ or if } p = 0 \text{ and } s > 0 \\ \ln \frac{1}{p} & \text{if } s = 1 \text{ and } p\in(0,1] \\ (1-s)\ln[(1-s)/(1-p)]+s\ln[s/p] & \text{if } 0 < p < s < 1 \\ 0 & \text{if } s \le p \text{ and } p\in[0,1]. \end{cases}$$

Denote

$$(1.5)\qquad h^{+}(s) = \inf\{h(F(x),s(1+\alpha f(x))+F(x)):\ x\in R\},$$

$$(1.6)\qquad h^{-}(s) = \inf\{h(1-F(y),s(1+\alpha f(y))+1-F(y)):\ y\in R\},$$

$$(1.7)\qquad h_0(s) = \min\{h^{+}(s),h^{-}(s)\}.$$

Theorem 1. If $X_1,\ldots,X_n$ are iid F and F meets Condition A, then

$$\limsup_{n\to\infty} P(L_n > s_n)^{1/n} \le \exp[-h_0(s-Bs^2)]$$

and

$$\liminf_{n\to\infty} P(L_n > s_n)^{1/n} \ge \exp[-h_0(s+Bs^2)],$$

where s_n is a sequence converging to $s\in(0,1/B)$.

Corollary 1. For df's F satisfying Condition A we have

$$h_0(s) = \frac{1}{2}\,\gamma(F,\alpha)s^2+o(s^2),$$

where

(1.8) $$\gamma(F,\alpha) = \sup\left\{\frac{(1+\alpha f(x))^2}{F(x)(1-F(x))}: x\in R\right\}.$$

Thus, for n large $-\frac{1}{n}\log P(L_n>s_n) \approx \frac{1}{2}\gamma(F,\alpha)s_n^2+o\left(s_n^2\right)$, $s_n\to s$, $s\in(0,1/B)$.

2. PROOFS

Lemma 1. Let $g_X(t) = \ln Ee^{tX} < 0$ for $t\in[0,T_1)$.

a) If X is not degenerate, then $g_X(t)$ is strictly convex and differentiable on $[0,T_1)$.

b) If $Y = aX+b$, then $g_Y(t) = bt+g_X(at)$.

c) If X and Y are independent, then $g_{X+Y}(t) = g_X(t)+g_Y(t)$.

Proof. See Bahadur (1971). □

Lemma 2. If $g(t)$ is a convex function on $[0,T_1)$ and $g_1(t) = g(at)+bt$, $a>0$, then the conjugate $g_1^*(s) = \sup\{st-g_1(t): t\geq 0\}$ is of the form

$$g_1^*(s) = g^*((s-b)/a),$$

where g^* is the conjugate of g.

Proof. See Rockafellar (1970). □

Lemma 3. a) The function h given by (1.4) is convex for $(p,s)\in[0,1]\times R$ and strictly convex on set $\{(p,s): 0<p<s<1\}$. Moreover

b) $\frac{\partial h}{\partial s}(p,s) = \ln\frac{s(1-p)}{(1-s)p} > 0$ for $0<p<s<1$, hence $h(p,\cdot)$ is increasing on $(p,1)$,

c) $\frac{\partial h}{\partial p}(p,s) = \frac{p-s}{p(1-p)} < 0$ for $0<p<s<1$, hence $h(\cdot,s)$ is decreasing on $(0,s)$.

Proof. Since $\frac{\partial^2 h}{\partial s^2}(p,s) = (s(1-s))^{-1}$, $\frac{\partial^2 h}{\partial p^2}(p,s) = \frac{(p-s)^2+s(1-s)}{(p(1-p))^2}$ and $\frac{\partial^2 h}{\partial s\partial p}(p,s) = -(p(1-p))^{-1}$ hold, the matrix of the second derivatives of h is positive definite for $0<p<s<1$. Hence part a) follows. Parts b) and c) are obvious. □

Our proof of Theorem 1 is based on the following theorem

Theorem 2 (Plachky-Steinbach (1975)). Let W_n, $n = 1, 2,\ldots$, be a sequence of random variables such that

$$g_n(t) = \ln Ee^{tW_n} < \infty \qquad \text{for } t\in[0,T_1)$$

and

$$g(t) = \lim_{n\to\infty} g_n(t)/n \qquad \text{for } t\in[0,T_1).$$

If $g(t)$ is differentiable and strictly convex on $[0,T_1)$ then

$$\lim_{n\to\infty} P(W_n > ns_n)^{1/n} = \exp[-g^*(s)]$$

holds, where $s_n \to s = g'(t_0)$, $t_0 \in [0,T_1)$ and g^* is the conjugate of g on $[0,T_1)$. □

Let

$$r_n = \max\{r_n^+, r_n^-\}, \tag{2.1}$$

where

$$r_n^+ = \sup\{(F_n(x)-F(x))/(1+\alpha f(x)) \colon x \in R\}, \tag{2.2}$$

$$r_n^- = \sup\{(F(y)-{}_1F_n(y))/(1+\alpha f(y)) \colon y \in R\} \tag{2.3}$$

and ${}_1F_n$ stands for the left continuous version of F_n.

Theorem 3. We have

$$\lim_{n\to\infty} P(r_n > s_n)^{1/n} = \exp(-h_0(s)),$$

where h_0 is given by (1.7).

Proof. Let

$$g_n^+(x,t) = \ln E \exp\left[t \sum_{i=1}^{n} (1_{(-\infty,x]}(X_i)-F(x))/(1+\alpha f(x))\right].$$

From Lemma 1 it follows that

$$g^+(x,t) = \frac{1}{n} g_n^+(x,t) = g(F(x), t/(1+\alpha f(x))) - tF(x)/(1+\alpha f(x)),$$

where g given by (1.3) is strictly convex and differentiable for $t \geq 0$. The derivative $\frac{d}{dt} g^+(x,t)$ takes values in $[0,(1-F(x))/(1+\alpha f(x)))$. Thus, by Lemma 2 and Theorem 2 we get

$$\lim_{n\to\infty} P((F_n(x)-F(x))/(1+\alpha f(x)) > s_n)^{1/n} = \exp[-h(F(x), s(1+\alpha f(x))+F(x))], \tag{2.4}$$

where $s_n \to s$, $s \in (0,(1-F(x))/(1+\alpha f(x)))$.

Similarly, we obtain that

$$\lim_{n\to\infty} P((F(y)-{}_1F_n(y))/(1+\alpha f(y)) > s_n)^{1/n} = \exp[-h(1-F(y), s(1+\alpha f(y))+1-F(y))], \tag{2.5}$$

where $s_n \to s$, $s \in (0, F(y)/(1+\alpha f(y)))$.

Now let us fix $s \in (0,1)$. From (2.4), (2.5) and (1.5) - (1.7) it follows that for $s_n \to s$

$$\underline{\lim}\, P(r_n > s_n)^{1/n} \geq \exp[-h_0(s)]. \tag{2.6}$$

We now turn to the other inequality. Since $f(x) \geq 0$ and $\lim_{|x|\to\infty} f(x) = 0$, Lemma 3 and (1.4) imply that the infimum in (1.6) is attained for some $x \in R$. Moreover $h^+(\cdot)$ and $h^-(\cdot)$ are finite, continuous and increasing for $s \in (0,1)$ and $\lim_{s\to 0} h^+(s) = 0$, $\lim_{s\to 0} h^-(s) = 0$ hold.

Let N and C be integers such that $s > 3/N$ and $\sup\{\alpha f(x): x \in R\} \leq C$, respectively. The sets $F^{-1}\left[\frac{i}{N}, \frac{i+1}{N}\right]$, $i = 0,\ldots, N-1$ are intervals, whereas the sets $(\alpha f)^{-1}\left[\frac{j}{N}, \frac{j+1}{N}\right]$, $j = 0,\ldots, CN-1$ are unions of at most K intervals. Thus, there exist at most KCN^2 points $x_1,\ldots, x_m$, $m \leq KCN^2$, such that the intervals $I_0 = (-\infty, x_1]$, $I_j = [x_j, x_{j+1}]$, $j = 1,\ldots, m-1$, $I_m = [x_m, \infty)$ have the following properties

$$|F(x')-F(x'')| \leq 1/N \qquad \forall x', x'' \in I_j,\ j = 0,\ldots, m,$$

$$|\alpha f(x')-\alpha f(x'')| \leq 1/N \qquad \forall x', x'' \in I_j,\ j = 0,\ldots, m.$$

Theorem 2 implies that for $j = 0,\ldots, m-1$ we have

$$\overline{\lim_{n\to\infty}}\, P\left(\sup_{x\in I_j} \frac{F_n(x)-F(x)}{1+\alpha f(x)} \geq s_n\right)^{1/n} \leq$$

$$\leq \lim_n P\left(\frac{F_n(x_{j+1})-F(x_{j+1})}{1+\alpha f(x_{j+1})} \geq \frac{1+\alpha f(x_{j+1})-1/N}{1+\alpha f(x_{j+1})} s_n - \frac{F(x_{j+1})-F(x_j)}{1+\alpha f(x_{j+1})}\right)^{1/N}$$

$$= \exp[-h(F(x_{j+1}), s(1+\alpha f(x_{j+1}))+F(x_{j+1})-s/N-(F(x_{j+1})-F(x_j))]$$

$$\leq \exp[-h(F(x_{j+1}), s(1+\alpha f(x_{j+1}))+F(x_{j+1})-2/N],$$

where $F(x_0) = 0$. Moreover

$$P\left(\sup_{x\geq x_m} \frac{F_n(x)-F(x)}{1+\alpha f(x)} \geq s_n\right) \leq P\left(\frac{1}{N} \geq s_n\right) = 0 \tag{2.7}$$

for n sufficiently large. Let n_0 be such that for all $n > n_0$ (2.7) holds and for $j = 1,\ldots, m$

$$P\left(\sup_{x\in I_j} \frac{F_n(x)-F(x)}{1+\alpha f(x)} \geq s_n\right) \leq \exp\left[-nh\left(F(x_j), \left(s-\frac{3}{N}\right)(1+\alpha f(x_j))+F(x_j)\right)\right].$$

Thus

$$P\left(\sup_{x\in R} \frac{F_n(x)-F(x)}{1+\alpha f(x)} \geq s_n\right) \leq KCN^2\exp\left[-nh^+\left(s-\frac{3}{N}\right)\right]$$

for $n > n_0$. A similar inequality may be proved for $\sup_{y\in R} (F(y)-{}_1F_n(y))/(1+\alpha f(x))$ so that

$$P(r_n \geq s_n) \leq 2KCN^2\exp[-n\min\{h^+(s-3/N),h^-(s-3/N)\}$$

is valid for $n > n_1$. Hence

(2.8) $$\overline{\lim}\, P(r_n \geq s_n)^{1/n} \leq \exp[-\min\{h^+(s-3/N),h^-(s-3/N)\}]$$

for every $N > 3/s$. Since h^+ and h^- are finite and continuous for $s\in(0,1)$ we infer from (2.6) and (2.8) that

$$\lim P(r_n \geq s_n)^{1/n} = \exp[-h_0(s)]. \quad \Box$$

Proof of Theorem 1. Let $\varepsilon_n^+(x)$ and $\varepsilon_n^-(y)$ be defined by the following equalities

(2.9) $$F_n(x) = F(x+\alpha\varepsilon_n^+(x))+\varepsilon_n^+(x),$$

(2.10) $${}_1F_n(y) = F(y-\alpha\varepsilon_n^-(y))-\varepsilon_n^-(y),$$

respectively. Let

(2.11) $$\varepsilon_n^+ = \sup\{\varepsilon_n^+(x)\colon x\in R\}, \qquad \varepsilon_n^- = \sup\{\varepsilon_n^-(y)\colon y\in R\}.$$

Then

(2.12) $$L_n = \rho_\alpha(F_n,F) = \max\{\varepsilon_n^+,\varepsilon_n^-\}.$$

From Taylor expansions for F in (2.9) and (2.10) we obtain

(2.13) $$\varepsilon_n^+(x) = \frac{F_n(x)-F(x)}{1+\alpha f(x)} - \frac{\alpha^2}{2}\,\frac{f'(\tilde{x})}{1+\alpha f(x)}\,(\varepsilon_n^+(x))^2$$

and

(2.14) $$\varepsilon_n^-(y) = \frac{F(y)-{}_1F_n(y)}{1+\alpha f(y)} + \frac{\alpha^2}{2}\,\frac{f'(\tilde{y})}{1+\alpha f(y)}\,(\varepsilon_n^-(y))^2,$$

where $\tilde{x}$, $\tilde{y}$ are points between x, $x+\alpha\varepsilon_n^+(x)$ and y, $y-\alpha\varepsilon_n^-(y)$, respectively. By Condition A we have $|\alpha^2 f'(x)| \leq B$. Thus, relations (2.9) - - (2.14) imply that

(2.15) $$L_n/(1+BL_n) \leq r_n \leq L_n+BL_n^2.$$

Let $l(x) = x/(1+Bx)$. Inequalities (2.15) imply that

$$P(L_n \geq s_n) = P(l(L_n) \geq l(s_n)) \leq P\left(r_n \geq s_n - Bs_n^2\right)$$

and

$$P\left(r_n \geq s_n + Bs_n^2\right) \leq P\left(L_n + BL_n^2 \geq s_n + Bs_n^2\right) = P(L_n \geq s_n).$$

Thus Theorem 3 implies the assertion of Theorem 1 provided $s_n + Bs_n^2 \in (0,1)$ and $s_n - Bs_n^2 \in (0,1)$. These conditions are satisfied for n large enough while $s_n \to s$ and $s \in (0,1/B)$, $B > 2$. □

Proof of Corollary 1. Given s, $s > 0$, denote

$$Z_s = \{(p,a): (p,a) \in [0,1]\times[0,1],\ p+s \leq a \leq p+(C+1)s\},$$

$$Z_1 = Z^l_{s,p_1} = \{(p,a) \in Z_s: p \leq p_1\}, \quad Z_2 = Z^u_{s,s_1} = \{(p,a) \in Z_s;\ a \geq s_1\}.$$

From Lemma 3 it follows that function $k(p) = h(p,s+p)$ is strictly convex and finite on $(0,1-s]$. Moreover $\lim_{p\to 0} k'(p) = -\infty$ and $\lim_{p\to 1-s} k'(p) = +\infty$. Let us choose p_1 such that $p_1 > 0$ and $k'(p_1) < 0$. Then

$$h(p_1, s+p_1) \leq h(p,a) \tag{2.16}$$

for every $(p,a) \in Z^l_{s,p_1}$. Now, let $k_1(p) = h(p,1-p)$, $p \in (0,1/2)$. By Lemma 3 $k_1(\cdot)$ is decreasing for $p \in (0,1/2)$. We shall show that for s small enough

$$k_1\left(\frac{1}{2} - (C+1)s\right) < h(1-s,1) \tag{2.17}$$

holds. Indeed, for $s < \frac{1}{4}(C+1)^{-2}$ we have

$$k_1\left(\frac{1}{2} - (C+1)s\right) = (C+1)s \ln\frac{1+(C+1)s}{1-(C+1)s} < 4(C+1)^2 s^2 < s < -\ln(1-s) = h(1-s,1).$$

Thus there exists p_2 such that $k'(p_2) > 0$ and

$$k_1(1/2-(C+1)s) < k(p_2) \tag{2.18}$$

holds. Let $s_1 = p_2 + s$. Then

$$h(p_2, s_1) \leq h(p,a) \tag{2.19}$$

for every $(p,a) \in Z^u_{s,s_1}$. Finally let us note that set

$$Z = \{(p,a): p = F(x),\ a = s(1+\alpha f(x))+F(x),\ x \in R\}$$

has the following properties: $Z \subset Z_s$, $Z \cap Z_1 \neq \emptyset$, $Z \cap Z_2 \neq \emptyset$ and $Z \cap (Z_s \setminus (Z_1 \cup Z_2)) \neq \emptyset$. Thus, (2.16) - (2.19) imply that the infimum in

(1.5) is attained for some $x \in R$ such that $(F(x), s(1+\alpha f(x))+F(x)) \in \in Z_s \setminus (Z_1 \cup Z_2)$. However on $Z_s \setminus (Z_1 \cup Z_2)$ all derivatives of h are uniformly continuous and the Taylor expansion

$$h(F(x), s(1+\alpha f(x))+F(x)) = \frac{s^2}{2} (1+\alpha f(x))^2/(F(x)(1-F(x)))+o(s^2)$$

is valid uniformly. Hence

$$h^+(s) = \frac{1}{2} \gamma(F,\alpha)s^2+o(s^2)$$

where $\gamma(F,\alpha)$ is given by (1.8). Similarly we obtain the expansion for $h^-(s)$ with the same $\gamma(F,\alpha)$. This concludes the proof of Corollary 1. □

REFERENCES

Abrahamson I.G. (1967): Exact Bahadur efficiencies for the Kolmogorov--Smirnov and Kuiper one- and two-sample statistics. Ann. Math. Statist. 38, 1475-1490.

Bahadur R.R. (1971): Some Limit Theorems in Statistics. SIAM Philadelphia.

Chernoff H. (1952): A measure of asymptotic efficiency for tests of a hypothesis based on the sum of observations. Ann. Math. Statist. 23, 493-507.

Groeneboom P., Shorack G.R. (1981): Large deviations of goodness of fit statistics and linear combinations of order statistics. Ann. Probability 9, 971-987.

Kozek A. (1982): The efficiency of tests based on the Lévy and Kolmogorov distances. University of Cologne. Preprints in Statistics No 74.

Plachky D., Steinbach J. (1975): A theorem about probabilities of large deviations with an application to queuing theory. Period. Math. Hungar. 6, 343-345.

Rockafellar R.T. (1970): Convex Analysis. Princeton University Press. Princeton, New Jersey.

Institute of Computer Science
University of Wrocław
Przesmycki Str. 20
51-151 Wrocław, Poland

KNOWLEDGE PREPROCESSING FOR INTENSIONAL-TYPE EXPERT SYSTEMS

O. Kříž

Prague

Key words: Expert systems, decision-making, finite diagnostic problem

Expert systems (ES) are usually expected to deal with decision making under uncertainty. They should involve some aspects of artificial intelligence (e. g. graph theory), they imitate the way the experts take decision deliberately avoiding direct statistical approach. The basic mode of communication with an ES is dialogue at some sort of terminal. To promote confidence in the gadget among potential users explanatory mechanism is required. At the same time it helps to debug knowledge base (KB) to improve decision abilities. There is not a unanimous consensus about all the mentioned points and there are attempts to expel uncertainty and incorporate strictly logical systems. As classical prototypes of ES are mentioned MYCIN (Shortliffe) and PROSPECTOR. KB of such systems consists in hundreds of decision rules of form IF A THEN B (w) interpreted: if event A happens (with certain degree of confidence) we may infer existence of B with confidence weight W. As a good common model for propagation of uncertain knowledge in these "extensional" ES (extensional rules for combining weights) may serve algebraic theory suggested by Hájek (1985). Intensional ES INES proposed by Perez and Jiroušek (1985) belongs to the models with probabilistic background. Infering here takes place in accordance with theory and it has clear interpretation. The basic idea, due to Perez (1977), is to construct an "optimal" approximation of the hypothetical joint distribution from its alleged marginals. The reconstruction is performed "intensionally" obeying only certain optimality principle. The final decision is taken on the basis of specific restriction of this approximation.

1. INPUT TYPES FOR INTENSIONAL EXPERT SYSTEM INES

We have partially adopted traditional ways to express KB but weights have strictly probabilistic interpretation. This may enable experts to formulate the pieces of knowledge in a more user-friendly way and may ease up the take over of KB from other ES. At the same time some other input types support the direct statistical input. Irrespective of the input types it was submitted in, all knowledge is finally transformed to marginal distributions kept in internal structures of INES.

1. IF A (w_1) THEN B (w_2) w_1 stands for measure P(A) of the antecedent A and may be omitted, w_2 stands for conditional probability $P(B|A) = w_2$.
2. "k FROM N": $P(B|\bigcap_{i=1}^{k} A_{j_i}, j_i \in \langle 1,w \rangle) = f(k)$, $f(n) = 1$ f is monotonic stepwise function, probability depends on number of fulfilled antecedents. Used e. g. in CODEX (Popper (1984)).
3. logical approach used in CADIAG I (Adlassnig, Kolarz, Grabner) symptoms are given certain attributes with respect to disease d:

OP = obligatory + proves	OP <=> d
FP = facultative + proves	$\overline{FP} \Rightarrow \overline{d}$
ON = obligatory but does not prove	$\overline{ON} \Rightarrow \overline{d}$
FN = facultative but does not prove	mentioned
E = precludes	$E = \overline{d}$
- = unknown relation	unmentioned

4. Direct input of marginal distributions in form of vectors or matrices.
5. Direct input of conditional probabilities in form of vector or matrices.

2. FORMAL MODEL OF AN EXPERT SYSTEM

As shape of an ES is still a bit fuzzy, we shall try to bring in a more formal description - a model of an ES based on probability background that could formally host most features of existing systems. Let $\{S_i | i \in I, |I| < +\infty; i,j \in I: i \neq j \Rightarrow S_i \cap S_j = \emptyset\}$ is a finite system of disjoint sets (symptom domains), D (diseases, diagnoses) is finite set disjoint of any S_i.

$D = \{d_m | m \in M, |M| < +\infty; m \in M, i \in I: d_m \notin S_i\}$. Let $S = \mathsf{X}_{i \in I} S_i$, $\mathcal{S} = \mathsf{X}_{i \in I} \mathcal{S}_i$ are cartesian products of S_i and their respective σ-algebras $\mathcal{S}_i$. Let us suppose there exists some relation between $\mathcal{S}$ and 2^D ($R \subset \mathcal{S} \times 2^D$) and we wish to infere from the fact that certain object (patient or state of organism) $x \in R$ "belongs" to $S^* \subset \mathcal{S}$ ($\pi_1(x) \in S^*$, π_i applied on product is projector onto its i-th component) the corresponding subset D' of D ($\pi_2(x) \in D' \subset D$). We shall choose the standard approach where the link between $\mathcal{S}$ and D is established by supposed existence of measurable functions over the same abstract space with measure $(X,\mathcal{X},P) \xrightarrow{\xi_i} (S_i,\mathcal{S}_i)$ $i \in I$ $(X,\mathcal{X},P) \xrightarrow{\eta} (D,2^D)$ and we define $\xi : (X,\mathcal{X},P) \to (S,\mathcal{S})$ by $\xi = \xi_1 \times \xi_2 \ldots \xi_{|I|}$. If $\mathcal{F}(S)$ denotes system of all finite algebras of subsets of S and $A(K)$ denotes minimal set algebra generated by finite system K of subsets of S ($K \subset \mathcal{S}$, $|K| < +\infty$) we may formulate some desirable properties of relation R:

D is discernible by S iff $\exists_{S' \in F(S)} \forall_{A \in 2^D} \eta^{-1}(A) \in \xi^{-1}(S')$ (1)

or given a measure μ on $(X,\mathcal{X})$ and $\varepsilon \in R^+$, $\mu \ll P$

D is (μ,ε)*-discernible by* S iff

$$\exists_{S' \in F(S)} \mu(\{x \in X | \eta^{-1}(\eta(x)) \notin \xi^{-1}(S')\}) \le \varepsilon \quad (2)$$

or in a more specific way: given "risk function" $m: D \to R^+$, $\ell, \delta \in R^+$

D is (m,ℓ,δ)*-discernible by* S

$$\exists_{S' \in F(S)} P(\{x \in X | m(\eta(x)) > \ell \ \& \ \eta^{-1}(\eta(x)) \notin \xi^{-1}(S')\}) < \delta \quad (3)$$

To characterize ability of K to discern (if D is discernible by S) and denoting by $F_0(S)$ all finite algebras satisfying (1) we may define

D is discernible by K iff $\exists_{S_0 \in F_0(S)} S_0 \subset A(K)$ (4)

After requirements on a potential KB let us turn attention to decision procedures, supposing D is discernible by S and setting $\mathcal{B} = A(\xi^{-1}(K))$. Let us define a partial decision function $\rho: \mathcal{B} \to 2^D$.

We may again request for ρ some special properties:

1. Resultativness: ρ decides for each patient

   $\forall_{x \in X} \exists_{B \in \mathcal{B}} x \in B \ \& \ B \in \mathcal{D}(\rho)$ ($\mathcal{D}(\rho)$ is a definition domain of ρ)

2. Consistence: ρ decides about each patient in a unique way

   $\forall_{x \in X} \exists !_{A \in 2^D} \forall_{B \in \mathcal{B}} x \in B \ \& \ B \in \mathcal{D}(\rho) \Rightarrow \rho(B) = A.$

3. Correctness: consistent ρ decides about each patient correctly

$$\bigvee_{x \in X} \bigvee_{B \in \mathcal{B}} x \in B \ \& \ B \in \mathcal{D}(\rho) \Rightarrow \rho(B) = \eta(x)$$

Similarly we may define (μ,ε) or (m,ℓ,δ)-correctness.

Our asking about the symptom $S \in \mathcal{S}$ of an patient $x \in X$ is not free but restricted by finite set $\mathcal{T}$ of tests/questions T_k

$\mathcal{T} = \{T_k | k \in K, |K| +\infty, T_k: X \to \mathcal{R}_k(S) \subset \mathcal{S}\}$, $\mathcal{R}_k(S)$ is a partition of S according to $\mathcal{T}$ we may ask

4. $\mathcal{T}$-realizability of ρ: $\mathcal{T}$ is rich enough for ρ to decide correctly

$$\bigvee_{x \in X} \ \exists_{C \in \mathcal{D}(\rho)} \ \exists_{k < +\infty} \ \bigcap_{i=1}^{k} T_{ji}(x) \subseteq C \text{ for } T_{ji} \in \mathcal{T} \tag{4}$$

Let further $\phi: \mathcal{T} \to R^+$ is a function reflecting the cost of tests (or even more generally $\phi: \mathcal{T} \times X \to R^+$ may express even contraindication of some tests for certain type of patients). As we want the ES not only to decide but to ask the questions about patient, we define mapping $\psi: \mathcal{S} \setminus \xi(\mathcal{D}(\rho)) \to \mathcal{T}$ such that

$$\bigvee_{S' \in \mathcal{S} \setminus \xi(\mathcal{D}(\rho))} \ \bigvee_{x \in X} \ \exists_{k < +\infty} \ S_k^* \in \xi(\mathcal{D}(\rho)) \tag{5}$$

where S_k^* is defined recursively $S_k^* = \psi(\bigcap_{i=0}^{k-1} S_i^*)(x)$

$S_0^* = S'$, $\xi(x) \in S^*$, $i = 1,\ldots,k$ or we may require the "cheapest" ψ_0

$$\psi_0 = \arg\min_{\psi:(5)} \sum_{\ell=1}^{k} \phi(\psi(\bigcap_{i=0}^{\ell-1} S_i^*)) \tag{6}$$

Finite diagnostic problem may be then formulated as:
Given D find S, $\mathcal{S}$, K, $\mathcal{T}$, ϕ, ρ, ψ satisfying some of above mentioned requirements (1) - (6) and then (D, S, $\mathcal{S}$, K, $\mathcal{T}$, ϕ, ρ, ψ) stands for formal description of an ES.

Surprisingly enough decision function is almost never realized directly. Traditional approach in ES is to split ρ into two parts. The first part consists in numerization i. e. ascribing numbers (confidence weights) to different $B \in \mathcal{B}$ (or $K \in K \subset \mathcal{S}$). The grounds for numerization lie mainly in possibility to derive the weights for $K \in \mathcal{S} \setminus K$ via arithmetic operations and it is assumed explanatory mechanism and interpretation is more transparent. The second part, which is often taciturnly bypassed, when describing an ES, consists in a simple rule that maps numbers into diagnoses D (or 2^D). It is usually based on the natural ordering of real numbers.

We take decision $A \in 2^D$ with the greatest value of confidence weight or benefit from trespassing certain threshold etc. Not only weights and their combining but exact formulation of this rule must be explicitely mentioned when defining an ES. In probabilistic ES we follow the line of splitting $\rho = \rho^* \tau$. τ stands for conditional probability $\tau: \mathcal{B} \to \{\mu: 2^D \to \langle 0,1\rangle,\ \mu$ - prob. measure$\}$

$$B \in \mathcal{B} \to \mu_B(.) : A \in 2^D \to \mu_B(A) = P(\eta^{-1}(A) \cap B)$$

The way τ is constructed is the gist of ES INES and is described in Perez, Jiroušek (1985). The similar split holds for $\psi = \psi^* \tau$. Let us mention 2 possible examples of decision functions ρ^* and ψ^* where $p = (p_1, \ldots, p_{|2^D|})^T$.

$$\rho_1^*(p) = d_{i^*(p)} \quad \text{where} \quad i^*(p) = \arg\max_{i \in \langle 1..2^D\rangle} \{\pi_i(p)\}$$

$$\mathcal{D}(\rho_1^*) = \{p \mid I(p) < \varepsilon |D|\} \text{ where } I(p) = -\sum_{i=1}^{|2^D|} p_i \log p_i$$

is entropy of discrete distribution p

and $\psi_1^*(p,S') = \arg\max_{T_k \in \mathcal{T}} I[\eta; T_k | S']$ where conditional mutual information

$$I[\eta; T_k | S'] = \sum_{i=1}^{|2^D|} \sum_{jk}^{|R_k|} P(d_i, R_j^k, S') \log \frac{P(d_i | R_{ji}^k, S')}{P(d_i | S')}$$

or if Λ is diagonal matrix with "danger weights" for combinations of various diseases; $r, s \in R^+$

$$i^*(p) = \arg\max_{i \in \langle 1,..,|2^D|\rangle} \{\pi_i(\Lambda p)\} \qquad \rho_2^*(p,B) = d_{i^*(p)}$$

$$\mathcal{D}(\rho_2^*) = \{(p,B) \mid \pi_{i^*(p)}(\Lambda p) > r \quad \& \quad P(B) < s\}$$ where condition $P(B) < s$ protects against dominancy of high apriori probabilities of diagnoses $d \in 2^D$

$$\psi_2(p,S') = \arg\max_{T_k \in \mathcal{T}} \frac{1}{\phi(T_k)} \sum_{i=1}^{|2^D|} \lambda_i \sum_{j=1}^{|R_k|} P(d_i, R_j^k, S') \log \frac{P(d_i | R_j^k, S')}{p(d_i | S')}$$

where $\lambda_i = [\Lambda]_{ii}$, $P(d_i, R_j^k, S) = P(\eta^{-1}(d_i) \cap \xi^{-1}(R_j^k) \cap \xi^{-1}(S'))$.

Strong dependency of ρ^* and ψ^* on values of p justifies the assumption that invariants of knowledge preprocessing should be conditional probabilities supplied explicitly by experts. Having in view this "numerization split" we may understand under KB certain relation

$$KB \subset K \times \{S \vee 2^D \vee S \times 2^D\} \times \{R \vee R^+ \vee \langle 0,1\rangle\} \times \{\langle 0,1\rangle \vee \emptyset\}$$

where V is exclusive alternative and ρ, ψ may then represent inference mechanism.

3. KNOWLEDGE PREPROCESSING PROCEDURE

Intensional expert system INES accepts knowledge only in form of a set of distributions (supposed marginals of global P). Transition phase from a knowledge base submitted in input types 1. - 5. to input joint distributions involves *probabilization* (1.,2.,3.,5.), *completness* problem (e. g. one rule of type 1. is only single entry in a table and we need further rules or generate unspecified probabilities automatically) and *consistence* in sense of marginal problem. Sufficient conditions were given by Kellerer (1964), some new results are achieved by Studený (1986) and algorithm suggested by Jiroušek, Perez (1985) has the sufficient conditions built in intrinsically. Necessary conditions (weak consistence) may be defined in the following way:

For a given index set N let us suppose existence of sample spaces $(X_i, \mathcal{X}_i)$ for $i \in N$. For nonempty $I \subseteq N$ $(X_I, \mathcal{X}_I)$ denotes product space $(X_I, \mathcal{X}_I) = (\prod_{i \in I} X_i, \prod_{i \in I} \mathcal{X}_i)$. $\mathcal{A}$ denotes a given system of indices covering all $N: \mathcal{A} = \{I \mid I \subsetneq N\} : \cup\mathcal{A} = \bigcup_{I \in \mathcal{A}} I = N$ and $\mathcal{P}$ is a set of measures $\mathcal{P} = \{P_I \mid I \in \mathcal{A},\ P_I$ prob. measure on $(X_I, \mathcal{X}_I)\}$; $\mathcal{P}$ is *weakly consistent* iff

$$\bigvee_{I,J \in \mathcal{A}} \quad I \cap J \neq 0 \Rightarrow P_I^{I \cap J} = P_J^{I \cap J} \tag{7}$$

where for $K \subset I \subseteq N$ P_I^K is the restriction of P_I on subspace $(X_K, \mathcal{X}_K)$.

For brevity sake we drop formal description and give only outlines of preprocessing algorithm:

1. Knowledge base given by input types 1. - 5. is transformed to distributions or their fragments
2. All conflicting probabilities are replaced by their weighted average (weights are expressing our confidence in the source of knowledge)
3. All missing (undefined) probabilities are completed to comply with lower-order marginals
4. Not all input distributions are selected by algorithm reconstructing global distribution. We may either transform only selected ones or all inputs claiming better use of all knowledge. Weak consistence algorithm may be formulated as a mathematical programming problem. For better insight let us suppose all input distributions P_ℓ, $\ell = 1,\dots,L$ are tridimensional marginals (they may

be in fact of different dimensionality up to 10). I, J, K, Z denotes functions: $\langle 1,\dots,L\rangle \to N$.

Distribution P_ℓ is tridimensional matrix then

$$P_\ell = [[[a^\ell(I(\ell) = i,\ J(\ell) = j,\ K(\ell) = k)]_{i=1}^{|I(\ell)|}]_{j=1}^{|J(\ell)|}]_{k=1}^{|K(\ell)|}$$

general element will be abbreviated to a_{ijk} and primed symbols denote objects after the transformation.

Find set $\{P'_\ell,\ \ell = 1,\dots,L\}$ satisfying

$$\sum_{\ell=1}^{L} V_\ell \rho(P_\ell, P'_\ell) \to \min$$

where ρ stands for some metrics on distributions (e. g. I-divergence, total variation, X^α-divergence,...); V_ℓ are confidence weights for sources; subject to

$$\left.\begin{array}{l} 0 \le a'^{\ell}_{ijk} \le 1 \quad \ell = 1,\dots,L \\ \sum_{ijk} a'^{\ell}_{ijk} = 1 \end{array}\right\} \text{ distribution conditions}$$

$$\forall_{a'^{\ell}_{IJK}} \ \forall_{a'^{m}_{IJZ}} \quad \sum_{k=1}^{|K(\ell)|} a'^{\ell}_{IJk} = \sum_{z=1}^{|Z(m)|} a'^{m}_{IJz} \quad \} \text{ weak consistence}$$

where $I(\ell) = I(m)$ and $J(\ell) = J(m)$ and sums represent restrictions to twodimensional matrices

$$a'^{\ell}_{i_r j_r k_r} = a^{\ell}_{i_r j_r k_r} \quad \begin{array}{l} r = 1,\dots,R(\ell) \\ \ell = 1,\dots,L \end{array} \quad \} \text{ fixed probabilities}$$

$$\frac{a'^{\ell}_{i_q j_q k_q}}{a'^{m}_{i_q j_q}} = \frac{a^{\ell}_{i_q j_q k_q}}{a^{m}_{i_q j_q}} \quad \begin{array}{l} q = 1,\dots,Q(\ell) \\ \ell = 1,\dots,L \end{array} \quad \} \text{ fixed conditional probabilities}$$

To obtain an versatile mean for describing knowledge base K a compiler for a special language EXLAN covering all mentioned input types was coded (Kříž (1986)).

4. REFERENCES

Perez A. (1977): ε-admissible Simplifications of the Dependence Structure of a Set of Random Variables. Kybernetika - vol. 13, Number 6.

Perez A., Jiroušek R. (1985): Constructing an intensional expert system INES Medical Decision Making: Diagnostic Strategies and Expert Systems, J. H. van Bemel, F. Grémy and J. Zvárová (eds.) North Holland, Amsterdam, pp. 307-315.

Kellerer H. G. (1964): Masstheoretische Marginalprobleme. Math. Annalen 168-198.

Studený M. (1986): The existence problem for joint probability measure with given marginal measures. Research Report ÚTIA ČSAV, Prague.

Shortliffe E. H.: Computer-based medical consultations: Mycin, New York Elsevier/North Holland.

Popper M. (1984): Expert system CODEX - some characteristics and principles (in Slovak). In: Expert Systems: Principles, realization, applications, ČSVTS FEL ČVUT, Praha, pp. 119-143.

Adlassnig K.-P., Kolarz G., Grabner G. (Wien): CADIAG I: A computer assisted diagnostic system using symbolic logic.

Hájek P. (1985): Combining functions for certainty degrees in consulting systems. Int. J. Man-Machine Studies 22, 59-76.

Kříž O. (1986): EXLAN - a knowledge base language, Research Report ÚTIA ČSAV, Prague.

Czechoslovak Academy of Sciences
Institute of Information Theory and Automation
182 08 Prague 8
Pod vodárenskou věží 4
Czechoslovakia

SOME RESULTS AND COUNTEREXAMPLES
ON RANDOM FIXED POINTS

Anna Kucia, Andrzej Nowak

Katowice

Key words: Random operator, stochastic domain, random fixed point

ABSTRACT

We present two general random fixed point theorems for random operators with stochastic domains. Then we give some examples of continuous random operators with wide-sense fixed points and no random fixed points.

INTRODUCTION AND PRELIMINARIES

Throughout this paper (X,d) denotes a metric space, $(T,\mathcal{T})$ a measurable space, D a subset of $T\times X$ with nonempty t-sections

$$D_t=\{x: (t,x) \in D\},\ t \in T,$$

and F a mapping from D to X. A function $f:T\to X$ is a <u>random fixed point</u> of F if it is measurable and

(1) $$f(t) \in D_t \text{ and } F(t,f(t))=f(t)$$

for all $t \in T$.

The study of random fixed points was initiated by the Prague school of probabilists in the Fifties. Bharucha-Reid(1976) gave a survey of results in this area and put some questions. After his paper there has been a revived interest in random fixed point theorems and their applications (see e.g. Engl(1978), Itoh(1979), Nowak(1981)).

Define the set-valued mapping

(2) $$\varphi(t)=\{x\in D_t:\ F(t,x)=x\},\quad t\in T,$$
i.e. $\varphi(t)$ is the set of all fixed points of $F(t,\cdot)$ in D_t. Deterministic fixed point theorems give sufficient conditions for $\varphi(t)\neq\emptyset$. A function $f:T\to X$ is a selector of φ if $f(t)\in\varphi(t)$ for $t\in T$. Note that f is a random fixed point of F iff it is a measurable selector for φ.

In this paper we study the existence of random fixed points under assumption $\varphi(t)\neq\emptyset$, $t\in T$. Our proofs are based on measurable selection theorems. In order to state these theorems we need some additional notation and definitions.

A metric space X is called a Souslin space if it is a continuous image of a Polish space. By $\mathcal{B}_X$ we denote the Borel σ-field on X, and by $\mathcal{T}\times\mathcal{B}_X$ the product σ-field on $T\times X$. Let $\psi:T\to 2^X$ be a set-valued mapping; 2^X stands for the family of all nonempty subsets of X. The graph of ψ is defined as follows
$$\mathrm{Gr}\,\psi=\{(t,x):\ x\in\psi(t)\}.$$
We say ψ is measurable if for each open $G\subset X$, $\{t\in T:\psi(t)\cap G\neq\emptyset\}\in\mathcal{T}$. A Castaing representation of ψ is a countable family U of measurable selectors for ψ such that $\{u(t):\ u\in U\}$ is dense in $\psi(t)$ for $t\in T$.

We shall use the following measurable selection theorems:

Theorem 1 (Wagner (1977), Th. 4.2). Suppose X is separable and $\psi:T\to 2^X$ has complete values. Then ψ is measurable iff ψ has a Castaing representation.

Theorem 2 (Wagner (1977), Th. 5.10). Suppose X is a Souslin space, $\mathcal{T}$ is closed under the Souslin operation, and $\psi:T\to 2^X$ is a set-valued map with $\mathrm{Gr}\,\psi\in\mathcal{T}\times\mathcal{B}_X$. Then ψ has a Castaing representation.

We say $F:D\to X$ is product-measurable if for any open $G\subset X$, $F^{-1}(G)\in\mathcal{T}\times\mathcal{B}_X$, i.e. $D\in\mathcal{T}\times\mathcal{B}_X$ and F is measurable with respect to the trace σ-field $D\cap(\mathcal{T}\times\mathcal{B}_X)$. F is called a Carathéodory map if for each $t\in T$, $F(t,\cdot)$ is continuous on D_t, and for each $x\in X$ and each open $G\subset X$, $\{t:\ x\in D_t$ and $F(t,x)\in G\}\in\mathcal{T}$, i.e. $D^x=\{t:(t,x)\in D\}\in\mathcal{T}$ and $F(\cdot,x)$ is measurable with respect to the trace σ-field $D^x\cap\mathcal{T}$.

If F is a Carathéodory map and $D=T\times A$, where $A\in\mathcal{B}_X$, then F is product-measurable. It is not the case when the domain of $F(t,\cdot)$ depends on t, even if $D\in\mathcal{T}\times\mathcal{B}_X$ (see Example 1).

In the study of the existence of random fixed points we shall use the following condition introduced by Bocşan (1979):

(C) There exists a Castaing representation U for the set-valued map $t \to D_t$, $t \in T$, such that the function $t \to F(t,u(t))$ is measurable for each $u \in U$.

Remarks. 1. If $(T,\mathcal{T},P)$ is a probability space, then the random fixed point may be alternatively defined as a measurable mapping $f:T \to X$ such that the condition (1) holds almost surely. In this case the probabilistic terminology is justified.
2. A Carathéodory map $F:D \to X$ is called a continuous random operator with a stochastic domain D. The notion of the stochastic domain was introduced by Engl.
3. If $\mathcal{T}$ is complete, then it is closed under the Souslin operation.

EXISTENCE OF RANDOM FIXED POINTS

In this section we present two general random fixed point theorems and discuss their assumptions.

Theorem 3. Suppose X is a Souslin space, $\mathcal{T}$ is closed under the Souslin operation and $D \in \mathcal{T} \times \mathcal{B}_X$. If $F:D \to X$ is measurable and for each $t \in T$, $F(t,\cdot)$ has a fixed point in D_t, then F has a random fixed point. Moreover, the set-valued map φ defined by (2) has a Castaing representation.

Proof. We show that φ has a measurable graph. Observe that

$$\mathrm{Gr}\,\varphi = \{(t,x) \in D: F(t,x) = x\} = v^{-1}(0),$$

where $v(t,x) = d(x,F(t,x))$. Since F is measurable, v is also measurable and, consequently, $\mathrm{Gr}\,\varphi \in \mathcal{T} \times \mathcal{B}_X$. The application of Theorem 2 completes the proof.

Now we shall discuss the condition (C).

Proposition 1. Suppose X is separable, D has closed t-sections D_t, and $F(t,\cdot)$ is continuous on D_t, $t \in T$. If the condition (C) is satisfied then F is measurable.

Proof. Let A be a closed subset of X, and let $A_k = \{x: d(x,A) < 1/k\}$, $k \in N$. If U is such as in the condition (C), then

$$F^{-1}(A) = \bigcap_{k \in N} \bigcup_{u \in U} \Big[(\{t: F(t,u(t)) \in A_k\} \times X) \cap \{(t,x): d(x,u(t)) < 1/k\} \Big].$$

Because of the condition (C) and the measurability of the mapping

$(t,x)\to d(x,u(t))$, $F^{-1}(A)\in \mathcal{T}\times\mathcal{B}_X$. Hence, F is product-measurable.

Proposition 2 (cf. Bocşan (1979)). The condition (C) is satisfied in the following two cases:

(i) X is separable, the set-valued map $t\to D_t$ is measurable, D_t are complete for $t\in T$, and F is the restriction to D of a Carathéodory map on $T\times X$.

(ii) There exists a countable subset $E\subset X$ such that $E\cap D_t$ is dense in D_t for $t\in T$, $D^e\in\mathcal{T}$ and $F(\cdot,e)$ is measurable for $e\in E$.

Proof. (i) By the measurability of $t\to D_t$, $D\in\mathcal{T}\times\mathcal{B}_X$. If $u:T\to X$ is measurable and $u(t)\in D_t$, then $t\to F(t,u(t))$ is also measurable, F being the restriction to D of a product-measurable map. Now (C) follows by Theorem 1.

(ii) Since $D^x\in\mathcal{T}$, $x\in E$, for each $e\in E$ there exists a measurable map $p_e:T\to E$ such that $p_e(t)\in D_t$ for $t\in T$, and $p_e(t)=e$ iff $e\in D_t$. Hence, $\{p_e(t): e\in E\}=E\cap D_t$ (cf. Schäl (1974)). Because the range of p_e is countable, the map $t\to F(t,p_e(t))$ is measurable. It completes the proof.

Without any assumption on $\mathcal{T}$ in Theorem 3 we also obtain $\mathrm{Gr}\,\varphi\in\mathcal{T}\times\mathcal{B}_X$, but this condition does not imply the existence of a random fixed point (see Example 2). However in the most interesting cases we need not assume that $\mathcal{T}$ is closed under the Souslin operation.

Denote by (S) the following condition:

(S) For each $t\in T$, each sequence $\{x_n: n\in N\}\subset D_t$ with $d(x_n,F(t,x_n))$ converging to 0 has a convergent subsequence.

This condition was introduced by Sehgal and Waters (1983) for the case X is a Banach space and $D=T\times A$. Note that if $F(t,\cdot)$ is continuous and D_t is closed, then the condition (S) implies the compactness of $\varphi(t)$, $t\in T$.

The following proposition is a slight generalization of a result of Sehgal and Waters (1983), therefore we omit its proof.

Proposition 3. The condition (S) holds if one of the following assumptions is satisfied:

(i) For each $t\in T$, D_t is complete and there is $0<k_t<1$ such that

$$d(F(t,x),F(t,y))\leqslant k_t\cdot d(x,y),\quad x,\ y\in D_t.$$

(ii) For each $t \in T$, cl D_t is compact.
(iii) X is a Banach space, for each $t \in T$, D_t is closed and convex, $F(t,\cdot)$ is condensing and $F(t,D_t)$ is bounded.

The next theorem is a generalization of Th. 5 from Sehgal and Waters (1983) for random operators with stochastic domains.

Theorem 4. Let X be separable, $D \subset T \times X$ has closed t-sections D_t and $F:D \to X$. If, for each $t \in T$, $F(t,\cdot)$ is continuous and has a fixed point in D_t, and conditions (C) and (S) are satisfied, then F has a random fixed point. Moreover, the set-valued map φ defined by (2) has a Castaing representation.

Proof. Let U be such as in (C). Let $A \subset X$ be closed, and let $A_k = \{x: d(x,A_k) < 1/k\}$, $k \in N$. Then

$$(3) \quad \{t: \varphi(t) \cap A \neq \emptyset\} = \bigcap_{k \in N} \bigcup_{u \in U} (\{t: d(u(t),F(t,u(t))) < 1/k\} \cap u^{-1}(A_k)).$$

In fact, denote by L and R, respectively, left and right hand side of (3). Since $F(t,\cdot)$ is continuous and $\{u(t): u \in U\}$ is dense in D_t, we obtain $L \subset R$. Now for $t \in R$ let $u_k \in U$ be such that

$$d(u_k(t),F(t,u_k(t))) < 1/k \quad \text{and} \quad d(u_k(t),A) < 1/k.$$

By (S), $\{u_k(t): k \in N\}$ has a subsequence which converges to some $x \in D_t$. Hence, $x \in A$ and $d(x,F(t,x)) = 0$. It means that $x \in \varphi(t) \cap A$, i.e. $t \in L$.

The condition (C) implies the measurability of the mapping $t \to d(u(t),F(t,u(t)))$. Because of (3), $\{t: \varphi(t) \cap A \neq \emptyset\} \in \mathcal{T}$. Since any open subset of X is a countable union of closed sets, φ is measurable. The application of Theorem 1 completes the proof.

By use of Theorems 3 and 4, for each fixed point theorem we can obtain its random analogue. As an example we give a probabilistic version of the Schauder-Tychonoff theorem.

Corrolary. Suppose X is a locally convex metric space, $D \subset T \times X$ has compact convex t-sections D_t, and $F:D \to X$. If the condition (C) is satisfied, X is separable, and for each $t \in T$, $F(t,\cdot)$ is continuous and $F(t,D_t) \subset D_t$, then F has a random fixed point.

Proof. Apply Theorem 4.

Our Corrolary is a generalization of Th. 10 from Bharucha-

Reid (1976).

Remark 4. If all assumptions of Theorem 4 except (S) are fulfilled, then (3) holds for any compact $A \subset X$. So in this case we obtain $\mathrm{Gr}\,\varphi \in \mathcal{T} \times \mathcal{B}_X$ (cf. Prop. 1 and Th. 3) and $\{t: \varphi(t) \cap A \neq \emptyset\} \in \mathcal{T}$ for compact $A \subset X$. In particular, if X is σ-compact, then φ is measurable.

EXAMPLES

In this section we present two examples of Carathéodory maps F such that, for each $t \in T$, $F(t,\cdot)$ has a fixed point and there is no random fixed point.

Example 1. Let $T=X=[0,1]\times[0,1]$, and let $\mathcal{T}$ be the Lebesgue σ-field on T. Let $D=\{((t,s),(x,y)) \in T\times X : t=x\}$, and let $g:[0,1]\to[0,1]$ be not Lebesgue-measurable. Note $D \in \mathcal{T}\times\mathcal{B}_X$, D being a closed subset of $T\times X$. We define

$$F((t,s),(x,y)) = (t,g(s)), \quad ((t,s),(x,y)) \in D.$$

$F((t,s),\cdot)$ is constant, hence continuous. If $G \subset X$ is open and $(x,y) \in X$, then the set

$$\{(t,s): ((t,s),(x,y)) \in D \text{ and } F((t,s),(x,y)) \in G\} \subset \{(t,s): t=x\}$$

is negligible and, consequently, Lebesgue-measurable. Thus F is a Carathéodory map. For each $(t,s) \in T$, $F((t,s),\cdot)$ has a unique fixed point $(t,g(s))$ in $D_{(t,s)}$, and this fixed point does not depend measurably on (t,s). This example shows that in Theorem 3 we cannot weaken the product-measurability of F to the assumption that F is a Carathéodory map.

The second example proves that the assumption that $\mathcal{T}$ is closed under the Souslin operation is essential for Theorem 3. It is suggested by Example 4 of Himmelberg, Parthasarathy and Van Vleck (1981).

Example 2. Let $T=R$, $\mathcal{T}=\mathcal{B}_R$, $X=R\times Z$ and $D=T\times X$, where R is the real line and Z is the set of all irrationals. Let A_1, A_2 be analytic subsets of T such that $A_1 \cup A_2 = T$ and $T\setminus A_1$, $T\setminus A_2$ cannot be separated by Borel sets. Denote by Z_1 and Z_2, respectively, the sets of negative positive irrationals (both Z_1, Z_2 are clopen and homeomorphic to Z). Let h_i be a continuous map of Z_i onto A_i, and let $h(z)=h_i(z)$ for $z \in Z_i$, $i=1, 2$. Note that h is a continuous map of Z onto R. Define $F:D\to X$ as $F(t,(s,z))=(h(z)-t+s,z)$. It is immediate that F is continuous, and the set $\varphi(t)$ of all fixed points of $F(t,\cdot)$ is closed nonempty and consists of $(s,z) \in X$ such that $h(z)=t$. We shall prove that φ has no measurable selector.

On the contrary, suppose there is a measurable map $f:T \to X$ such that $f(t)=(s,z)$ and $h(z)=t$. Then $g=\pi_Z \circ f$, where π_Z is the projection from X to Z, is Borel-measurable. Note that if $g(t)=z$ then $h(z)=t$, i.e. $\mathrm{Gr}\, g \subset \mathrm{Gr}\, h$. Consequently, $T \setminus A_1$, $T \setminus A_2$ are separated by Borel sets $g^{-1}(Z_2)$, $g^{-1}(Z_1)$. This contradiction completes the proof that F has no random fixed point.

In connection with these examples the following question arises: Suppose $(T, \mathcal{T})$ is a measurable space, X is a Polish space, and $F:T \times X \to X$ is a Carathéodory map such that $F(t,\cdot)$ has a unique fixed point for each $t \in T$. Does this fixed point depend measurably on t ? The answer is "Yes" in situations described by Theorems 3 and 4, and if T is a Polish space and $\mathcal{T}=\mathcal{B}_T$. This last case is related to the following more general fact: If T and X are Polish, $\mathcal{T}=\mathcal{B}_T$, $D \subset T \times X$ has σ-compact t-sections D_t, and $F:D \to X$ is product-measurable and such that for each $t \in T$, $F(t,\cdot)$ is continuous and has a fixed point in D_t, then φ is measurable. The proof of this statement is based on the Kunugui-Arsenin theorem (cf. Remark 4).

REFERENCES

Bharucha-Reid A. (1976): Fixed point theorems in probabilistic analysis. Bull. Amer. Math. Soc. 82, No. 5, 641-657.

Bocşan G. (1979): Some properties of continuous random functions on random domains with applications to continuous extensions and random fixed point theorems. Semin. Teor. Functiilor Mat. Apl. A45.

Engl H. W. (1978): A general stochastic fixed point theorem for continuous random operators on stochastic domains. J. Math. Anal. Appl. 66, No. 1, 220-231.

Himmelberg C. J., Parthasarathy T. and Van Vleck F. S. (1981): On measurable relations. Fund. Math. 111, No. 2, 161-167.

Itoh S. (1979): Random fixed point theorems with an application to random differential equations in Banach spaces. J. Math. Anal. Appl. 67, No. 2, 261-273.

Nowak A. (1981): A note on random fixed point theorems. Prace Nauk.

Uniw. Śląsk. Katowice. Prace Mat. 11, 33-35.

Sehgal V. M. and Waters Ch. (1983): Some random fixed point theorems. In: Topological Methods in Nonlinear Functional Analysis, Contemp. Math. 21, AMS, 215-218.

Shäl M. (1974): A selection theorem for optimization problems. Arch. Math. 25, No. 2, 219-224.

Wagner D. H. (1977): Survey of measurable selection theorems. SIAM J. Control 15, No. 5, 859-903.

Institute of Mathematics
Silesian University
40-007 Katowice
Poland

ON THE OPTIMALITY OF (σ,S)-STRATEGIES IN AN INVENTORY MODEL WITH UNCOMPLETELY KNOWN DEMAND DISTRIBUTION

Heinz-Uwe Küenle

Leipzig

Key words: Multi-item inventory models, Markov games

ABSTRACT

It is considered a periodic rewiev multi-item inventory problem, in which the demand distribution is dependent on an unknown parameter. The problem is interpreted as a game against nature. In this sense the optimality of (σ,S)-strategies can be shown.

INTRODUCTION

The optimality of (σ,S)-strategies in a multi-product inventory problem was shown first by Johnson (1967) in the case of completely known discrete demand distribution using Howards improvement. Küenle (1976) carried over Johnson's method to models with continuous demand. Kalin (1980) has given an optimality proof for a model with general demand distribution.

Dynamic minimax inventory models was studied by Dietzsch (1977), Jagannathan (1978), Girlich and Küenle (1983) and Küenle (1986) for example.

In this paper a Markov game approach is used for the investigation of a multi-product inventory problem with uncompletely known demand distribution. Proofs of our results can be found in Küenle (1986).

THE INVENTORY PROBLEM

A warehouse is considered in which m different products are held. We denote by x_n^j the stock of product number j at the beginning of period number n, $x_n:=(x_n^1,\dots,x_n^m)\in\mathbb{R}^m$. After observing the state x_n the manager chooses an action, that means he orders quantities $z_n^{(j)}$ ($z_n^{(j)}\geq 0$) or he sells quantities $-z_n^j$ ($z_n^j<0$), respective. We suppose that the ordered quantities arrive the warehouse immediately after the order.

The demand in each period is a random variable $\hat{\xi}_n=(\hat{\xi}_n^1,\dots,\hat{\xi}_n^m)$ with a distribution function $\Psi^{(n)}$ from a parametric set $\{\Psi_b: b\in D\}$ of distribution functions. Unfilled demand is backlogged. Then, it holds

$$x_{n+1}=x_n+z_n-\xi_n\ ,$$

where ξ_n is a realisation of $\hat{\xi}_n$.

We consider in each period costs of the following types:

$l(x_n+z_n-\xi_n)$ - cost of stocking or shortage

K - fixed cost per order/selling

$c\cdot z_n=\sum_{i=1}^{n} c^i z_n^i$ - linear ordering cost or selling reward

$(c:=(c^1,\dots,c^m)$ with $c^i\geq 0)$.

For a moment, let D be a set which contains only one element, that means the distribution of the demand is the same in every period. The objective is then to find a reorder/selling-strategy which minimizes the expected disconted costs over an infinite horizon. For $m\geq 2$ we have then that problem which was studied first by Johnson (1967) in the case of discrete demands and without selling.

A MARKOV GAME WITH COMPLETE INFORMATION

Markov games in the sense of this paper are mathematical models for the following situation:
The state of a dynamic system is periodically controlled. In every control point the first player chooses an action a from an action set $\mathcal{A}(x)$, where x is the observed state. After first players choice the second one chooses an action from an action set $\mathcal{B}(x,a)$. Then the system moves to the state in the next control point according to a probability $q(\cdot/x,a,b)$, and costs $k(x,a,b)$ are assumed.

Let us consider the expected total discounted costs over infinitely many periods. The aim of the first player is then to minimize these costs, but the aim of the second player is to maximize they. (Hence, we have a two-person zero-sum game.) A mathematical model for such a situation is given by a tupel

$\mathcal{M} = (X, A, \mathcal{A}, B, \mathcal{B}, q, k, \alpha)$

where the symbols have the following meaning:

- X, A and B are Borel sets in polish spaces; we denote by $\mathfrak{X}$, $\mathbb{A}$, $\mathbb{B}$ the corresponding Borel σ-algebras.
  X is called state space,
  A action space of the first player, and
  B action space of the second player.
- $\mathcal{A}$ is a map $\mathcal{A}: X \to \mathbb{A}$ with the property, that $gr\mathcal{A} = \{(x,a): a \in \mathcal{A}(x)\}$ is a Borel set and contains the graph of an $\mathfrak{X}$-$\mathbb{A}$-measurable function.
- $\mathcal{B}$ is a map $\mathcal{B}: X \otimes A \to \mathbb{B}$ with the property, that $gr\mathcal{B} = \{(x,a,b): b \in \mathcal{B}(x,a)\}$ is a Borel set and contains the graph of an $\mathfrak{X} \otimes \mathbb{A}$-$\mathbb{B}$-measurable function.
  $\mathcal{A}(\cdot)$ and $\mathcal{B}(\cdot)$ are called action set of the first player or of the second player, resp.
- q is a transition probability from $(X \times A \times B, \mathfrak{X} \otimes \mathbb{A} \otimes \mathbb{B})$ to $(X, \mathfrak{X})$, called transition law.
- k is an $\mathfrak{X} \otimes \mathbb{A} \otimes \mathbb{B}$-measurable real function, called cost function.
- $\alpha \in [0,1)$ is called discount factor.

We consider only Markov deterministic strategies in this paper.

DEFINITION 1

An $\mathfrak{X}$-$\mathbb{A}$-measurable function e with the property $e(x) \in \mathcal{A}(x)$ for each $x \in X$ is called decision rule of the first player, an $\mathfrak{X} \otimes \mathbb{A}$-$\mathbb{B}$-measurable function f with the property $f(x,a) \in \mathcal{B}(x,a)$ for all $x \in X$, $a \in \mathcal{A}(x)$ is called decision rule of the second player. A sequence $\pi = (e_n)$ of decision rules of the first player or $\rho = (f_n)$ of decision rules of the second player is called strategy of the first or second player, resp. A strategy $\pi = (e_n)$ of the first player is called stationary, if $e_n \equiv e$ for all $n \in \mathbb{N}$. Such a strategy is also denoted by e^∞. The set of all strategies of the first or second player we denote by $\mathfrak{E}$ or $\mathfrak{F}$, resp.

If an initial state $x_0 \in X$, a strategy π of the first player and a strategy ρ of the second player are given, then it is possible to construct a suitable probability space by means of the Ionescu-Tulcea-theorem. In this paper the cost function k shall be bounded from below. Then the expected total discounted costs $V_{\pi\rho}(x_0)$ exist.

DEFINITION 2

A strategy π^* of the first player is called optimal, if

$$\inf_{\rho\in\mathcal{F}} V_{\pi^*\rho}(x_0) = \inf_{\pi\in\mathcal{C}} \sup_{\rho\in\mathcal{F}} V_{\pi\rho}(x_0) \text{ for every } x_0 \in X.$$

We put $V_\pi := \sup_{\rho\in\mathcal{F}} V_{\pi\rho}$ for all $\pi\in\mathcal{C}$.

We consider now the inventory problem described in the introduction as a special case of the Markov game with complete information. Let be $\bar{S} = (\bar{S}^1,\dots,\bar{S}^m)$ the vector of the capacities of the several products in the inventory. Then we put

$$X:=(-\infty,\bar{S}].$$

An action of the first player (which is the manager) may be the stock after order/selling. With the notation of the introduction we have then $a_n=x_n+z_n$ for the action in period number n. Let be

$\bar{Z} = (\bar{Z}^1,\dots,\bar{Z}^m)$ the vector of the maximal selling quantities. Then we have

$$\mathcal{A}(x)=[x-\bar{Z},\bar{S}] \text{ for every } x\in X \text{ and } A=X=(-\infty,\bar{S}].$$

The action of the second player (nature) is the parameter of the demand distribution. Hence we have

$$B=\mathcal{B}(x,a)=D \text{ for all } x\in X,\ a\in A.$$

The transition law q is given by

$$q([y,\infty)/x,a,b) = \Psi_b(a-y) \text{ for each } x,a,y\in X,\ b\in B.$$

We put $L(y,b) := \int_{\mathbb{R}^m} l(y-\xi)\,d\Psi_b(\xi)$ and

$$k(x,a,b):=K\delta(a-x)+c\cdot(a-x)+L(a,b) \text{ for each } x,a\in X,\ b\in B$$

$$(\delta(y)=1 \text{ for } y\neq 0,\ \delta(0)=0).$$

The Markov game $\mathcal{M}$ with this special assumptions is called minimax inventory model.

We assume $\sup_{b\in B} L(a,b)<\infty$ for each $a\in A$. Let be

$$\mu_b^i:= \int_{\mathbb{R}^m} \chi^i(\xi)\,d\Psi_b(\xi) \text{ with } \chi^i(\xi^1,\dots,\xi^i,\dots,\xi^m)=\xi^i \text{ for}$$

$$\text{each } (\xi^1,\dots,\xi^m)\in\mathbb{R}^m \text{ and } i\in\{1,\dots,m\}.$$

$$\mu_b := (\mu_b^1, \ldots, \mu_b^m),$$

$$L'(x,b) := L(x,b) + (1-\alpha)c\cdot x + \alpha c\cdot \mu_b,$$

$$L'(x) := \sup_{b\in B} L'(x,b) \quad \text{for all } x\in X.$$

For the rest of this paper we propose, that the following assumption is fulfilled.

ASSUMPTION 1

(a) The distribution function Ψ_b has a density ψ_b (with respect to the Lebesgue measure λ in $\mathbb{R}^m$), which is bounded with the same bound for all b,
and it holds

$$\mu^i := \max_{b\in B} \mu_b^i < \infty .$$

(b) l is uniformly continuous.

(c) There exists an $x^0 \leq S$ such that

$$\bar{L}'(x) \geq L'(x^0) \quad \text{for all } x \leq \bar{S}$$

and

$$\bar{L}'(x) \geq \bar{L}'(x^0) + K + \alpha \max\{0, c\cdot(x^0 - x)\} \quad \text{for all } x \in \bar{S} - \bar{Z} =: \underline{s},\ x \in X.$$

THE OPTIMALITY OF (σ,S)-STRATEGIES

We direct our attention now to the class of strategies given in the following definition.

DEFINITION 3

A stationary strategy e^∞ is called (σ,S)-strategy, if

$$e(x) = \begin{cases} S & \text{for } x\in\sigma \\ x & \text{for } x \in X\setminus\sigma \end{cases}$$

where $\sigma \in \mathfrak{X}$ and $S \in X\setminus\sigma$.

We denote e^∞ by $\langle\sigma,S\rangle$ and put $\bar{\sigma} := X\setminus\sigma$

In the following one it is $J := [\underline{s}, \bar{S}]$ and $\bar{L}(x) = \sup_{b\in B} L(x,b)$ for each $x\in X$, and we consider only the case "c=0". But the case "c$\geq$0" can be reduced to that case (with L' for L) by the transformation

$$V'_{\langle\sigma,S\rangle}(x) := V_{\langle\sigma,S\rangle}(x) + c\cdot x \quad \text{for each } x\in X.$$

The next theorem bases on results for Markov games with complete information which can be found in Küenle (1986).

THEOREM 1

Let be $\bar{\sigma} \subseteq J$, $S \in \bar{\sigma}$.

(a) $V_{\langle\sigma,S\rangle}$ is the unique solution of the functional equation

$$v(x)=\sup_{b\in B}\left\{L_b(x)+\alpha\int_X v(\xi)\,\psi_b(x-\xi)\lambda(d\xi)\right\} \quad \text{for } x\in\bar{\sigma}$$

$$v(x)=v(S)+K \quad \text{for } x\in\sigma$$

(b) If

$V_{\langle\sigma,S\rangle}(x)\le K+V_{\langle\sigma,S\rangle}(y)$ for all $x\in X$, $y\in\bar{\sigma}$,

$V_{\langle\sigma,S\rangle}(x)\le \sup_{b\in B}\left\{L_b(x)+\alpha\int_X V_{\langle\sigma,S\rangle}(\xi)\,\psi_B(x-\xi)\lambda(d\xi)\right\}$ for each $x\in\sigma$,

then $\langle\sigma,S\rangle$ is optimal.

(c) If $V_{\langle\sigma,S\rangle}(S')<V_{\langle\sigma,S\rangle}(S)$,

then $V_{\langle\sigma,S'\rangle}\lneq V_{\langle\sigma,S\rangle}$, $V_{\langle\sigma,S'\rangle}(x)<V_{\langle\sigma,S\rangle}(x)$ for $x\in\sigma$.

(d) If $\{x:V_{\langle\sigma,S\rangle}(x)>K+V_{\langle\sigma,S\rangle}(S)\}=:M\neq\emptyset$, then for a strategy $\langle\sigma',S'\rangle$ with the property $\sigma\subseteq\sigma'\subseteq\{x:V_{\langle\sigma,S\rangle}(x)\geq K+V_{\langle\sigma,S\rangle}(S)\}$ it holds $V_{\langle\sigma',S\rangle}\lneq V_{\langle\sigma,S\rangle}$, $V_{\langle\sigma',S\rangle}(x)<V_{\langle\sigma,S\rangle}(x)$ for $x\in\sigma'\cap M$.

We consider now a special class of (σ,S)-strategies, in which the sets σ are defined by a functional equation.

We put $r_0:=L(x^0)+(1-\alpha)K$

$r^0:=L(x^0)+K$.

For every real function u on X we put

$\sigma(u):=\{x:x\in X,\ u(x)>0\}$ and

$\bar{\sigma}(u):=X\setminus\sigma(u)$.

THEOREM 2

For every $r\in[r_0,r^0]$ there is one and only one $\mathfrak{X}$-measurable function $u=u_r$, bounded on J, which is a solution of the functional equation

$$u(x)=\sup_{b\in B}\left\{L_b(x)-r+\alpha\int_{\sigma(u)} u(\xi)\,\psi_b(x-\xi)\lambda(d\xi)\right\} \quad \text{for all } x\in X.$$

In the next remark we state some properties of u_r.

REMARK

Let be $r\in[r_0,r^0]$

(a) If $\bar{L}(x)\le r$ then it follows $x\in\bar{\sigma}_r$

(b) If $\bar{L}(y)>r$ for every $y\le x$, then it follows $x\in\sigma_r$.

Using Theorem 1 (c) and (d) and some continuity properties of $V_{\langle\sigma,S\rangle}$ for $\sigma=\sigma(u_r)$ it is possible to construct sequences (r_n) and (S_n) with $r_n\in[r_0,r^0]$, $S_n\in\sigma(u_{r_n})$, such that

$\lim r_n=r^*$ and $\lim S_n=S^*$ exist and the (σ,S)-strategy with $\sigma=\sigma_r$, $S=S^*$ fulfils Theorem 1(b), and hence it is optimal.

REFERENCES

Dietzsch V. (1977): Dynamische Minimax-Entscheidungsmodelle. Diss.(A), Karl-Marx-Universität Leipzig.

Girlich H.-J. and H.-U. Küenle (1983): On dynamic min-max-decision models. Trans. of the Ninth Prague Conference, 1982, Academia, 257-262.

Jagannathan R. (1978): A minimax ordering policy for the infinite stage dynamic inventory problem. Management Science 24, 1138-1149.

Johnson E.L. (1967): Optimality and computation of (σ,S) policies in the multi-item infinite horizon inventory problem. Management Science 13, 475-491.

Kalin D. (1980): On the optimality of (σ,S) policies. Math. Oper. Res. 5, 293-307.

Küenle H.-U. (1976): Markovsche Entscheidungsmodelle mit allgemeinem Zustands- und Aktionsraum und ihre Anwendung in der Lagerhaltungstheorie. Diss.(A), Karl-Marx-Universität Leipzig.

(1986): Stochastische Spiele und Entscheidungsmodelle. B.S.B. Teubner Verlagsgesellschaft, Leipzig (in preparation)

Karl-Marx-Universität Leipzig
Sektion Mathematik
DDR-7010 Leipzig
Karl-Marx-Platz 10
G.D.R.

4. Using Theorem 3 (c) and (d) and some continuity properties of $V_{(\sigma,S)}$ for $\sigma = \sigma(u_r)$ it is possible to construct sequences (r_n) and (S_n) with $r_n \in [r_0, r^0]$, $S_n \in Q(u_{r_n})$, such that $\lim r_n = r^*$ and $\lim S_n = S^*$ exists and the (σ, S)-strategy with $\sigma = \sigma_{r^*}$, $S = S^*$ fulfils Theorem 4(b), and hence it is optimal.

References

[illegible]

[illegible] Spieltheorie und [illegible] B.G. Teubner Verlagsgesellschaft, Leipzig (in preparation)

Karl-Marx-Universität Leipzig
Sektion Mathematik
DDR-7010 Leipzig
Karl-Marx-Platz 10
G.D.R.

ON A DERIVATION OF PROBABILITY DENSITY FUNCTIONS. THE PDF OF GREENWOOD'S STATISTIC.

Murat Kumganbayev, Vasilij Voinov

Alma-Ata

Key words: Probability density functions, Greenwood's statistic

ABSTRACT

An expression for limits of integration or summation useful in deriving a probability density function (pdf) is considered. A way of deriving the pdf of Greenwood's statistic for arbitrary number of spacings has been given.

Let $X_1,\dots,X_n$ be random variables the joint pdf of which is $f(x_1,\dots,x_n)$. Consider a statistic $U=F(X_1,\dots,X_n)$. Its pdf can be written as

$$(1)\quad q(u)=\sum_{i=1}^{m} \int\limits_{x_1\in Q(u)} \cdots\cdots \int\limits_{x_{n-1}\in Q(x_1,\dots,x_{n-2},u)} \left|\frac{dx_n}{du}\right|\Bigg|_{x_n=x_{ni}} p(x_1,\dots,x_{n-1},u)\,dx_1\dots dx_{n-1},$$

where $p(x_1,\dots,x_{n-1},u)=f(x_1,\dots,x_{n-1},x_n(x_1,\dots,x_{n-1},u))$. The summation is performed over all singlevalued branches of solutions $x_n=x_{ni}(x_1,\dots,x_{n-1},u)$, $i=1,2,\dots,m$, of equation $u=F(x_1,\dots,x_n)$ and

$$(2)\quad \begin{aligned} Q(u)&=W[x_1]\cap W[F^{-1}(u)],\\ Q(x_1,u)&=W[x_2]\cap W[F^{-1}(x_1,u)],\\ &\dots\dots\dots\dots\dots\dots\dots\dots\\ Q(x_1,\dots,x_{n-2},u)&=W[x_{n-1}]\cap W[F^{-1}(x_1,\dots,x_{n-2},u)]. \end{aligned}$$

$W[x]$ means a set of values of x, $F^{-1}(u)$ is the solution of $u=F(x_1,\ldots,x_n)$ with respect to x_1, u being fixed, $F^{-1}(x_1,u)$ is that of $u=F(x_1,\ldots,x_n)$ with respect to x_2, u and x_1 being fixed, etc. The representation (2) proves to be very useful in deriving a pdf.

Consider for example the well-known Greenwood's statistic

$$(3) \qquad U=\sum_{i=0}^{n}\left[X_{(i+1)}-X_{(i)}\right]^2, X_{(0)}=0,\ X_{(n+1)}=b,$$

where $X_{(i)}$ (i=1,2,...,n) are order statistics, X_i being points taken at random in (0,b). In the sequel we shall denote values of $X_{(1)},X_{(2)},\ldots,X_{(n)}$ by $x_1,x_2,\ldots,x_n$.

This statistic was proposed by Greenwood (1946) for goodness-of-fit tests of uniformity on the unit interval. Many authors have been studied the pdf of (3) and its percentage points (Moran (1947),(1953),(1981), Gardner (1952), Burrows (1979), Currie (1981), Stephens (1981), Does, Helmers and Klaassen (1984)). However, the exact pdf of U is not known for n>3. We shall proceed now to show how one can write down the pdf of U for any n.

In accordance with (1) the pdf q(u) of U is

$$(4) \quad q(u)=\frac{n!}{b^n}\int\limits_{x_1\in Q(u)}\cdots\int\limits_{x_{n-1}\in Q(x_1,\ldots,x_{n-2},u)}\frac{\sqrt{3}\,dx_1\ldots dx_{n-1}}{\sqrt{6u_{n-1}-2(b-x_{n-2})^2-(3x_{n-1}-b-2x_{n-2})^2}}$$

where $u_{k+1}=u_k-(x_k-x_{k-1})^2$, $u_1=u$, $x_0=0$, $k\geq 1$. To derive Q(u) note that

$$F^{-1}(u)=\frac{1}{2}\left(x_2\pm\sqrt{2u-x_2^2-2(x_3-x_2)^2-\ldots-2(b-x_n)^2}\right).$$

This function possesses two branches - $F_1^{-1}(u)$ if sign is plus and $F_2^{-1}(u)$ if that is minus. Evidently, the maximum of $F^{-1}(u)$ coincides with that of $F_1^{-1}(u)$. To find it note that a value of $\sqrt{2u-x_2^2-2(x_3-x_2)^2-\ldots-2(b-x_n)^2}$ decreases with increasing of a value of $2(x_3-x_2)^2+\ldots+2(b-x_n)^2$ the minimum of which under the condition $x_2\leq x_3\leq x_4\leq\ldots\leq x_n=b$ can be shown to be

$$\min\left[2(x_3-x_2)^2+\ldots+2(b-x_n)^2\right]=\frac{2(b-x_2)^2}{n-1}.$$

Hence, the maximum of $F_1^{-1}(u)$ coincides with that of the expression

$$\frac{1}{2}(x_2+\sqrt{2u-x_2^2-\frac{2(b-x_2)^2}{n-1}})$$

and is, evidently, equal to

$$\max\left[F^{-1}(u)\right]=\frac{b+\sqrt{n(n+1)u-nb^2}}{n+1}.$$

The same consideration shows that

$$\min\left[F^{-1}(u)\right]=\frac{b-\sqrt{n(n+1)u-nb^2}}{n+1}.$$

Using the fact that $\frac{b-\sqrt{n(n+1)u-nb^2}}{n+1}\geqslant 0$, if $u\leqslant\frac{b^2}{n}$, and $\frac{b-\sqrt{n(n+1)u-nb^2}}{n+1}\leqslant 0$, if $u\geqslant\frac{b^2}{n}$, and that $Q(u)=W[x_1]\cap W\left[F^{-1}(u)\right]$, we find

$$(5)\quad Q(u)=\begin{cases}\left[\frac{b-\sqrt{n(n+1)u-nb^2}}{n+1},\ \frac{b+\sqrt{n(n+1)u-nb^2}}{n+1}\right], & \frac{b^2}{n+1}\leqslant u<\frac{b^2}{n};\\ \left[0,\frac{b+\sqrt{n(n+1)u-nb^2}}{n+1}\right], & \frac{b^2}{n}\leqslant u<\frac{b^2}{2};\\ \left[0,\frac{b-\sqrt{2u-b^2}}{2}\right]\cup\left[\frac{b+\sqrt{2u-b^2}}{2},\ \frac{b+\sqrt{n(n+1)u-nb^2}}{n+1}\right], & \frac{b^2}{2}\leqslant u\leqslant b^2.\end{cases}$$

To find $Q(x_1,u)$ we use the function $F^{-1}(x_1,u)=\frac{1}{2}(x_1+x_3\pm\sqrt{2u-2x_1^2-(x_3-x_1)^2-2(x_4-x_3)^2-\ldots-2(b-x_n)^2})$, x_1 being fixed. Change the variables by formulae $x_3-x_1=x_3'$, $x_4-x_1=x_4',\ldots,x_n-x_1=x_n'$, $u-x_1^2=u_2$ and $b-x_1=b_2$ we get

$$F^{-1}(x_1,u)=x_1+\frac{1}{2}(x_3'\pm\sqrt{2u_2-x_3'^2-2(x_4'-x_3')^2-\ldots-2(b_2-x_n')^2}).$$

It is seen that $F^{-1}(x_1,u)-x_1$ possesses the same form as $F^{-1}(u)$ does, with the exclusion that the number of variables under the radical sign is less by one. Generalizing this conclusion we see that a set $Q(x_1,\ldots,x_k,u)$, $k=1,2,\ldots,n-2$, of values of x_{k+1} is

$$(6)\quad Q(x_1,\dots,x_k,u)=\begin{cases} B_k=\left[x_k+\dfrac{b_{k+1}-\sqrt{A_{k,n}}}{n-k+1},\ x_k+\dfrac{b_{k+1}+\sqrt{A_{k,n}}}{n-k+1}\right], \\ \qquad\qquad \text{a) } \dfrac{b_{k+1}^2}{n-k+1}\leqslant u_{k+1}\leqslant\dfrac{b_{k+1}^2}{n-k}; \\ C_k=\left[x_k,x_k+\dfrac{b_{k+1}+\sqrt{A_{k,n}}}{n-k+1}\right], \\ \qquad\qquad \text{b) } \dfrac{b_{k+1}^2}{n-k}<u_{k+1}\leqslant\dfrac{b_{k+1}^2}{2}; \\ D_k=\left[x_k,x_k+\dfrac{b_{k+1}-\sqrt{2u_{k+1}-b_{k+1}^2}}{2}\right]\cup \\ \cup\left[x_k+\dfrac{b_{k+1}+\sqrt{2u_{k+1}-b_{k+1}^2}}{2},x_k+\dfrac{b_{k+1}+\sqrt{A_{k,n}}}{n-k+1}\right], \\ \qquad\qquad \text{c) } \dfrac{b_{k+1}^2}{2}<u_{k+1}\leqslant b_{k+1}^2. \end{cases}$$

where $A_{k,n}=(n-k+1)(n-k)u_{k+1}-(n-k)b_{k+1}^2$, $b_{k+1}=b_k-(x_k-x_{k-1})$, $k\geqslant 1$, $b_1=b$. If $k=0$, then $Q(x_1,\dots,x_k,u)=Q(u)$ (see (5)). Performing the integration with respect to x_{k+1} we consider x_k as being fixed. In view of this the set $Q(x_1,\dots,x_{k-1},u)$ can be divided into subsets so that in each of them only one of the conditions a), b) or c) is true.

If $u_k\leqslant\dfrac{b_k^2}{n-k+1}$ then

$$x_k\in B_{k-1}\equiv\left[x_{k-1}+\frac{b_k-\sqrt{A_{k-1,n}}}{n-k+2},\ x_{k-1}+\frac{b_k+\sqrt{A_{k-1,n}}}{n-k+2}\right].$$

It is not difficult to show that in this case only condition a) is valid and

$$(7)\qquad x_{k+1}\in B_k\equiv\left[x_k+\frac{b_{k+1}-\sqrt{A_{k,n}}}{n-k+1},\ x_k+\frac{b_{k+1}+\sqrt{A_{k,n}}}{n-k+1}\right].$$

If $\dfrac{b_k^2}{n-k+1}<u_k\leqslant\dfrac{b_k^2}{2}$ then $x_k\in C_{k-1}$. This interval is subdivided into subsets

$$(8)\quad\begin{cases}\left[x_{k-1},x_{k-1}+\dfrac{b_k-\sqrt{A_{k-1,n-1}}}{n-k+1}\right]\cup\left[x_{k-1}+\dfrac{b_k+\sqrt{A_{k-1,n-1}}}{n-k+1},\ x_{k-1}+\right. \\ \left.+\dfrac{b_k+\sqrt{A_{k-1,n}}}{n-k+2}\right]\ \text{for } u_{k+1}\leqslant\dfrac{b_{k+1}^2}{n-k}\text{, where } x_{k+1}\in B_k;\end{cases}$$

$$\left\{\begin{array}{l}\left[x_{k-1}+\frac{b_k-\sqrt{A_{k-1,n-1}}}{n-k+1},\, x_{k-1}+\frac{b_k-\sqrt{6u_k-2b_k^2}}{3}\right]\cup\left[x_{k-1}+\frac{b_k+\sqrt{6u_k-2b_k^2}}{3},\right.\\ \left. x_{k-1}+\frac{b_k+\sqrt{A_{k-1,n-1}}}{n-k+1}\right] \text{ for } \frac{b_{k+1}^2}{n-k}<u_{k+1}\leqslant\frac{b_{k+1}^2}{2}, \text{ where } x_{k+1}\in C_k \text{ and}\end{array}\right. \tag{9}$$

$$\left\{\begin{array}{l}\left[x_{k-1}+\frac{b_k-\sqrt{6u_k-2b_k^2}}{3},\, x_{k-1}+\frac{b_k+\sqrt{6u_k-2b_k^2}}{3}\right] \text{ for } u_{k+1}>\frac{b_{k+1}^2}{2},\\ \text{where } x_{k+1}\in D_k.\end{array}\right. \tag{10}$$

If $u_k>\frac{b_k^2}{2}$ then $x_k\in D_{k-1}$. This set is subdivided into subsets

$$\left\{\begin{array}{l}\left[x_{k-1},\, x_{k-1}+\frac{b_k-\sqrt{A_{k-1,n-1}}}{n-k+1}\right]\cup\left[x_{k-1}+\frac{b_k+\sqrt{A_{k-1,n-1}}}{n-k+1},\, x_{k-1}+\right.\\ \left.+\frac{b_k+\sqrt{A_{k-1,n}}}{n-k+2}\right] \text{ for } u_{k+1}\leqslant\frac{b_{k+1}^2}{n-k}, \text{ where } x_{k+1}\in B_k;\end{array}\right. \tag{11}$$

$$\left\{\begin{array}{l}\left[x_{k-1}+\frac{b_k-\sqrt{A_{k-1,n-1}}}{n-k+1},\, x_{k-1}+\frac{b_k-\sqrt{6u_k-2b_k^2}}{3}\right]\cup\left[x_{k-1}+\frac{b_k+\sqrt{6u_k-2b_k^2}}{3},\right.\\ \left. x_{k-1}+\frac{b_k+\sqrt{A_{k-1,n-1}}}{n-k+1}\right] \text{ for } \frac{b_{k+1}^2}{n-k}<u_{k+1}\leqslant\frac{b_{k+1}^2}{2}, \text{ where } x_{k+1}\in C_k \text{ and}\end{array}\right. \tag{12}$$

$$\left\{\begin{array}{l}\left[x_{k-1}+\frac{b_k-\sqrt{6u_k-2b_k^2}}{3},\, x_{k-1}+\frac{b_k-\sqrt{2u_k-b_k^2}}{2}\right]\cup\left[x_{k-1}+\frac{b_k+\sqrt{2u_k-b_k^2}}{2},\right.\\ \left. x_{k-1}+\frac{b_k+\sqrt{6u_k-2b_k^2}}{3}\right] \text{ for } u_{k+1}>\frac{b_{k+1}^2}{2}, \text{ where } x_{k+1}\in D_k.\end{array}\right. \tag{13}$$

To illustrate the utilization of these formulae consider for example the case n=3. In accordance with (4)

$$q(u)=\frac{6}{b^3}\int\limits_{x_1\in Q(u)}\int\limits_{x_2\in Q(x_1,u)} Q(x_1,x_2)\,dx_1dx_2,$$

where $Q(x_1,x_2)=\sqrt{3}/\sqrt{6u_2-2b_2^2-(3x_2-3x_1-b_2)^2}$. From (6) we obtain three nonintersecting sets of values of u: a)$b^2/4\leqslant u\leqslant b^2/3$, b)$b^2/3<u\leqslant b^2/2$ and c)$b^2/2<u\leqslant b^2$. If $b^2/4\leqslant u\leqslant b^2/3$ then $x_1\in B_0\equiv\left[\frac{b-\sqrt{12u-3b^2}}{4},\, \frac{b+\sqrt{12u-3b^2}}{4}\right]$ and in view of (7) $x_2\in B_1\equiv$ $\equiv\left[x_1+\frac{b_2-\sqrt{6u_2-2b_2^2}}{3},\, x_1+\frac{b_2+\sqrt{6u_2-2b_2^2}}{3}\right]$.

If $b^2/3<u\leqslant b^2/2$ then $x_1\in C_0\equiv\left[0,\frac{b+\sqrt{12u-3b^2}}{4}\right]$. This interval is sub-

divided into subsets

$$\left[0,\frac{b-\sqrt{6u-2b^2}}{3}\right]\cup\left[\frac{b+\sqrt{6u-2b^2}}{3},\frac{b+\sqrt{12u-3b^2}}{4}\right],\ \text{where}$$

$$x_2\in B_1\equiv\left[x_1+\frac{b_2-\sqrt{6u_2-2b_2^2}}{3},x_1+\frac{b_2+\sqrt{6u_2-2b_2^2}}{3}\right]\text{and}\left[\frac{b-\sqrt{6u-2b^2}}{3},\frac{b+\sqrt{6u-2b^2}}{3}\right],$$

$$\text{where } x_2\in D_1\equiv\left[x_1,x_1+\frac{b_2-\sqrt{2u_2-b_2^2}}{2}\right]\cup\left[x_1+\frac{b_2+\sqrt{2u_2-b_2^2}}{2},x_1+\frac{b_2+\sqrt{6u_2-2b_2^2}}{3}\right].$$

If $\frac{b^2}{2}<u\leqslant b^2$ then $x_1\in D_o\equiv\left[0,\frac{b-\sqrt{2u-b^2}}{2}\right]\cup\left[\frac{b+\sqrt{2u-b^2}}{2},\frac{b+\sqrt{12u-3b^2}}{4}\right]$.

This set is subdivided into subsets

$$\left[0,\frac{b-\sqrt{6u-2b^2}}{3}\right]\cup\left[\frac{b+\sqrt{6u-2b^2}}{3},\frac{b+\sqrt{12u-3b^2}}{4}\right],\ \text{where } x_2\in B_1\equiv\left[x_1+\frac{b_2-\sqrt{6u_2-2b_2^2}}{3},x_1+\frac{b_2+\sqrt{6u_2-2b_2^2}}{3}\right]$$

$$\text{and }\left[\frac{b-\sqrt{6u-2b^2}}{3},\frac{b-\sqrt{2u-b^2}}{2}\right]\cup\left[\frac{b+\sqrt{2u-b^2}}{2},\frac{b+\sqrt{6u-2b^2}}{3}\right],\ \text{where } x_2\in D_1\equiv\left[x_1,x_1+\frac{b_2-\sqrt{2u_2-b_2^2}}{2}\right]\cup$$

$$\cup\left[x_1+\frac{b_2-\sqrt{2u_2-b_2^2}}{2},\ x_1+\frac{b_2+\sqrt{6u_2-2b_2^2}}{3}\right].$$

Denoting $x_2'=x_2-x_1$ we have

$$(14)\quad q(u)=\begin{cases}\dfrac{6}{b^3}\displaystyle\int_{a_4^-}^{a_4^+}dx_1\int_{b_3^-}^{b_3^+}Q(x_1,x_1+x_2')dx_2', & \dfrac{b^2}{4}\leqslant u\leqslant\dfrac{b^2}{3};\\[2ex] \dfrac{6}{b^3}\left[\displaystyle\int_0^{a_3^-}dx_1\int_{b_3^-}^{b_3^+}Q(x_1,x_1+x_2')dx_2'+\int_{a_3^-}^{a_4^+}dx_1\int_{b_3^-}^{b_3^+}Q(x_1,x_1+x_2')dx_2'\right]+ & \\[2ex] +\dfrac{6}{b^3}\displaystyle\int_{a_3^-}^{a_3^+}dx_1\left[\int_0^{b_2^-}Q(x_1,x_1+x_2')dx_2'+\int_{b_2^+}^{b_3^+}Q(x_1,x_1+x_2')dx_2'\right], & \dfrac{b^2}{3}<u\leqslant\dfrac{b^2}{2};\end{cases}$$

$$\left| \begin{array}{l} \frac{6}{b^3}\left[\int_0^{a_3^-} dx_1 \int_{b_3^-}^{b_3^+} Q(x_1,x_1+x_2')dx_2' + \int_{a_3^+}^{a_4^+} dx_1 \int_{b_3^-}^{b_3^+} Q(x_1,x_1+x_2')dx_2'\right] + \\ + \frac{6}{b^3}\int_{a_3^-}^{a_2^-} dx_1 \left[\int_0^{b_2^-} Q(x_1,x_1+x_2')dx_2' + \int_{b_2^+}^{b_3^+} Q(x_1,x_1+x_2')dx_2'\right] + \\ + \frac{6}{b^3}\int_{a_2^+}^{a_3^+} dx_1 \left[\int_0^{b_2^-} Q(x_1,x_1+x_2')dx_2' + \int_{b_2^+}^{b_3^+} Q(x_1,x_1+x_2')dx_2'\right], \quad u > \frac{b^2}{2}, \end{array} \right.$$

where $a_2^{\pm}=\frac{b\pm\sqrt{2u-b^2}}{2}$, $a_3^{\pm}=\frac{b\pm\sqrt{6u-2b^2}}{3}$, $a_4^{\pm}=\frac{b\pm\sqrt{12u-3b^2}}{4}$, $b_2^{\pm}=\frac{b_2\pm\sqrt{2u_2-b_2^2}}{2}$ and $b_3^{\pm}=\frac{b_2\pm\sqrt{6u_2-2b_2^2}}{3}$.

Performing the integration in (14) and substituting for simplisity b=1 we get

$$(15)\quad q(u)=\begin{cases} 6\pi\sqrt{u-1/4}, \quad 1/4\leqslant u<1/3, \\ 2\sqrt{3}\pi\left[\frac{1+\sqrt{12u-3}}{4}-\frac{2\sqrt{6u-2}}{3}\right]+2\sqrt{3}\int_{\frac{1-\sqrt{6u-2}}{3}}^{\frac{1+\sqrt{6u-2}}{3}} R(u,x)dx, \quad 1/3\leqslant u<1/2; \\ 2\sqrt{3}\pi\left[\frac{1+\sqrt{12u-3}}{4}-\frac{1+\sqrt{6u-2}}{3}\right]+ \\ +2\sqrt{3}\int_0^{\frac{1-\sqrt{2u-1}}{2}} R(u,x)dx+2\sqrt{3}\int_{\frac{1+\sqrt{2u-1}}{2}}^{\frac{1+\sqrt{6u-2}}{3}} R(u,x)dx, \quad 1/2\leqslant u\leqslant 1, \end{cases}$$

where $R(u,x)=\arcsin\left[\frac{1-x-3\sqrt{2u-2x^2-(1-x)^2}}{2\sqrt{6u-6x^2-2(1-x)^2}}\right]+\arcsin\left[\frac{1-x}{\sqrt{6u-6x^2-2(1-x)^2}}\right]$ $+\arccos\left[\frac{1-x+3\sqrt{2u-2x^2-(1-x)^2}}{2\sqrt{6u-6x^2-2(1-x)^2}}\right]$. It can be easily shown that this expression is identical to that obtained by Gardner (1952) who used a geometrical approach. The above given technique may be utilized in deriving the pdf of Greenwood's statistic for any n. Unfortunately, we have not found some common expression for this pdf.

REFERENCES

Burrows P. (1979): Selected percentage points of Greenwood's statistic. J.R.Statist.Soc.A,142,Part 2,256-258.

Currie I. (1981): Further percentage points of Greenwood's statistic. J.R.Statist.Soc.A,144,Part 3,360-363.

Does R., Helmers R., Klaassen C. (1984): Approximating the percentage points of Greenwood's statistic with Cornish-Fisher expansion. Amsterdam, - 6p. (Report/Centre for mathematics and Computer Sciences: MS-R8405).

Gardner A. (1952): Greenwood's "problem of intervals": An exact solution for N=3. J.R.Statist.Soc.B,14,135-139

Greenwood M. (1946): The statistical study of infectious diseases. J.R.Statist.Soc.A, 109, 85-110.

Moran P. (1947): The random division of an interval. J.R.Statist. Soc.B, 9, 92-98.

(1953): The random division of an interval-Part III. J.R.Statist.Soc.B, 15, 77-80.

(1981): Corrigendum: The random division of an interval. J.R.Statist.Soc.A, 144, 388.

Stephens M. (1981): Further percentage points for Greenwood's statistic. J.R.Statist.Soc.A, 144, 364-366.

High Energy Physics Institute
of Kazakh Academy of Sciences
480082, Alma-Ata-82
USSR

ON THE DIFFUSION APPROXIMATION OF THE GENERALIZED STEIN´S NEURONAL MODEL WITH SYNAPTIC REVERSAL POTENTIALS

Věra Lánská

Prague

Key words: stochastic processes, weak convergence, neuronal model

ABSTRACT

Diffusion approximation to the generalized Stein´s neuronal model which includes exponential decay, random excitation and inhibition and synaptic reversal potential is investigated. Using the weak convergence of probability measures the non-Gaussian diffusion model is derived.

1. INTRODUCTION

A diffusion process that has often been used to model the subthreshold depolarization of a nerve cell membrane is the Ornstein-Uhlenbeck process, e.g. Kallianpur (1983), Lánský (1984). This model has been derived from Stein´s neuronal model (Stein (1965)), and both the models are characterized by the state-independence of the depolarization. However, it is well known fact that these changes of the nerve cell potential are state-dependent in such a way that they depend on the difference between the prior voltage and the synaptic reversal potential characteristic of the activated synapses. Therefore a modification of the Stein´s model which incorporates a more complete characterization of the process of synaptic transmission by inclusion of the reversal potentials has been proposed by Tuckwell (1979).

Under the assumption of the model with reversal potentials

the subthreshold values of the membrane potential $V(t)$ are given by stochastic equation

$$(1) \quad dV(t)=-1/\tau\, V(t)dt+a(V_E-V(t))dN^+(t)+i(V(t)-V_I)dN^-(t),$$

$$V(0)=v_0,$$

where $-1<i<0<a<1$, $V_I<0<V_E$, $V_I<v_0<V_E$ are constants, $\tau>0$ is a constant intensity of decay. The value a , resp. i, represents excitatory, resp. inhibitory, postsynaptic amplitude, V_E and V_I are the postsynaptical reversal potentials, $N^+(t)$ and $N^-(t)$ are two independent Poisson processes representing the input of the neuron, $N^+(0)=N^-(0)=0$. This model postulates that each event of the excitatory process $N^+(t)$ depolarizes the membrane potential by $a(V_E-V(t-))$ and analogously the inhibition process $N^-(t)$ produces hyperpolarization of size $i(V(t-)-V_I)$. So, aV_E and iV_I are the values of the jump size under the condition that the membrane potential is equal to zero at the instant of its generation. Between events of input process $V(t)$ decays exponentially to zero.

The model (1) has discontinuous trajectories and therefore it is too complicated for a mathematical treatment. Thus, the scheme of "usual" diffusion approximation was used by Hanson and Tuckwell (1983). The aim of this paper is a formulation of the diffusion approximation for the generalized stochastic neuronal model with spontaneous decay and synaptic reversal potentials via weak convergence of probability measures.

2. DISCONTINUOUS MODEL

Let us consider a generalization of the Tuckwell´s model (1) including also nondeterministic postsynaptic potential amplitudes. Moreover, let us assume that these stochastic parts of postsynaptic potential amplitudes depend on both reversal potentials in the following way

$$dV(t)=-1/\tau V(t)dt+a(V_E-V(t))dN^+(t)+i(V(t)-V_I)dN^-(t)+$$

$$(2)\qquad +A_t(V_E-V(t))^p(V(t)-V_I)^q dN^+(t)+I_t(V_E-V(t))^q(V(t)-V_I)^p\cdot$$

$$dN^-(t),$$

$$V(0)=v_0,$$

where A_t, I_t are mutually independent random variables with $EA_t=EI_t=0$, and p,q are constants, $p>0$. The interpretation of the equation (2) is the same as for (1). For $V(t)$ given by (2) the first and the second infinitesimal moments are

$$(3)\qquad M_1(v)=\lim_{\Delta t\to 0}\frac{E[\Delta V(t)\mid V(t)=v]}{\Delta t}=-1/\tau\, v+\lambda a(V_E-v)+\omega i(v-V_I),$$

$$(4)\qquad M_2(v)=\lim_{\Delta t\to 0}\frac{E[\Delta V(t)^2\mid V(t)=v]}{\Delta t}=\lambda a^2(V_E-v)^2+i\omega^2(v-V_I)^2$$

$$+\lambda EA^2(V_E-v)^{2p}(v-V_I)^{2q}+\omega EI^2(V_E-v)^{2q}(v-V_I)^{2p},$$

where λ and ω are intensities of the processes $N^+(t)$ and $N^-(t)$. The above summarized models attempt to characterize the stochastic variability of interspike intervals. This is achieved proposing that the neuron fires at the moment when the membrane potential exceeds a fixed threshold $S>v_0$ for the first time. So, the most important quantity is the random variable $T_S(V)$ defined as the first passage time of the process $V(t)$ across the threshold S, $T_S(V)=\inf(t\geq 0, V(t)\geq S)$. It is worth to note that the relation $S<V_E$ can be fairly assumed and thus the behaviour of the membrane potential above S is of no interest to us. To find at least the moments of interspike intervals $T_S(V)$ one has to solve difference-differential equations. The solution of such equations is

very difficult and it is therefore useful to turn to diffusion approximation. Then instead of $T_S(V)$ the random variable $T_S(X)$ is studied, where $X(t)$ is a diffusion process substituted for the process $V(t)$.

The weak convergence of the probability distributions of $V(t)$ to probability distribution of $X(t)$ induces the convergence in law of the first passage times in question. For the diffusion process $X(t)$ with the transition density function $f(x,t;x_0)$ the equation

$$f(S,t;x_0)=\int_0^t g(S,\Theta;x_0)f(S,t-\Theta;S)d\Theta$$

for the unknown first passage time density $g(S,t;x_0)$ can be used. Similarly, Siegert's recursion formula for the first passage time moments can be applied.

3. DIFFUSION APPROXIMATION

The method of obtaining a diffusion approximation $X(t)$ to the process $V(t)$ defined by (2) is to construct a sequence of processes with the probability distributions converging weakly to the probability distribution of the diffusion process $X(t)$. To prove the weak convergence the martingale approach used also in Lánská (1983) is applied.

Let us define the sequence of $V_n(t)$ analogously to (2) in the form

$$V_n(t)=-1/\tau \int_0^t V_n(s)ds+ \sum_{j=1}^{N_n^+(t)} a_n(V_E-V_n(t_{nj}^+))+A_{nj}(V_E-V_n(t_{nj}^+))^p$$

$$(V_n(t_{nj}^-)-V_I)^q+ \sum_{j=1}^{N_n^-(t)} i_n(V_n(t_{nj}^-)-V_I)+ \tag{5}$$

$$+I_{nj}(V_E-V_n(t_{nj}^-))^q(V_n(t_{nj}^-)-V_I)^p,$$

where for each $n=1,2,\ldots,$ $\{A_{nj}\}_{j=1}^{\infty},\{I_{nj}\}_{j=1}^{\infty}$ are two sequences of i.i.d. random variables with the distribution functions

$F_A,\ F_I,\ EA_{nj}=0=EI_{nj},\ \{t^+_{nj}\}^{\infty}_{j=1},\{t^-_{nj}\}^{\infty}_{j=1}$

are time moments of jumps in $N^+_n(t),\ N^-_n(t)$.

For the sake of simplicity $V_n(0)=0$.

Next, assumptions guaranteeing the weak convergence will be stated.

Assumption 1. The input intensities λ_n, ω_n tend to infinity, as $n\to\infty$, in such a way that

$$\lim_{n\to\infty} \lambda_n a_n=\alpha>0, \qquad \lim_{n\to\infty} \omega_n i_n=\beta<0,$$

$$\lim_{n\to\infty} \lambda_n EA^2_{nj}=\sigma^2_E,\ \lim \omega_n EI^2_{nj}=\sigma^2_I>0.$$

Assumption 2.

$$EA^4_{nj}=O[(EA^2_{nj})^2],\ EI^4_{nj}=O[(EI^2_{nj})^2], n\to\infty.$$

Under Assumption 1 there exist limits of the first two infinitesimal moments (3) and (4) for $v\in[V_I,V_E]$, as $n\to\infty$,

$$\{M_1(v)\}_n\to -v/\tau+\alpha(V_E-v)+\beta(v-V_I)=\mu(v),$$

$$\{M_2(v)\}_n\to \sigma^2_E(V_E-v)^{2p}(v-V_I)^{2q}+\sigma^2_I(V_E-v)^{2q}(v-V_I)^{2p}=\sigma^2(v).$$

Theorem: The sequence of processes V_n defined by (5) under Assumptions 1,2 converges weakly to the diffusion process $X(t)$ given by stochastic differential equation

$$(6) \qquad dX(t)=\mu(X(t))dt+\sigma(X(t))dW_t,\ X(0)=0,$$

where $W=\{W_t, t\geq 0\}$ is a standard Wiener process.

The proof of this Theorem can be done in the same way as in Lánská (1983). This means that the assumptions of Corollary III.4 in Rebolledo (1979), concerning the weak convergence of a sequence of probability measures for quasimartingales to the probability distribution of a diffusion process, are checked for martingale

$$M_n(t)=V_n(t)-1/\tau \int_0^t V_n(s)ds-\lambda_n a_n \int_0^t (V_E-V_n(s))ds-\omega_n i_n \int_0^t (V_n(s)-V_I)ds.$$

The Assumption 2 guarantees that the jumps in $M_n(t)$ are "small", as $n\to\infty$. It should be noted that the discontinuous process $V_n(t)$ can reach or exceed the boundaries V_I, V_E, however, under Assumption 2

$$P(V_n(t+\Delta)\notin(V_I,V_E)\mid V_n(t)\in(V_I,V_E))\to 0, \text{as} \quad n\to\infty.$$

Let us briefly mention the boundary behaviour of the derived model. (See Mandl (1968) for more details about the boundary behaviour of diffusion processes.)

For $\sigma_E^2>0$ and $min(p,q)>\frac{1}{2}$ both boundaries V_I, V_E are inaccessible for the diffusion process *(6)* and for $0\leq min<(p,q)<\frac{1}{2}$, $\sigma_E^2>0$ both boundaries are regular. In case $\sigma_E^2=0$ we investigate mainly the behaviour of the process *(6)* near the boundary V_I, as this one is the only important for us. For $p>\frac{1}{2}$ V_I is inaccessible, for $p<\frac{1}{2}$ it is regular and for $p=\frac{1}{2}$ the behaviour depends on the constant $\dfrac{-V_I/\tau+\alpha(V_E-V_I)}{\sigma_I^2(V_E-V_I)^{2q}}$.

One of the most interesting cases occurs for $q=p=\frac{1}{2}$. We have

$$dX_t=-1/\tau X_t dt+\alpha(V_E-X_t)dt+\beta(X_t-V_I)dt+\sigma\cdot[(X_t-V_I)(V_E-X_t)]^{\frac{1}{2}}dW_t,$$

$$\sigma^2=\sigma_E^2+\sigma_I^2>0.$$

For this case the stationary distribution of X_t exists and therefore proper asymptotic properties of maximum likelihood estimate for unknown parameters α,β are guaranteed. The boundary V_I is reflecting, resp. entrance,

if $0<\frac{-\frac{V_I}{\tau}+\alpha(V_E-V_I)}{\sigma^2(V_E-V_I)}<1$ resp. $\frac{-\frac{V_I}{\tau}+\alpha(V_E-V_I)}{\sigma^2(V_E-V_I)}>1$.

REFERENCES

Hanson, F.B., Tuckwell, H.O. (1983) : Diffusion approximations for neuronal activity including synaptic reversal potentials. J. Theor. Neurobiol. 2, 127-153.

Kallianpur, G. (1983) : On the diffusion approximation to a discontinuous model for a single neuron. In: Sen, P.K. Ed. : Contributions to statistics. Amsterdam, North Holland.

Lánská, V. (1983) : An application of the diffusion approximation to a controlled service system. In: Trans. of the Ninth Prague Conference on Information Theory, Statistical decision Functions, Random Processes, 1982, Academia Praha, 43-47.

Lánský, P. (1984) : On approximations of Stein´s neuronal model. J. Theor. Biol. 107, 631-647.

Mandl, P. (1968) : Analytical treatment of one-dimensional Markov processes, Academia Praha.

Rebolledo, R. (1979) : Martingales at convergence étroite de measures de probabilité. Kybernetika 15, 1-7.

Stein, R.B. (1965) : A theoretical analysis of neuronal variability. Biophys. J. 5, 173-194.

Tuckwell, H.C. (1979) : Synaptic transmission in a model for stochastic neural activity. J. Theor. Biol. 77, 65-81.

Institute of Clinical and Experimental Medicine
Vídeňská 800, 146 29 Prague 4

BOUNDS FOR THE r-th ABSOLUTE MOMENT OF A DIFFERENCE OF TWO RANDOM VARIABLES

Gabriele Laue

Leipzig

Key words: Non-negative random variables, fractional moments

ABSTRACT

Let X and Y be independent, identically distributed random variables (i.i.d.r.v.). We denote their distribution function (d.f.) and the corresponding characteristic function (c.f.) by F and f, respectively. The aim of the present paper is to estimate $E/X-Y/^r$ if $r > 0$ is an arbitrary real number. We present a simple method to estimate $E/X-Y/^r$ provided that X and Y are non-negative r.v. It is based on the fact that under the above assumptions $E/X-Y/^r$ with $r > 2$ can be expressed by $E/X_1-Y_1/^{r-2}$ where X_1 and Y_1 are independent r.v. with the common d.f.

(0.1) $$F_1(x) := \frac{1}{m_{F,1}} \cdot \int_0^x (1-F(u))\, du.$$

Therefore, we need only good estimates for $0 < r < 2$ and can reduce all the other cases to them. To give such estimates we follow an idea of C.G. Esseen (see /3/). Finally we give some analogous results concerning arbitrary r.v.

1. A general method to estimate $E/X-Y/^r$ if X and Y are nonnegative r.v.

In the following we give a basic formula to estimate moments of /X-Y/ of order $r := k+1+\lambda$ where k is a non-negative integer and $-1 < \lambda < 1$. It permits us to estimate not only moments of integer order but also of arbitrary order greater than zero.

Theorem 1.1

$$E/X-Y/^r \lesseqgtr 2\int_0^\infty x^{1+\lambda}\left[\int_0^x (x-v)^k dF(v) - \int_x^\infty (v-x)^k dF(v)\right]dF(x)$$

(1.1) $$\pm 2\left|\cot\frac{(1+\lambda)\pi}{2}\right|\left|\int_0^\infty x^{1+\lambda}\left[\int_0^\infty /x-v/^k dF(v)\right]dF(x)\right.$$

For $\lambda=0$ we have the equality sign.

Proof: We have to distinguish the three cases $\lambda=0$, $-1<\lambda<0$ and $0<\lambda<1$. At first let $\lambda=0$. It follows from the definition

(1.2) $$E/X-Y/^r = \int_0^\infty\int_0^\infty /x-v/^r dF(v)\, dF(x)$$

that $$E/X-Y/^{k+1} = 2\int_0^\infty dF(x)\int_x^\infty (v-x)^{k+1}dF(v)$$

$$= 2\int_0^\infty dF(x)\left[\int_x^\infty v(v-x)^k dF(v) - \int_x^\infty x(v-x)^k dF(v)\right].$$

Changing the order of integrations in the first double integral we get the assertion. Let now $-1<\lambda<0$. Putting

$$C := \frac{1}{1+\lambda}\Gamma(-\lambda)\sin\frac{(1+\lambda)\pi}{2} = \int_0^\infty \frac{\sin t}{t^{2+\lambda}}dt$$ (see /4/)

we can express $(v-x)^{1+\lambda}$ for $v\geq x$ by

$$(v-x)^{1+\lambda} = \frac{1}{C}\int_0^\infty \frac{\sin (v-x)u}{u^{2+\lambda}}du.$$ Therefore we get

$$E/X-Y/^r = 2\int_0^\infty dF(x)\int_x^\infty (v-x)^{k+1+\lambda}\, dF(v)$$

$$= \frac{2}{C}\int_0^\infty dF(x)\int_x^\infty (v-x)^k dF(v)\int_0^\infty \frac{\sin vu\cdot\cos xu - \sin xu\cdot\cos vu}{u^{2+\lambda}}du$$

$$= \frac{2}{C}\int_0^\infty dF(x)\int_x^\infty (v-x)^k dF(v)\left[\int_0^\infty \frac{\sin vu}{u^{2+\lambda}}du - \int_0^\infty \frac{\sin xu}{u^{2+\lambda}}du\right.$$

$$\left. + \int_0^\infty \frac{1-\cos vu}{u^{2+\lambda}}\sin xu\, du - \int_0^\infty \frac{1-\cos xu}{u^{2+\lambda}}\sin vu\, du\right]$$

$$= 2\int_0^\infty dF(x)\int_x^\infty (v-x)^k(v^{1+\lambda} - x^{1+\lambda})dF(v) + R_k^-$$

where R_k^- is given by

$$R_k^- := \frac{2}{C}\int_0^\infty dF(x)\int_x^\infty (v-x)^k dF(v)\left[\int_0^\infty \frac{1-\cos vu}{u^{2+\lambda}}\sin xu\, du - \int_0^\infty \frac{1-\cos xu}{u^{2+\lambda}}\sin vu\, du\right]$$

It is easy to see that R_k^- satisfies the following inequality

$$\left|R_k^-\right| \leq \frac{2K}{C}\int_0^\infty dF(x)\int_x^\infty (v-x)^k(v^{1+\lambda} + x^{1+\lambda})\, dF(v).$$

Here the constant K is defined by

$$K:=\frac{1}{1+\lambda}\Gamma(-\lambda)\cos\frac{(1+\lambda)\pi}{2}=\int_0^\infty\frac{1-\cos t}{t^{2+\lambda}}\,dt \qquad (\text{ see } /4/).$$

This yields (1.1). Finally let $0<\lambda<1$. Using the constant

$$K':=\frac{1}{\lambda}\Gamma(1-\lambda)\cos\frac{\lambda\pi}{2}=(1+\lambda)\int_0^\infty\frac{t-\sin t}{t^{2+\lambda}}\,dt$$

and representing $(v-x)^{1+\lambda}$ for $v\geqslant x$ by

$$(v-x)^{1+\lambda}=\frac{1+\lambda}{K'}\int_0^\infty\frac{(v-x)u-\sin(v-x)u}{u^{2+\lambda}}\,du$$

after similar computations as above we see that (1.1) is valid also in this case.

In the following we need only the cases k=0 and k=1. For these cases the expressions which occur in (1.1) can be essentially simplified. We obtain

Corollary 1.2: Let $-1<\lambda<1$. Then we have the estimates

$$(1.3)\quad E/X-Y/^{1+\lambda}\lesseqgtr 2\left\{m_{F,1+\lambda}-(1+\lambda)\int_0^\infty x^\lambda(1-F(x))^2dx\pm\left|\cot\frac{(1+\lambda)\pi}{2}\right|m_{F,1+\lambda}\right\}$$

$$(1.4)\quad E/X-Y/^{2+\lambda}\lesseqgtr 2\left\{m_{F,2+\lambda}-m_{F,1}m_{F,1+\lambda}\pm(m_{F,2+\lambda}+m_{F,1}m_{F,1+\lambda})\cdot\left|\cot\frac{(1+\lambda)\pi}{2}\right|\right\}$$

Proof: Setting k=0 in (1.1) and integrating by parts we get (1.3). To prove (1.4) we have to simplify the expressions

$$A_{\mp}:=\int_0^\infty x^{1+\lambda}\,dF(x)\left[\int_0^x(x-v)dF(v)\mp\int_x^\infty(v-x)dF(v)\right].$$

It is easy to see that $A_-=m_{F,2+\lambda}-m_{F,1}m_{F,1+\lambda}$ and $/A_+/\leqslant m_{F,2+\lambda}+m_{F,1}m_{F,1+\lambda}$ from which (1.4) follows.

The two formulas (1.3) and (1.4.) express a general property of the estimate (1.1). Namely, if k is odd, the bounds of $E/X-Y/^r$ can be expressed only by certain moments of F. But, if k is even, this is not so: The given bounds contain the d.f. F. To obtain useful estimates further considerations are necessary (see §3). Remark that we must consider even and odd k, especially k=0 and k=1, because of the structure of our estimates. They are not applicable if $/\lambda/>\frac{1}{2}$.

2. Reduction of $E/X-Y/^{k+1+\lambda}$ if X and Y are non-negative r.v. and $k \geq 2$

Before giving final estimates for $E/X-Y/^{1+\lambda}$ we prove that the cases with $k \geq 2$ can be reduced to those of k=0 and k=1. Namely, introducing the i.i.d.r.v. X_1 and Y_1 subject to the d.f. F_1 (see (0.1.)) we get the following result.

Lemma 2.1: Let $r > 1$. $m_{F,r}$ exists if and only if $m_{F_1,r-1}$ exists. In either case we have the relation

(2.1) $m_{F,r} = r\, m_{F,1}\, m_{F_1,r-1}$.

Proof: If $m_{F,r} < \infty$ then

$$\infty > m_{F,r} = r \int_0^\infty x^{r-1}(1-F(x))\, dx = r\, m_{F,1}\, m_{F_1,r-1} .$$

The computations can be reversed so that the assertion follows.

Theorem 2.2: Let $r > 2$ and $m_{F,r} < \infty$. Then we have

(2.2) $E/X-Y/^r = 2m_{F,r} - r(r-1)\, m_{F,1}^2\, E/X_1-Y_1/^{r-2}$.

Proof: Starting from (1.2) and integrating by parts in the inner integral we get

$$E/X-Y/^r = 2r \int_0^\infty dF(x) \int_x^\infty (v-x)^{r-1}(1-F(v))dv.$$

Now we integrate by parts in the outer integral and obtain

$$E/X-Y/^r = 2r \left\{ \int_x^\infty (v-x)^{r-1}(1-F(v))\, dv\, (F(x)-1)\Big|_0^\infty - (r-1)\int_0^\infty (1-F(x)) \int_x^\infty (v-x)^{r-2}(1-F(v))\, dv\, dx \right\}$$

$$= 2m_{F,r} - 2r(r-1)\, m_{F,1}^2 \int_0^\infty dF_1(x) \int_x^\infty (v-x)^{r-2} dF_1(v).$$

Using again (1.2) we get (2.2).

Setting in (2.2) $r=3+\lambda$, using (1.) to estimate $E/X_1-Y_1/^{r-2}$ and applying (2.1) to express the moments of F_1 by moments of F we obtain an estimate for $E/X-Y/^{3+\lambda}$. Setting $r=4+\lambda$ and using (1.4) and again (2.1) we get an estimate for $E/X-Y/^{4+\lambda}$. We summarize these results in

Corollary 2.3: Let $-1 < \lambda < 1$. Then we have

(2.3) $E/X-Y/^{3+\lambda} \lesseqgtr 2m_{F,3+\lambda} - 2(3+\lambda)m_{F,1}m_{F,2+\lambda}$

$$+ 2(3+\lambda)(2+\lambda)(1+\lambda)\int_0^\infty x^\lambda \left\{ \int_x^\infty (1-F(u))\, du \right\}^2 dx$$

$$\pm 2(3+\lambda) \left|\cot \frac{(1+\lambda)\pi}{2}\right| m_{F,1}m_{F,2+\lambda}$$

$$(2.4)\quad E/X-Y/^{4+\lambda} \lesseqgtr 2m_{F,4+\lambda} - 2(4+\lambda)m_{F,1}m_{F,3+\lambda} + \frac{(4+\lambda)(3+\lambda)}{2+\lambda} m_{F,2}m_{F,2+\lambda}$$

$$\pm 2(4+\lambda)\left|\cot\frac{(1+\lambda)\pi}{2}\right|(m_{F,1}m_{F,3+\lambda} + \frac{3+\lambda}{2(2+\lambda)} m_{F,2}m_{F,2+\lambda})$$

Remark: (2.3) is a generalization of the formula which was used by Esseen /3/ and Daley /2/; it is the basis of our further considerations.

Applying this corollary and formula (2.2) we get estimates for $E/X-Y/^{5+\lambda}$ and $E/X-Y/^{6+\lambda}$ and so on.

3. Further estimations of (1.3) and (2.3)

We will now estimate the integrals occuring in (1.3) and (2.3). For this reason we generalize a trick which was used by Esseen for r=3. First we give lower bounds for the integral

$$(3.1)\quad I := \int_0^\infty x^\lambda \left\{ \int_x^\infty [1-F(v)]\, dv \right\}^2 dx \quad \text{occuring in} \quad (2.3)$$

Lemma 3.1: Let $m_{F,3+\lambda} < \infty$, $-1 < \lambda < 1$. Then we have the estimates

$$(3.2)\quad I \geq \frac{9}{(3+\lambda)^3}\left(m_{F,\frac{1}{3}(3+\lambda)} \right)^3 \qquad \text{if } -1<\lambda<0,$$

$$(3.3)\quad I \geq \frac{16}{3(3+\lambda)^3}\left(m_{F,\frac{1}{4}(3+\lambda)} \right)^4 \qquad \text{if } 0\leq\lambda<1.$$

Proof: Let $-1 < \lambda < 0$ and $a > 0$. Then the integral I can be written as

$$I = \int_0^\infty x^\lambda \left[\int_x^\infty \frac{u^a}{u^a} (1-F(u))\, du \right]^2 dx$$

$$\geq \int_0^\infty x^{\lambda+2a}(1-F(x)) \left[\int_x^\infty u^{-a}(1-F(u))\, du \right]^2 dx.$$

Choosing $a = -\frac{1}{3}\lambda$ which is possible by the assumptions we can substitute $v = \int_x^\infty u^{-a}(1-F(u))\, du$ and obtain

$$I \geq \int_0^{v_0} v^2\, dv = \frac{1}{3} v_0{}^3, \text{ where } v_0 := \int_0^\infty u^{\frac{1}{3}\lambda} (1-F(u))du = \frac{m_{F,\frac{1}{3}(3+\lambda)}}{\frac{1}{3}(3+\lambda)}$$

Let now $0 \leq \lambda < 1$ and $0 < b < 1$. Writing the integral I in the form

$$I = \int_0^\infty x^{\lambda+b}\, x^{-b} \left[\int_x^\infty (1-F(u))du \right]^2 dx$$

and integrating by parts we get

$$I = (\lambda+b) \int_0^\infty x^{\lambda+b-1} \int_x^\infty v^{-b} \left[\int_v^\infty (1-F(u))du \right]^2 dv\, dx.$$

Considering the inner integral $I' := \int_x^\infty v^{-b} \left\{ \int_v^\infty (1-F(u))du \right\}^2 dv$

we see that it can be estimated by the method used above. We obtain

$$I' \geq \frac{1}{3}\left[\int_x^\infty u^{-\frac{1}{3}b}(1-F(u))du\right]^3 \text{, so that}$$

$$I \geq \frac{\lambda+b}{3}\int_0^\infty x^{\lambda+b-1}\left[\int_x^\infty u^{-\frac{1}{3}b}(1-F(u))\,du\right]^3 dx .$$

We are still free to choose b. Setting $b = \frac{3}{4}(1-\lambda)$ which is positive by the assumptionswe can again apply the above method. Our result will be the inequality (3.3).

Next we give upper bounds for the expression I defined by (3.1).

Lemma 3.2: Let $m_{F,3+\lambda} < \infty$, $-1 < \lambda < 1$. Upper bounds for the integral I are given by

$$(3.4)\quad I \leq \frac{2}{(1+\lambda)(2+\lambda)(3+\lambda)} m_{F,2+\lambda}\, m_{F,1} .$$

Proof: Integrating (3.1) by parts we get

$$I = \frac{2}{1+\lambda}\int_0^\infty x^{1+\lambda}(1-F(x))\int_x^\infty(1-F(u))\,du\,dx .$$

It follows from $x(1-F(x)) \leq \int_0^x (1-F(u))\,du$ that

$$I \leq \frac{2}{1+\lambda}\int_0^\infty x^\lambda\left[\int_0^x(1-F(u))du\right]\left[\int_x^\infty(1-F(v))dv\right]dx$$

$$= \frac{2}{1+\lambda}\int_0^\infty x^\lambda\left[m_{F,1} - \int_x^\infty(1-F(u))du\right]\left[\int_x^\infty(1-F(v))dv\right]dx$$

$$= \frac{2}{(1+\lambda)^2(2+\lambda)} m_{F,1}\, m_{F,2+\lambda} - \frac{2}{1+\lambda} I .$$

Hence we conclude (3.4).

Under certain assumptions we can estimate the integral

$$(3.5)\quad J := \int_0^\infty x^\lambda(1-F(x))^2dx \quad \text{occuring in (1.3)}$$

in an analogous manner. Namely, we have the following result.

Corollary 3.3: Let F be a d.f. of a non-negative r.v. with a monotone and bounded density p_F. Further on, let $m_{F,1+\lambda} < \infty$, where $-1 < \lambda < 1$.

a) Lower bounds for J are given by

$$(3.6)\quad J \geq \frac{1}{3}\frac{1}{p_F(0)}\left(m_{F,\frac{1}{3}\lambda}\right)^3 , \qquad \text{if } -1 < \lambda < 0, \text{ and by}$$

$$(3.7)\quad J \geq \frac{3+\lambda}{48}\frac{1}{p_F^2(0)}\left(m_{F,-\frac{1}{4}(1-\lambda)}\right)^4 , \qquad \text{if } 0 < \lambda < 1.$$

b) An upper bound for J is given by

(3.8) $J \leq \frac{2}{(1+\lambda)(3+\lambda)} m_{F,1+\lambda}$.

Proof: Obviously, a monotone and bounded density function p_F can be written as: $p_F(x) = p_F(0)(1-F_o(x))$, where F_o is the d.f. of a non-negative r.v. Therefore, the integral J defined by (3.5) gets the form

$$J = (p_F(0))^2 \int_0^\infty x^\lambda \left[\int_x^\infty (1-F_o(u))du \right]^2 dx .$$

Applying lemma 3.1 and lemma 3.2 we obtain estimates for J by moments of F_o. Using further the relation

$$m_{F_o,r+1} = \frac{r+1}{p_F(0)} m_{F,r} \qquad \text{(see (2.1))}$$

we get the assertions.

Setting the estimates (3.6) - (3.8) into (1.3) and the estimates (3.2) - (3.4) into (2.3), respectively, we obtain final estimates for $E/X-Y/^{1+\lambda}$ and $E/X-Y/^{3+\lambda}$, respectively. We will summarize these results in the next two theorems.

<u>Theorem 3.4</u>: Suppose that the d.f. F has a monotone and bounded density function p_F and that $m_{F,1+\lambda} < \infty$, where λ is an arbitrary real number satisfying $-1 < \lambda < 1$.

a) Upper bounds for $E/X-Y/^{1+\lambda}$ are given by

$$(3.9)\quad E/X-Y/^{1+\lambda} \leq 2(m_{F,1+\lambda} - \frac{1+\lambda}{3}\frac{1}{p_F(0)} (m_{F,\frac{1}{3}\lambda})^3) + $$
$$+ 2m_{F,1+\lambda} \left| \cot\frac{(1+\lambda)\pi}{2} \right| \qquad \text{if } -1 < \lambda < 0$$

$$(3.10)\quad E/X-Y/^{1+\lambda} \leq 2(m_{F,1+\lambda} - \frac{(1+\lambda)(3+\lambda)}{48(p_F(0))^2} (m_{F,\frac{1}{4}(\lambda-1)})^4) + $$
$$+ 2m_{F,1+\lambda} \left| \cot\frac{(1+\lambda)\pi}{2} \right| \qquad \text{if } 0 \leq \lambda < 1.$$

b) A lower bound for $E/X-Y/^{1+\lambda}$ is given by

$$(3.11)\quad E/X-Y/^{1+\lambda} \geq \frac{2(1+\lambda)}{3+\lambda} m_{F,1+\lambda} - 2\left|\cot\frac{(1+\lambda)\pi}{2}\right| m_{F,1+\lambda} .$$

<u>Theorem 3.5</u>: Let $m_{F,3+\lambda} < \infty$, $-1 < \lambda < 1$.

a) An upper bound for $E/X-Y/^{3+\lambda}$ is given by

$$(3.12)\quad E/X-Y/^{3+\lambda} \leq 2(m_{F,3+\lambda} - (1+\lambda)m_{F,1}m_{F,2+\lambda}) + 2(3+\lambda)m_{F,1}m_{F,2+\lambda}\left|\cot\frac{(1+\lambda)\pi}{2}\right|$$

b) Lower bounds for $E/X-Y/^{3+\lambda}$ are given by

$$(3.13)\quad E/X-Y/^{3+\lambda} \geq 2(m_{F,3+\lambda} - (3+\lambda)m_{F,1}m_{F,2+\lambda}) + + 18\frac{(1+\lambda)\ (2+\lambda)}{(3+\lambda)^2}\left(m_{F,\frac{1}{3}(3+\lambda)}\right)^3 - - 2(3+\lambda)m_{F,1}m_{F,2+\lambda}\left|\cot\frac{(1+\lambda)\pi}{2}\right| \quad \text{if } -1<\lambda<0$$

$$(3.14)\quad E/X-Y/^{3+\lambda} \geq 2(m_{F,3+\lambda} - (3+\lambda)m_{F,1}m_{F,2+\lambda}) + + \frac{4^3}{6}\frac{(1+\lambda)\ (2+\lambda)}{(3+\lambda)^2}\left(m_{F,\frac{1}{4}(3+\lambda)}\right)^4 - - 2(3+\lambda)m_{F,1}m_{F,2+\lambda}\left|\cot\frac{(1+\lambda)\pi}{2}\right| \quad \text{if } 0\leq\lambda<1.$$

Remarks: (3.11) is similar to estimates of von Bahr and Esseen /1/. Setting $\lambda=0$ we can compare our results with those of Esseen /3/ and Daley /2/. Namely, the upper bound (3.12) coincides with that given by Daley while the lower bound (3.14) is somewhat weaker than that of Esseen.

4. Extension of the results to moments of a difference of two arbitrary r.v.

Let now X and Y be arbitrary i.i.d.r.v. with the d.f. F. Our basic result, theorem 1.1, can be carried over to this case. Because the proof is similar, we omit it.

Theorem 4.1: Let k be a non-negative integer, $-1<\lambda<1$. Then we have

$$E/X-Y/^{k+1+\lambda} \lesseqgtr 2\int_0^\infty x^{1+\lambda}dF(x)\left[\int_{-\infty}^{x}(x-v)^k dF(v) - \int_x^\infty (v-x)^k dF(v)\right]$$

$$(4.1)\qquad + 2\int_{-\infty}^{0}/x/^{1+\lambda}dF(x)\left[\int_x^\infty (v-x)^k dF(v) - \int_{-\infty}^{x}(x-v)^k dF(v)\right]$$

$$\pm 2\left|\cot\frac{(1+\lambda)\pi}{2}\right|\int_{-\infty}^{\infty}/x/^{1+\lambda}dF(x)\int_{-\infty}^{\infty}/x-v/^k dF(v)$$

We discuss only the cases k=0, 1 and 2. If k=0 or 1 we use the method described in §3 for non-negative r.v. But, if $k>2$ we cannot apply the idea of reducing moments of higher order to moments of lower order. Nevertheless, we can evaluate (4.1) and obtain results analogous to corollary 1.2. These computations are the more complicated the greater is k. Again, if k is even further estimates are necessary and we proceed as in §3. For instance, we get the following result.

Theorem 4.2: Let $-1 < \lambda < 1$ and $m_{F,3+\lambda} < \infty$.
An upper bound for $E/X-Y/^{3+\lambda}$ is given by

$$E/X-Y/^{3+\lambda} \leq 2\hat{m}_{F,3+\lambda} - 4m_{F,1}(m^+_{F,2+\lambda} - m^-_{F,2+\lambda}) - 2(m^+_{F,1+\lambda} - m^-_{F,1+\lambda})(m^+_{F,2} - m^-_{F,2})$$
$$+ 4\frac{(1+\lambda)(4+\lambda)+2}{(2+\lambda)(3+\lambda)}(m^+_{F,1}m^+_{F,2+\lambda} + m^-_{F,1}m^-_{F,2+\lambda})$$
$$+ 2\left|\cot\frac{(1+\lambda)\pi}{2}\right|(\hat{m}_{F,3+\lambda} + 2\hat{m}_{F,2+\lambda}\hat{m}_{F,1} + \hat{m}_{F,1+\lambda}m_{F,2})$$

where $\hat{m}_{F,r} = \int_{-\infty}^{0} |x|^r \, dF(x) + \int_0^{\infty} x^r dF(x) =: m^-_{F,r} + m^+_{F,r}$

REFERENCES

/1/ Von Bahr, B.; Esseen, C.G. (1965): Inequalities for the r-th absolute moment of a sum of random variables, $1 < r < 2$
Ann. Math. Statist. 36, 299 - 303

/2/ Daley, D.J. (1977): Tighter bounds for the absolute third moment
Scand J Statist 4, 183 - 184

/3/ Esseen, C.G. (1975): Bounds for the absolute third moment
Scand J Statist 2, No. 3, 149 - 152

/4/ Ryshik, I.M.; Gradstein, I.S. (1957): Summen-, Produkt- und Integraltafeln
Deutscher Verlag der Wissenschaften, Berlin

Karl-Marx-Universität
Sektion Mathematik
Karl-Marx-Platz 10
7010 Leipzig
DDR

ON COOPERATIVE GAMES ARISING FROM
A PROBLEM OF FAIR DIVISION

Jerzy Legut

Wrocław

Key words: cooperative game, problem of fair division

ABSTRACT

We introduce a cooperative game of fair division which turns out to be totally balanced. We propose to consider the core of the game as a cooperative solution of the well-known problem of fair division.

I. INTRODUCTION

1. COOPERATIVE GAMES

Let $I:=\{1,2,\dots,n\}$ denote the set of players, 2^I the set of coalitions. By n-person cooperative game (in characteristic form) we mean any function $v: 2^I \to R_+$, where $v(\emptyset) = 0$.

Given a game v as above, the core of v is the set of outcomes (division of the amount v(I)) to which no coalition can object. The core C(v) of the game v is defined formally by

$$C(v) = \{x \in R^n,\ x(I) = v(I)\ ,\ x(S) \geqslant v(S)\ \forall\ S \subset 2^I\}$$

where $x(S) = \sum_{i \in S} x_i$.

For each $T \in 2^I$ $(T \neq \emptyset)$ let B(T) denote the set of all function $\delta: 2^T \to R_+$ such that $\sum_{S \ni i} \delta(S) = 1$ whenever $i \in T$ and $\delta(\emptyset) = 0$.

A game v is said to be balanced (see Shapley (1967)) if for

each $\delta \in B(I)$

$$\sum_{S \subset I} \delta(S)\ v(S) \leqslant v(I)$$

The set of all balanced games is denoted by V_b.

A game v is said to be totally balanced (see Shapley and Shubik (1969)) if for each $T \in 2^I$ and $\delta \in B(T)$

$$\sum_{S \subset T} \delta(S)\ v(S) \leqslant v(T)$$

The set of all totally balanced games is denoted by V_{tb}. It is obvious that $V_{tb} \subset V_b$.

Bondareva (1962) and independently Shapley (1967) proved

THEOREM. $v \in V_b$ iff $C(v) \neq \emptyset$

2.PROBLEM OF FAIR DIVISION

Let $(X,\mathcal{B})$ be a measurable space, where X is an object (e.g. a cake), which has to be divided among n players. By an ordered partition $P = \{A_i\}_{i=1}^n$ of the measurable space $(X,\mathcal{B})$ is meant a collection of disjoint subsets $A_1, A_2, \ldots, A_n$ of X satisfying $A_i \in \mathcal{B}$ for all $i \in I = \{1,2,\ldots,n\}$ and $\bigcup_{i=1}^n A_i = X$. Let $\mathcal{P}$ denote the set of all measurable partitions $P = \{A_i\}_{i=1}^n$ of X. Suppose we are given n non-atomic probability measures $\{\mu_i\}_{i=1}^n$ defined on the same σ-algebra $\mathcal{B}$. Each μ_i represents the individual evaluation of sets from $\mathcal{B}$. In the problem of fair division we are interested in giving to each person i a part A_i of X so that $\mu_i(A_i) \geqslant 1/n$, $i \in I$. A simple and well-known method of realizing a fair division for two players is "for one to cut, the other to choose".In this method each person can ensure that he receives at least a half of the cake according to his own evaluation measure , independently of what the other does. Banach and Knaster (cf. Knaster (1946)) showed that this method can be extended to arbitrary n.

II. COOPERATIVE GAME OF FAIR DIVISION

Let $P = \{A_i\}_{i=1}^n \in \mathcal{P}$ be a partition obtained by mean of certain procedure of fair division (cf. Knaster (1946), Kuhn(1973) and Steinhaus (1949)). Now, we are interested in a secondary division: the players form coalitions to improve the initial partition $P = \{A_i\}_{i=1}^{n}$. For this reason we introduce a cooperative game v_P (associated with P) as follows

$$v_P(S) = \sup \left\{ \sum_{i \in S} \mu_i(E_i),\ \bigcup_{i \in S} E_i = \bigcup_{i \in S} A_i,\ E_i \cap E_j = \emptyset \text{ for } i \neq j,\ E_i \in \mathcal{B},\ i \in I \right\},\ S \in 2^I.$$

It is easy to see that the cooperative game v_P is a special case of a market game with a continuum of commodities which as it was proved by Legut (1986) is totally balanced. Then the game v_P has the non-empty core which can be considered as a cooperative solution of the problem of fair division or more precisely the secondary fair division.

The following example illustrates the nature of the core of a two-person cooperative game of fair division.

EXAMPLE. Let $I = \{1,2\}$ be the set of players and let $\mathcal{P}$ denote the set off all measurable partitions $P = (A, X-A)$. The individual evaluations of player 1 and 2 are represented by probability non-atomic measures μ_1 and μ_2, respectively. It follows from the Lyapunov convexity theorem that the set of all points $\bar{\mu}(P) = (\mu_1(A), \mu_2(X-A))$ is convex and compact in R^2. Figure 1 shows the range $\bar{\mu}(\mathcal{P})$ for some measures μ_1 and μ_2.

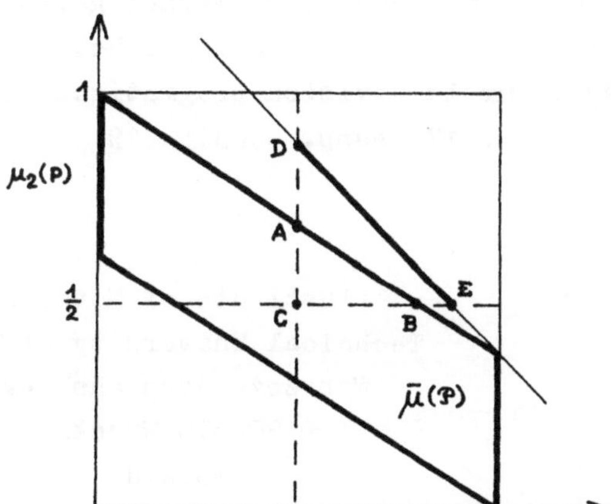

FIGURE 1

Let $P_o = (A_o, X-A_o)$ be a division obtained by using the scheme "divide-and-choose" such that $\mu_1(A_o) = \mu_2(X-A_o) = 1/2$ (the point C in fig. 1) . The core of the cooperative game v_{P_o} is represented by the segment DE, while the segment AB is the Pareto solution. Without a cooperation the players can receive the payoffs from the segment AC or CB only but with a cooperation and sidepayments they can obtain payoffs from segment DE which clearly dominates each noncooperative solution.

REFERENCES

Bondareva O.(1962) : The core of an n-person game. Vestnik Leningrad University, No.13, 141-142.

Knaster B.(1946) : Sur le probléme du partage pragmatique de H. Steinhaus. Ann. Soc. Polon. Math. No. 19 228-230.

Kuhn H.W.(1973) : On Games of Fair Division. In: Esseys in Modern Math. Economics, ed. Shubik M. Princeton, 1973, 29-37.

Legut J.(1986) : Market Games with a Continuum of Indivisible Commodities. Int. J. Game Theory , No. 15 Issue 1, 1-7.

Shapley L.S.(1967) : On balanced sets and cores. Naval Research Logistics Quarterly, No.14, 453-460.

Shapley L.S. and Shubik M.(1969) : On market games. J. Economic Theory , No.1, 9-25.

Steinhaus H.(1949) : Sur la division pragmatique. Econometrica, No. 17 supp. , 315-319.

Institute of Mathematics
Technical University of Wrocław
Wybrzeże Wyspiańskiego 27
50-370 Wrocław
Poland

A CRAMER-RAO TYPE INEQUALITY FOR A CONVEX LOSS FUNCTION

Friedrich Liese

Rostock

Key words: Rao-Cramer inequality, f-divergences, χ^{α}-divergences parameter estimation

ABSTRACT

For an unbiased estimator T for $\tau(\theta)$ lower bounds for the risk $\int g(|T-\tau(\theta)|)dP_{\theta}$ are established. Some generalizations of Rao-Cramer inequality are included as special cases.

RESULTS

Our generalization of the Cramer-Rao inequality is based on the Young inequality which is fundamental in the theory of Orlicz spaces (see A. Kufner, O. John, S. Fučík (1977), M. A. Krasnoselskij, J. B. Rutickij (1961)).

The functions $f,g:[0,\infty)\longrightarrow(-\infty,\infty)$ are said to be associated if they fulfil Young's inequality, i.e.

$$(1) \qquad uv \leq f(u) + g(v) \quad \text{for every } u,v\in[0,\infty).$$

A continuous convex function $f : [0,\infty)\longrightarrow[0,\infty)$ is called a Young function if it is strictly increasing, $f(0) = 0$

$$(2) \qquad \lim_{u\downarrow 0} f(u)/u = 0, \quad \lim_{u\to\infty} f(u)/u = \infty.$$

Put

$$g(v) = \sup\{uv - f(u),\quad 0 \leq u < \infty\}$$

for a Young function f. It is well-known (see M. A. Krasnoselskij, J. B. Rutickij (1961) that g is also a Young function. It is clear from the definition that g is the minimal function which is associated with f. In this case the functions f and g are called complementary. The functions

$$(3)\qquad f(u) = \frac{1}{\alpha} u^{\alpha},\quad g(v) = \frac{1}{\beta} v^{\beta},\quad \frac{1}{\alpha} + \frac{1}{\beta} = 1,\quad 1 < \alpha < \infty$$

are an example of complementary Young functions.

Let $f : (0,\infty) \to (-\infty,\infty)$ be a continuous convex function. We set

$$(4)\qquad f(\tfrac{0}{0})0 = 0,\quad f(0) = \lim_{u \downarrow 0} f(u),\quad f(\tfrac{u}{0})0 = u \cdot \lim_{t \to \infty} \frac{f(t)}{t},\quad u > 0.$$

The existence of the limits in the extended topology of $(-\infty, \infty]$ follows from the convexity of f.

Let P,Q be probability measures on the measurable space $(X, \mathfrak{A})$. μ denotes an σ-finite dominating measure for P, Q. Write p,q for the corresponding densities and put

$$I_f(P,Q) = \int (f(\tfrac{p}{q})q)d\mu .$$

The integral above exists, takes values in $(-\infty, \infty]$ and is independent of the choice of the dominating measure μ. $I_f(P,Q)$ is the f-divergence introduced by Csiszár (1963).

As every Young function f is increasing and convex, the function f_c defined by

$$f_c(u) = f(c|u-1|)$$

is also convex for $c > 0$. To each Young function we associate a family of f-divergences in the following way. Set for $c > 0$

$$J_f^c(P,Q) = \int (f_c(\tfrac{p}{q})q)d\mu .$$

Let for example $f(u) = u^{\alpha}$, $\alpha > 1$. Then

$$J_f^c(P,Q) = c^{\alpha} \int |p - q|^{\alpha} q^{1-\alpha} d\mu.$$

Hence $c^{-\alpha} J_f^c(P,Q) =: \chi^{\alpha}(P,Q)$ is the well-known χ^{α}-divergence (see Vajda (1971)).

Important for all further considerations are properties of the function $\varphi(c) = J_f^c(P,Q)$ for fixed P,Q and f. As every Young function is increasing the function $\varphi(c)$ is non-decreasing. Consequently, if there exists $c_o > 0$ so that

$J_f^{c_o}(P,Q) < \infty$ then $J_f^c(P,Q) < \infty$ for every $0 \leq c \leq c_o$.

Let $\mathcal{P}_f$ denote the class of all pairs of probability measures (P,Q) for which there exists $c_o = c_o(P,Q) > 0$ with $J_f^{c_o}(P,Q) < \infty$.

Some basic properties of the function $\varphi(c)$ will be established in the following lemma.

LEMMA: Let $r : X \to [0,\infty)$ be an $\mathfrak{A}$-measurable function and f a Young function. If there exists $c_o > 0$ such that

$0 < \int (f(\frac{c_o r}{q})q) d\mu < \infty$ then

$$G^f(t) = \inf_{c > 0} \frac{1}{c}(t + \int (f(\frac{cr}{q})q) d\mu)$$

is a continuous strictly increasing concave function on $[0,\infty)$ with

$$G^f(0) = 0 \quad \text{and} \quad \lim_{t \to \infty} G^f(t) = \infty.$$

If $r = |p-q|$ then we write $G^f_{P,Q}(t)$ instead of $G^f(t)$. In other words

$$G^f_{P,Q}(t) = \inf_{c > 0} \frac{1}{c}(t + J_f^c(P,Q)).$$

If $P \neq Q$ then $\mu(p \neq q) > 0$. As f is non-negative and strictly increasing, we get $\mu(f_c(\frac{p}{q})q > 0) > 0$ for every $c > 0$.

Hence for $P \neq Q$ and $(P,Q) \in \mathcal{P}_f$ it holds $0 < J_f^{c_0}(P,Q) < \infty$. The LEMMA yields that in this case the function $G^f_{P,Q}$ has an inverse function which is denoted by $H^f_{P,Q}$. It is an increasing convex function with

$$H^f_{P,Q}(0) = 0, \quad \lim_{t\to\infty} H^f_{P,Q}(t) = \infty.$$

EXAMPLE: Put $f(u) = \frac{1}{\alpha}u^\alpha$ for $\alpha > 1$. A simple calculation shows that in this case

$$H^f_{P,Q}(t) = \frac{1}{\beta}\, t^\beta \,(\chi^\alpha(P,Q))^{-\beta/\alpha}$$

if $0 < \chi^\alpha(P,Q) < \infty$ and $\chi^\alpha(P,Q) = \int |p-q|^\alpha q^{1-\alpha}\, d\mu$.

Let $(P_\theta)_{\theta\in\Theta}$ be a family of probability measures on $(X,\mathfrak{A})$ such that $P_{\theta_1} \neq P_{\theta_2}$ for $\theta_1 \neq \theta_2$. Let $\tau : \Theta \to (-\infty, +\infty)$ be a function and suppose $T : X \to (-\infty, +\infty)$ is an $\mathfrak{A}$-measurable function. We assume that T is an unbiased estimator for the function $\tau(\theta)$, i.e. T is integrable with respect to P_θ and it holds

$$\int T\, dP_\theta = \tau(\theta) \quad \text{for every } \theta \in \Theta .$$

In order to characterize the quality of the estimator T we use a non-negative measurable function g associated with the Young function f as a loss function. Similarly as in the classical Cramer-Rao inequality the following theorem shows that

$$\int g(|T - \tau(\theta)|)\, dP_\theta$$

is bounded below by a positive constant which is independent of the unbiased estimator T.

We write $H^f_{\theta_1,\theta_2}$ instead of $H^f_{P_{\theta_1},P_{\theta_2}}$.

THEOREM 1: Let f be a Young function and g a non-negative measurable function associated with f. If T is an unbiased estimator for $\tau(\theta)$ then for every $\theta_1, \theta_2 \in \Theta$ $\theta_1 \neq \theta_2$ with $(P_{\theta_1}, P_{\theta_2}) \in \mathcal{P}_f$ the following inequality holds

$$\int g(|T - \tau(\theta_2)|)\, dP_{\theta_2} \geq H^f_{\theta_1,\theta_2}(|\tau(\theta_2) - \tau(\theta_1)|).$$

COROLLARY: If $\beta > 1$, $\frac{1}{\alpha} + \frac{1}{\beta} = 1$, and T is an unbiased estimator for $\tau(\theta)$ then

$$\int |T - \tau(\theta)|^{\beta}\, dP_{\theta} \geq \sup_{\theta' \in \Theta,\ \theta' \neq \theta} \frac{|\tau(\theta') - \tau(\theta)|^{\beta}}{(\chi^{\alpha}(P_{\theta'}, P_{\theta}))^{\beta/\alpha}}.$$

The statement of the COROLLARY is due to Vajda (1971).

Now we turn to the case where Θ is an open interval (a,b) of the real line. We are iterested in the behavior of the term on the right hand side of the inequality in THEOREM 1 if θ_1 tends to θ_2. Similarlyas in the classical Cramer-Rao inequality we need some regularity conditions.

(5) $\tau: (a,b) \to (-\infty, +\infty)$ is differentiable.

(6) The family $(P_\theta)_{\theta \in \Theta}$ is dominated by an σ-finite measure μ and the density $p(\theta,x)$ is differentiable with respect to θ for every fixed $x \in X$.

(7) If f is a Young function then

$$\lim_{h \to 0} \int \left(f\left(\frac{c/|h|}{p(\theta,x)}|p(\theta+h,x) - p(\theta,x)|\right) p(\theta,x)\right) \mu(dx) =$$

$$\int \left(f\left(\frac{c|p'(\theta,x)|}{p(\theta,x)}\right) p(\theta,x)\right) \mu(dx) \quad \text{for every } c \geq 0,$$

where $p'(\theta,x) = \dfrac{\partial p(\theta,x)}{\partial \theta}$

and the conventions (4) are used.

We set

$$J^d_f(\theta) = \int \left(f\left(\frac{d}{p(\theta,x)}|p'(\theta,x)|\right) p(\theta,x)\right) \mu(dx)$$

and

$$G^f_{\theta}(t) = \inf_{d > 0} \frac{1}{d}(t + J^d_f(\theta)).$$

If there exists c_0 so that $0 < J_f^{c_0}(\theta) < \infty$ then the function G_θ^f is strictly increasing by the LEMMA. Hence the inverse function exists. It is denoted by H_θ^f.

THEOREM 2: Let T be an unbiased estimator for $\tau(\theta)$, $\Theta = (a,b)$ and f a Young function. If the regularity conditions (5), (6), (7) are fulfilled and there exists d_0 such that $0 < J_f^{d_0}(\theta) < \infty$ then for every non-negative measurable function g associated with f

$$\int g(|T - \tau(\theta)|)dP_\theta \geq H_\theta^f(|\tau'(\theta)|).$$

COROLLARY 1: Let T be an unbiased estimator for $\tau(\theta)$. If $\tau(\theta)$ is differentiable and the limit

$$I_\alpha(\theta) = \lim_{\theta' \to \theta} \frac{\chi^\alpha(P_{\theta'}, P_\theta)}{|\theta' - \theta|^\alpha}$$

exists for $\alpha > 1$, $1/\alpha + 1/\beta = 1$ then

$$(I_\alpha(\theta))^{1/\alpha} \left(\int |T - \tau(\theta)|^\beta dP_\theta\right)^{1/\beta} \geq |\tau'(\theta)|.$$

COROLLARY 2: Let T be an estimator for $\tau(\theta)$ with a bias $b(\theta)$. If both $\tau(\theta)$ and $b(\theta)$ are differentiable and the limit

$$I_2(\theta) = \lim_{\theta' \to \theta} \frac{\chi^2(P_{\theta'}, P_\theta)}{(\theta' - \theta)^2}$$

exists and is non-zero then

$$\int (T - \tau(\theta))^2 dP \geq b^2(\theta) + \frac{(\tau'(\theta) + b'(\theta))^2}{I_2(\theta)}.$$

REMARK: The statement of COROLLARY 1 is due to Vajda (1971). If the regularity condition (7) is fulfilled then it is easy to see that

$$I_2(\theta) = \int \left(\frac{p'(\theta,x)}{p(\theta,x)}\right)^2 p(\theta,x)\mu(dx).$$

This means $I_2(\theta)$ is the classical Fisher information and the statement of COROLLARY 2 is the familiar Cramer-Rao inequality.

PROOFS

The proof of the LEMMA requires more technical considerations. Therefore we omit it.

Proof of THEOREM 1:

Since $(P_{\theta_1}, P_{\theta_2}) \in \mathcal{P}_f$, we obtain from convention (4) that

$$\mu(\{x : p(\theta_2,x) = 0, |p(\theta_1,x) - p(\theta_2,x)| > 0\}) = 0.$$

Using the convention $\frac{0}{0} = 0$ we see that

$$\int c\left(\frac{p(\theta_1,x)}{p(\theta_2,x)} - 1\right)(T(x) - \tau(\theta_2))p(\theta_2,x)\mu(dx) = c\left(\int T dP_{\theta_1} - \int T dP_{\theta_2}\right)$$
$$= c(\tau(\theta_1) - \tau(\theta_2))$$

where the last relation follows from the assumption that T is an unbiased estimator for $\tau(\theta)$. As f and g are associated we may apply inequality (1) to the left hand side

$$(8) \quad c\,|\tau(\theta_1) - \tau(\theta_2)| \leq \int f\left(c\,\frac{|p(\theta_1,x)-p(\theta_2,x)|}{p(\theta_2,x)}\right)p(\theta_2,x)\mu(dx)$$
$$+ \int g(|T - \tau(\theta_2)|)dP_{\theta_2} = J_f^c(P_{\theta_1},P_{\theta_2}) + \int g(|T - \tau(\theta_2)|)dP_{\theta_2}.$$

This means

$$G^f_{P_{\theta_1},P_{\theta_2}}\left(\int g(|T - \tau(\theta_2)|)dP_{\theta_2}\right) \geq |\tau(\theta_1) - \tau(\theta_2)|$$

and

$$\int g(|T - \tau(\theta_2)|)dP_{\theta_2} \geq H^f_{\theta_1,\theta_2}(|\tau(\theta_1) - \tau(\theta_2)|)$$

as the inverse function $H^f_{\theta_1,\theta_2}$ is increasing.

The COROLLARY follows from the formula for $G^f_{P,Q}$ in the EXAMPLE.

Proof of THEOREM 2:

Inequality (8) yields for $\theta_1 = \theta + h$, $\theta_2 = \theta$, $c = \frac{d}{h}$, $d > 0$, as $h \to 0$ in view of the regularity condition (7)

$$d\,|\tau'(\theta)| \leq J_f^d(\theta) + \int g(|T-\tau(\theta)|)dP_\theta$$

and

$$G_\theta^f\left(\int g(|T-\tau(\theta)|dP_\theta\right) \geq |\tau'(\theta)|.$$

The assumption $0 < J_f^{d_o}(\theta) < \infty$ and the LEMMA guarantee the existence of a strictly increasing inverse function H_θ^f of G_θ^f. To end the proof we have only to apply H_θ^f to both sides of the above inequality.
COROLLARY 1 is an immediate consequence of THEOREM 2. For proving COROLLARY 2 one has only to put $\alpha = \beta = 2$ and $\tau(\theta) + b(\theta)$ instead of $\tau(\theta)$ in COROLLARY 1.

REFERENCES

Bouchon B., F. Pessou (1985): Logarithmic information of degree q linked with an extension of Fisher's information. Kybernetika 21, 346-359.

Boekee B. E. (1978): The D_f-information of order s. Trans. 8th Prague Conf. on Information Theory, ... Vol C, 55-56.

Csiszár I. (1963): Eine informationstheoretische Ungleichung und ihre Anwendung auf den Beweis der Ergodizität von Markoffschen Ketten. Publ. Inst. Hungar.Acad. Sci. 8, Ser. A, 85-108.

Kagan A. M. (1963): On the theory of Fisher's amount of information (In Russian). Dokl. Akad. Nauk SSSR, 151, 277-278.

Krasnoselskij M. A., J. B. Rutickij (1961): Convex functions and Orlicz Spaces. Noordhoff, Groningen

Kufner A., O. John, S. Fučík (1977): Function Spaces. Academia Praha.

Vajda I. (1973): χ^α-divergence and generalized Fisher's information. Trans. 6th. Prague Conf. on Inform. Theory 1971. Academia, Praha.

Wilhelm-Pieck-Universität Rostock
DDR-2500 Rostock,Universitätsplatz

DECISION MAKING IN IDENTIFICATION

OF RANDOM FIELDS

Antoni Ligęza, Małgorzata Więcek

Cracow

Key words: random field, regionalized variable, multicriteria optimization, identification, linear systems

ABSTRACT

A mathematical model for identification of the stationary random field is presented. Some multiobjective cases of the identification process are considered and possible procedures for aiding the multicriteria decision making are suggested. Remarks related to the application as well as to possible extension of presented methods are included .

INTRODUCTION

Decision making in the area of natural resources management calls for mathematical tools for the adequate description of natural phenomena as well as for the development of knowledge-based interactive systems for planning and decision support. Data observation and measurement methods, although accessible in great variety, are very expensive, and the importance of their accuracy renders the application of optimization techniques indispensable. Moreover, if the decision process involves choosing among multi-attributed alternatives it requires an initial search for a dominance stucture and ultimate identification of a set of nondominated alternatives.

The paper deals with a mathematical model for identification problems of the stationary random field. It is assumed that the

phenomenon of our concern is described by an unknown random function representing a system parameter of interest. Such a mathematical approach has to be applied to a large class of problems which are characterized by certain random variables and scarce data randomly scattered in space e.g. the quality or the quantity of geological resources, the water level in hydrology, soil parameters in pedology, the rainfall or the degree of pollution in environmental sciences, etc. The common feature of the phenomena above is that having given a number of values of a regionalized variable one is to choose such an additional data point that extremizes given multiple criteria.

The basis of the methods for locating a single data point /or multiple points/ is the theory of minimizing uncertainty in parameter estimation whereas consideration of the system in terms of multi-attribute decision theory elicits a process of multiple criteria alternative selection.

DEFINITIONS

Let x be a point in a one-, two- or three-dimensional space and $z(x)$ be a value of a function describing a phenomenon under consideration. The fundamental idea of the theory is to consider the function $z(x)$ as one realization of a random function $Z(x)$. The very obvious advantage of such an approach is that now we can make use of the statistical theory in order to analyse the character of the regionalized variable.

We assume that $Z(x)$, $x \in R^k$, $k \in \{1,2,3\}$, is a real-valued function whose deviations from the mean are purely nondeterministic and for which the following functions are defined:

- the mean value $E[Z(x)]$
- the spatial covariance

$$(1) \quad COV[Z(x), Z(x+h)] = E\{[Z(x) - E[Z(x)]][Z(x+h) - E[Z(x+h)]]\}, h \in R^k$$

- the spatial variogram defined as the semi-variance of random function increments

$$(2) \quad \gamma(x, x+h) = \frac{1}{2} VAR[Z(x+h) - Z(x)]$$

The variogram γ is a function of a vector h expressing how the values of a random function differ on average at points x and $x+h$ according to the mutual location of the points.

In order to consider the stationary random field let us make the following assumptions:

- the mean value of the regionalized variable $Z(x)$ is the same all over the field of interest

(3) $E[Z(x)] = m$

- the spatial covariance of the regionalized variable $Z(x)$ is the same all over the field of interest

(4) $COV[Z(x), Z(x+h)] = K(h) \quad , h \in R^k$

If a random function is second-order stationary it is also intrinsic and its variogram is

(5) $$\gamma(x,x+h) = \gamma(h) = \frac{1}{2}\left(VAR[Z(x+h)] + VAR[Z(x)] - 2COV[Z(x+h),Z(x)]\right)$$
$$= K(0) - K(h)$$

IDENTIFICATION MODEL - PROBLEM STATEMENT

Let V be the domain of our interest and let $z(V)$ denote the average value of the regionalized variable centred at point x in V :

(6) $$z(V) = \frac{1}{|V|} \int_V z(x)dx$$

Now, assume that the data from points $x_1, x_2, \dots, x_n$ are known, $n < N$. Let $\gamma(W,h)$ denote the variogram obtained in the usual manner, David (1977), on the basis of the experimental data, where $W = \{x_1, x_2, \dots, x_n\}$. Our primary goal is to estimate the value $z(V)$ as precisely as possible. It is presumed that the estimated value Z^* of $z(V)$ can be calculated as a weighted average of n available data points :

(7) $$Z^* = \sum_{i=1}^{n} a_i z(x_i)$$

The identification optimization problem is formulated as follows : in order to estimate the value $z(V)$ we have to find weights a_i, $i=1,2,\dots,n$, which will make the weighted average (7) the best estimation of $Z(V)$ - i.e. the following conditions must be fulfilled :

(8) $VAR[Z(V) - Z^*]$ must be minimal

The estimator must be unbiased - i.e. $E[Z(V) - Z^*] = 0$

Altogether what is desired is the best linear unbiased estimator BLUE. In the case of the considered random field the condition of unbiasedness leads to the following equation for

BLUE :

(9) $\sum_{i=1}^{n} a_i = 1$

Solving optimization problem (8) by means of the Lagrange function optimization method we obtain a linear system of n+1 equations and n+1 unknown coefficients a_i, i=1,...,n, μ, David (1977), Szidarovszky (1983b) :

$$(10)\quad \begin{aligned} &\sum_{j=1}^{n} a_j \gamma_{ij} + \mu = \gamma_{V,i} \qquad i=1,\dots,n \\ &\sum_{i=1}^{n} a_i = 1 \end{aligned}$$

where $\gamma_{ij} = \gamma(x_i - x_j)$ and $\gamma_{V,i} = \frac{1}{|V|} \int_V \gamma(x_i - x)dx$

The minimal value of the estimation variance VAR$[Z(V) - Z^*]$ denoted as var$[Z(V) - Z^*]$ can be written in the following form :

$$(11)\quad \mathrm{var}[Z(V) - Z^*] = \sum_{j=1}^{n} a_j \gamma_{V,j} + \mu - \gamma_{V,V}$$

where $\gamma_{V,V} = \frac{1}{|V|^2} \int_V \int_V \gamma(x - y)dxdy$

In matrix form system (10) is as follows :

(12) $G\, a = b$

where

$$G = \begin{bmatrix} 0 & 1 & \dots 1 \\ 1 & \gamma_{11} & \dots \gamma_{1n} \\ \cdot & & \\ \cdot & & \\ \cdot & & \\ 1 & \gamma_{n1} & \dots \gamma_{nn} \end{bmatrix}, \quad a = \begin{bmatrix} \mu \\ a_1 \\ a_2 \\ \vdots \\ a_n \end{bmatrix}, \quad b = \begin{bmatrix} 1 \\ \gamma_{V,1} \\ \cdot \\ \cdot \\ \cdot \\ \gamma_{V,n} \end{bmatrix}$$

and respectively

$$(13)\quad \mathrm{var}[Z(V) - Z^*] = b^T G^{-1} b - \gamma_{V,V} = b^T a - \gamma_{V,V}$$

MULTIOBJECTIVE OPTIMIZATION MODELS

Let us recall the problem. There are N possible observation point locations and among them n points already explored. In general, a multiobjective optimization problem consists in selecting from the p=N-n points left some additional data points in order to minimize given objective functions.

Simultaneous estimation of a random function on M domains.

This section is mainly based on the papers of Szidarovszky (1983a),(1983b) who has widely developed the idea of multiobjective approach in the case of a random field that has to be identified by means of its one characteristic feature on different domains belonging to this field. Let us consider two cases :

I. Optimal location of additional points $x_{n+1},\ldots,x_N$ should be determined so that the estimation variances of $Z(V_1),Z(V_2),\ldots,Z(V_M)$ are minimal.

On the basis of system (12) we obtain the following mathematical model consisting of m functions, each of them obtained from formula (13), in which the dimensions of matrix G and vector b have been extended in order to get a system of dimension N.

(14) $\min \mathrm{Var}_{N,m} = \min b_{N,m}^T G_N^{-1} b_{N,m}$, $m=1,\ldots,M$

where the only unknowns are point locations $x_{n+1},\ldots,x_N$.

Szidarovszky proposed two solution methods for problem (14): the method of weighting coefficients and the branch and bound algorithm based on the monotonic property of the estimation variance.

II. Optimal location of some additional measurement points chosen from among $p=N-n$ points left.

If all the additional points were to be included in the estimation process then we would have a system of large dimension:

(15) $G_{n+p}\, a_{n+p} = b_{n+p}$

And if several additional points from the set $x_{n+1},\ldots,x_N$ are not included, then we have either to drop the corresponding rows and columns of matrix G_{n+p} and the corresponding elements of vectors a_{n+p} and b_{n+p} to obtain a system which is based only on points that are really involved in the estimation process or to make them equal zero. This operation can be accomplished by introducing the diagonal matrix D, where its the first n+1 diagonal elements are equal to one, $d_i=1$, $i=1,2,\ldots,n+1$, and the remaining diagonal elements are

$$d_i = \begin{cases} 1 & \text{if the additional point is included in the system} \\ 0 & \text{else} \end{cases}$$

$i=n+2,\ldots,N+1$

Thus system (15) can be expressed as :

(16) $D\, G_{n+p}\, D\, a_{n+p} = D\, b_{n+p}$

where diagonal elements d_i, $i=n+2,\ldots,N+1$, and components of a_{n+p} are the unknowns.

The corresponding estimation variance is given by :

(17) $Var_{n+p} = b^T_{n+p} D a_{n+p} - \gamma_{V,V}$

and the multiobjective problem is formulated as follows :

$$\min Var_{n+p,m} = b^T_{n+p,m} D_m a_{n+p,m} \qquad m=1,\dots,M$$

(18) subject to

$$D_m G_{n+p} a_{n+p,m} = D_m b_{n+p,m}$$

This is a mixed integer 0-1 programming problem since diagonal elements $d_{i,m}$, $i=n+2,\dots,N+1$, of matrix D_m and components of $a_{n+p,m}$, $m=1,\dots,M$, are the unknowns.

There are different solution methods which could be applied to solving problem (18): the weighting method, the ε-constraint method, sequential optimization, the distance based methods, the branch and bound method /after transforming of the multiobjective problem into a special single objective problem, where the only objective function depends monotonically on the objectives of the original problem/.

Estimation of P random functions on a domain V.

Now we consider the identification process of different characteristic features of a random field over a given domain V. The optimization problem consists in selecting from $p=N-n$ points left some additional points in order to minimize the estimation variances of the random field. On the basis of the previous considerations the multiobjective problem can be formulated as follows :

$$\min Var_{n+p,r} = b^T_{n+p,r} D a_{n+p,r} \qquad r=1,\dots,P$$

(19) subject to

$$D G_{n+p,r} D a_{n+p,r} = D b_{n+p,r}$$

where matrix D has the same meaning as in model (18).

This is also a mixed integer 0-1 programming problem since diagonal elements d_i, $i=n+2,\dots,N+1$, of matrix D and components of $a_{n+p,r}$, $r=1,\dots,P$, are the unknow.

In the paper of Ligęza and Więcek (1985) a hierarchical multistage strategy for dealing with the optimal identification problem of the random field was presented. The strategy can be easily adapted to solving problem (19) extended by the additional limitation on the number of simultaneously explored points. It has to be emphasized that the strategy ensures maximal utilization of the experimental data being obtained while performing the identification process.

CONCLUDING REMARKS

A great number of real-life systems of different nature may be described by means of random fields and appropriately assumed for them statistical hypotheses. In practice, the identification process of a random field consists in performing of quantitative and qualitative analysis of its features and therefore in this case multiobjective models seem to be of great importance.

In this paper the multiobjective models for identification of the stationary random field have been discussed. The non-stationary random field, described in David (1977), Szidarovszky (1983a), Ligęza and Więcek (1985), may be also considered in terms of the multicriteria identification process. The non-stationary approach, which is often referred to as kriging, is based on the assumption that the mean value of the random function varies in a regular manner. This assumption leads to a more complicated mathematical model which, however, does not yield a different multiobjective interpretation. For the sake of simplicity, multiobjective models (14),(15),(18) and (19) have not been discussed in this paper in terms of the non-stationary case but they remain valid and could be easily extended to the form relevant to the universal kriging system.

On the basis of presented considerations one may conclude that there exist sufficient mathematical tools for constructing a decision support system that could perform the identification process of multifarious random fields. Existing optimization techniques may be easily engaged in the identification process. The choice of such a technique should ensure minimal computational effort as well as obtaining of results of high accuracy. The first exercises in implementing the random field identification were undertaken by David (1977). There already exist a few software packages e.g. the geostatistical BLUEPACK, the specialized in contour mapping KRIGEA/CARTO. Designing and developing of a decision support system for the purpose of random fields identification would create an effective tool in engineering practice.

REFERENCES

Bastin G.,Gevers M.(1981) :Identification and optimal estimation of random fields from scattered point--wise data. In: Automatica 21,No.2,139-155.

David M. (1977) :Geostatistical Ore Reserve Estimation. Elsevier, Amsterdam.

Dubrule O. (1983) :Two methods with different objectives: splines and kriging. In: Mathematical Geology 15,No.2,245-257.

Gonzalez P.L., Chami H. (1984) :Amélioration d'un réseau de surveillance de la pollution atmospherique. In: RAIRO Automatique 18,No.4,369-384.

Journel A.G. Huijbregts Ch.J. (1978) :Mining Geostatistics. Academic Press, New York.

Ligęza A., Więcek M. (1985) :Optimal identification problems of regionalized variables systems. In: Trans. of the 12th IFIP Conference on Modelling and Optimization, Budapest, in printing.

Szczepankiewicz E. (1985) :Zastosowanie pól losowych (Application of Random Fields). Państwowe Wydawnictwo Naukowe, Warszawa.

Szidarovszky F. (1983a) :Optimal observation network in geostatistics and underground hydrology. In: Applied Mathematical Modelling 7,25-32.

(1983b) :Multiobjective observation network design for regionalized variables. In: International Journal of Mining Eng. 1, 331-342.

Institute of Automatic Control
University of Mining and Metallurgy
al.Mickiewicza 30
30-059 Cracow
Poland

ZUR DYNAMISCHEN OPTIMIERUNG IN ZEITDISKRETEN STOCHASTISCHEN ENTSCHEIDUNGSMODELLEN MIT ERWEITERT - REKURSIVEN UND VEKTORWERTIGEN GEWINNFUNKTIONALEN

Wilfried Lipfert
Dresden

Schlüsselworte: Stochastische Entscheidungsmodelle, dynamische Optimierung, allgemeine (nicht-additive) Gewinnfunktionale, vektorwertige Gewinne.

ZUSAMMENFASSUNG

Es wird ein zeitdiskretes stochastisches Entscheidungsmodell mit erweitert-rekursiven Gewinnfunktionalen betrachtet und eine Aussage im Hinblick auf Lösungsmöglichkeiten mittels dynamischer Optimierung gemacht. Weiter wird gezeigt, wie dieses Entscheidungsmodell durch die Einführung vektorwertiger Gewinnfunktionale erweitert werden kann und unter welchen Voraussetzungen auf der Grundlage der Pareto-Optimalität eine Folge maximaler Gewinne ein System verallgemeinerter Optimalitätsgleichungen erfüllt.

VORBETRACHTUNGEN

Ausgehend von Bellman (1957) wurde die Theorie der dynamischen Programmierung über Blackwell (1965), Hinderer (1970), Müller (1984) weitgehend vollständig für den Fall additiver Gewinnstrukturen entwickelt. Gleichzeitig entstand zunehmend das

Bedürfnis z. B. für ökonomische, psychologische und andere Prozesse allgemeinere Ertragsfunktionale zu verwenden, die aber noch ein rekursives oder wenigstens erweitert rekursives Vorgehen im Sinne der dynamischen Optimierung zur Gewinnung von optimalen Lösungen gestatten. Mit dieser Problematik befaßten sich in den letzten Jahren eine Reihe von Autoren (z. B. Furukawa/Iwamoto (1976), Groenewegen/Wessels (1978), Lipfert (1980, 1985), Schäl (1981)). Es zeichnen sich dabei zwei Herangehensweisen ab. Während z. B. bei Furukawa/Iwamoto rekursive (u.a. rekursiv-additive) Gewinnstrukturen, die vom Fakt abhängig sind, zugrunde gelegt werden, verwendet man z. B. bei Groenewegen/Wessels und Schäl ein einziges Nutzenfunktional g für alle reduzierten Zeitabschnitte $[t,\infty) \subseteq \mathbb{N}$ und erklärt die Optimalität einer Strategie so, daß sie gleichzeitig für alle diese Zeitabschnitte gilt. Es kann dadurch auf topologische Strukturen bei der Angabe von Bedingungen für die Optimalität einer Strategie verzichtet werden.

Im folgenden wird die erstgenannte Linie betrachtet und von einem allgemeinen stochastischen Entscheidungsmodell (SEM), beschrieben durch das Tupel $((S,\mathcal{T}),(A,\mathfrak{A}),(E_t),(q_t),q_o,(g_{(t,N]}); t\in\mathbb{N})$, ausgegangen (vgl. Lipfert (1985)).

Die Räume $(S,\mathcal{T})$ und $(A,\mathfrak{A})$ (Zustandsraum bzw. Aktionsraum) sind standard-Borelsche Räume (vgl. Hinderer (1970)). (E_t) ist eine Folge von Abbildungen E_t von Mengen $V_t \subseteq \overline{V}_t = (S\times A)^{t-1}\times S$ zulässiger Vorgeschichten $v_t = (s_1,a_1,s_2,a_2,\ldots,s_t)$ in die Menge aller nichtleeren Untermengen von A. (q_t) ist eine Folge von Übergangswahrscheinlichkeiten q_t von $V_t\times A$ in $\mathcal{T}$ (Bewegungsgesetz im t-ten Takt), q_o bezeichnet eine Anfangsverteilung auf $\mathcal{T}$. $g_{(t,N]}$ ist ein reellwertiges meßbares Funktional, $g_{(t,N]} : V_{N-1}\times A \rightarrow R$, der (totale) Ertrag im Zeitintervall (t,N], t, $N\in\mathbb{N}$, $t<N$. Als bedingten erwarteten Gesamtertrag (Gewinn) im Intervall (t,N], falls unter der Strategie $e = (e_t)$ (die e_t sind meßbare Abbildungen von V_t in $E_t(v_t) \subseteq A$) die Vorgeschichte $v_t = (\underbrace{s_1,e_1(s_1),s_2}_{v_2}, e_2(v_2),\ldots,s_t)$ mit $e_i(v_i) = a_i$, $1 \leq i < t$, erschien, definieren wir

(1) $W^e_{(t,N]}(v_t) := E_e[g_{(t,N]}(\dot{V}_t, e_t(\dot{V}_t), X_{t+1}, \ldots, X_{N-1}, e_{N-1}(\dot{V}_{N-1})) \mid \dot{V}_t = v_t]$

mit $\dot{V}_k = (X_1, e_1(X_1), X_2, e_2(\dot{V}_2), \ldots, X_k)$ als Zufallsvektor der Vorgeschichte v_k.

Für die Anwendung der Grundgedanken der dynamischen Optimierung interessieren Ertragsfunktionale W^e_t, $t \in \mathbb{N}$, die trennbar sind und eine gewisse Rekursivität aufweisen.

Wir betrachten z. B. Ertragsfunktionale der Gestalt

(2) $W^e_{(t,N]}(v_t) = \bar{g}_{(t,N]}(v_t, e_t(v_t), \check{f}^e_{(t+1,N]}(v_t))$, $t \in \mathbb{N}$,

(3) mit $\check{f}^e_{(t+1,N]}(v_t) := f_{(t+1,N]}(W^e_{(t+1,N],t}(v_t), \bar{r}^e_{(t+1,N],t}(v_t))$

und $W^e_{(t+1,N],t}(v_t) := \int q_t(v_t, e_t(v_t), ds) W^e_{(t+1,N]}(v_t, e_t(v_t), s)$,

$\bar{r}^e_{(t+1,N],t}(v_t) := E_e[r^e_{(t+1,N]}(\dot{V}_t, e_t(\dot{V}_t), X_{t+1}, e_{t+1}(\dot{V}_{t+1}), \ldots, e_{N-1}(\dot{V}_{N-1})) \mid \dot{V}_t = v_t]$.

Dabei sind die Funktionale $\bar{g}_{(t,N]}, f_{(t+1,N]}, r^e_{(t+1,N]}$ meßbar und reellwertig. Eine Strategie e^o aus der Menge $\Delta(v_t)$ aller mit v_t 'verträglichen' Strategien e (für diese gilt $e_i(v_i) = a_i$, $1 \le i < t$ und damit $v_t = (s_1, e_1(s_1), s_2, e_2(v_2), \ldots, s_t)$) heißt optimal in (t,N], wenn

$$P_{e^o}\text{ - f.s. } W^{e^o}_{(t,N]}(v_t) = \sup_{e \in \Delta(v_t)} W^e_{(t,N]}(v_t) =: W_{(t,N]}(v_t), t \in \mathbb{N},$$

erfüllt ist. Das Wahrscheinlichkeitsmaß P_e ist bekanntlich nach einem Satz von Ionescu Tulcea eindeutig bestimmt.

EIN ENTSCHEIDUNGSMODELL MIT ERWEITERT-REKURSIVEN ERTRAGSFUNKTIONALEN

Satz 1: (vgl. Lipfert (1985)) In einem SEM seien die Ertragsfunktionale von der Gestalt (2). Die $g_{(t,N]}$ seien beschränkt und bez. des dritten Arguments linksseitig stetig und isoton. Dann genügt die Folge der maximalen bedingten erwarteten Gewinne $(W_{(t,N]})$, $t \in \mathbb{N}$, dem System der Gleichungen

(4) $W_{(t,N]}(v_t) = \sup_{a \in E_t(v_t)} \bar{g}_{(t,N]}(v_t, a, \sup_{e \in \Delta(v_t,a,s)} \check{f}^e_{(t+1,N]}(v_t))$, $t \in \mathbb{N}$,

$a = e_t(v_t)$, $s \in S$, die wir als verallgemeinerte Optimalitätsgleichungen auffassen können.

Im Hinblick auf Lösungsmöglichkeiten mittels dynamischer Optimierung ist die folgende Aussage von Interesse.

Satz 2: Sind in einem SEM mit Ertragsfunktionalen (2) die Voraussetzungen des Satzes 1 erfüllt und gilt für das Funktional (3) für alle $t \in \mathbb{N}$

$$W^{e}_{(t+1,N],t} \geq W^{e'}_{(t+1,N],t}, e,e' \in \Delta(v_t, e_t(v_t), s_{t+1}),$$

$$(5) \qquad \Longrightarrow \check{f}^{e}_{(t+1,N]} \geq \check{f}^{e'}_{(t+1,N]}, \text{ dann folgt für alle } t \in \mathbb{N}$$

$$\text{aus } W^{e^o}_{(t+1,N],t}(v_t) = \sup_{e \in \Delta(v_t, e_t(v_t), s_{t+1})} W^{e}_{(t+1,N],t}(v_t)$$

$$\text{stets } W^{e^o}_{(t,N)}(v_t) = \sup_{e \in \Delta(v_t)} W^{e}_{(t,N]}(v_t).$$

Bemerkung 1:
Die Voraussetzung (5) ist ersichtlich erfüllt, wenn die Funktionale $f_{(t+1,N]}$ isoton bez. beider Argumente ist, d. h. wenn aus

$$W^{e}_{(t+1,N],t} \geq W^{e'}_{(t+1,N],t} \wedge \bar{r}^{e}_{(t+1,N],t} \geq \bar{r}^{e'}_{(t+1,N],t} \quad \text{stets}$$

$\check{f}^{e}_{(t+1,N]} \geq \check{f}^{e'}_{(t+1,N]}$ folgt, und wenn z. B.

$$W^{e}_{(t+1,N],t} \geq W^{e'}_{(t+1,N],t} \quad \bar{r}^{e}_{(t+1,N],t} \geq \bar{r}^{e'}_{(t+1,N],t} \text{ impliziert.}$$

Beispiel 1: Gegeben sei ein SEM mit dem Ertragsfunktional

$$(6) \quad W^{e}_{(t,N]}(v_t) = \sum_{k=0}^{N-1} \bar{w}^{e}_{t+k,t}(v_t) - \bar{w}^{e}_{t+2,t}(v_t), t \in \{1,2,\ldots,N-1\},$$

$$t+k \leq N-1, \text{ wobei}$$

$\bar{w}^{e}_{t+k,t}(v_t) := E_e[w_{t+k}(\dot{V}_{t+k}, e_{t+k}(\dot{V}_{t+k})) \mid \dot{V}_t = v_t]$ der bedingte Erwartungswert des zufälligen Einschrittgewinnes $w_{t+k}(\dot{V}_{t+k}, e_{t+k}(\dot{V}_{t+k}))$ im Intervall $(t+k, t+k+1]$ ist.

Es liegt mit diesem gestört-additivem Ertragsfunktional ersichtlich ein spezielles erweitert-rekursives Ertragsfunktional vor:

$$\bar{r}^{e}_{(t+1,N],t}(v_t) := \bar{w}^{e}_{t+3,t}(v_t) - \bar{w}^{e}_{t+2,t+1}(v_t)$$

$$\check{f}^{e}_{(t+1,N]}(v_t) := W^{e}_{(t+1,N],t}(v_t) + \bar{r}^{e}_{(t+1,N],t}(v_t)$$

Die Voraussetzung (5) ist z. B. erfüllt (vgl. Bem. 1), wenn die Folgen der Einschritterträge$(\bar{w}^{e}_{t})$, $e \in \Delta$, die Bedingungen

$\overline{w}^{e}_{t+k,t} - \overline{w}^{e}_{t+k-1,t} \geq \overline{w}^{e'}_{t+k,t} - \overline{w}^{e'}_{t+k-1,t}$ erfüllen.

ERWEITERUNG DES SEM ZU EINEM ENTSCHEIDUNGSMODELL MIT VEKTORWERTIGEN ERTRAGSFUNKTIONALEN

Bei realen Entscheidungsproblemen liegen öfter nicht vollständig geordnete Mengen vor, so daß Fragen der Optimierung z. B. vektorwertiger Ertragsfunktionale von Bedeutung sind. Wird das SEM so erweitert, daß die allgemeinen Ertragsfunktionale vektorwertig sind, dann ergeben sich unter Beibehaltung der Grundgedanken inhaltlich wesentliche Unterschiede im Zusammenhang mit der Präferenz von erzielten Ergebnissen.
In Arbeiten von Klötzler (1978) werden für deterministische Entscheidungsmodelle rekursive vektorwertige Gewinnfunktionale eingeführt, Nollau (1985) untersucht ein stochastisches dynamisches Entscheidungsmodell mit vektorwertigen additiven Gewinnfunktionalen.
Im folgenden wird an das SEM angeknüpft.

Ein stochastisches Entscheidungsmodell mit vektorwertigem Gewinn wird durch das Tupel

(7) $((S,\mathcal{T}),(A,\mathfrak{A}), (E_t), (q_t), q_o, (\overline{g}_{(t,N]}); t<N, t\in\mathbb{N})$

beschrieben. Die $\overline{g}_{(t,N]}$, $t<N$ sind vektorwertige Funktionale des (totalen) Gewinns im Intervall (t,N],

$\overline{g}_{(t,N]} : V_{N-1} \times A \longrightarrow R^p$ (p-dimensionaler euklidischer Raum).
Die Räume S und A seien abzählbar.
Vom Modellansatz her sind die $\overline{g}_{(t,N]}$ Zufallsvektoren
$\overline{g}_{(t,N]} = ({}^1\overline{g}_{(t,N]},\ldots,{}^p\overline{g}_{(t,N]})$.
Es wird angenommen, daß die auftretenden Erwartungen existieren.
$\overline{W}^{e}_{(t,N]} := E_e\, \overline{g}_{(t,N]} = (E_e\, {}^1\overline{g}_{(t,N]},\ldots,E_e\, {}^p\overline{g}_{(t,N]}) =: ({}^1\overline{W}^{e}_{(t,N]},\ldots,{}^p\overline{W}^{e}_{(t,N]})$
Speziell betrachten wir trennbare Ertragsfunktionale der Form

$$W'^{e}_{(t,N]}(v_t) = E_e[\overline{g}_{(t,N]}\; \dot{V}_t, e_t(\dot{V}_t), X_{t+1},\ldots,X_{N-1}, e_{N-1}(\dot{V}_{N-1}) \mid \dot{V}_t = v_t]$$

(8)

$$= g'_{(t,N]}(v_t, e_t(v_t), f'^{e}_{(t+1,N]}(v_t)) \quad \text{mit}$$

$$f'^{e}_{(t+1,N]}(v_t) := \tilde{f}_{(t+1,N]}(W'^{e}_{(t+1,N],t}(v_t),\ r'^{e}_{(t+1,N],t}(v_t))$$

und $W'^{e}_{(t+1,N],t}(v_t) := \sum_{s\in S} q_t(v_t, e_t(v_t), s) W'^{e}_{(t+1,N]}(v_t, e_t(v_t), s)$,

$$r'^{e}_{(t+1,N],t}(v_t) = \sum_{s\in S} q_t(v_t, e_t(v_t), s) r'^{e}_{(t+1,N]}(v_t, e_t(v_t), s)$$

Dabei sind die Funktionale $g'_{(t,N]}$, $\tilde{f}_{(t+1,N]}$, $r'^{e}_{(t+1,N],t}$ vektorwertige Abbildungen in R^p und unter $\sum_{s\in S} \beta_s B_s$, $\beta_s \in R^+$, $s \in S$, $B_s \subset R^p$, verstehen wir die Menge $\left\{\sum_{s\in S} \beta_s b_s : b_s \in B_s,\ s \in S\right\}$.

Die Präferenz der Ertragsvektoren $\overline{W}^{e}_{(t,N]}(v_t)$, $e\in\Delta(v_t)$, wird wie folgt erklärt:

$$\overline{W}^{e}_{(t,N]}(v_t) \succ \overline{W}^{e'}_{(t,N]}(v_t) \Longleftrightarrow {}^{i}\overline{W}^{e}_{(t,N]}(v_t) \geq {}^{i}\overline{W}^{e'}_{(t,N]}(v_t) \forall\, i=1,\dots,p,$$

und für wenigstens ein i gilt echt die Relation '$>$'.
($a \succ b$ bedeutet: a vorzuziehen b (oder a besser als b))
Diese Ordnungsrelation ist nicht vollständig, ermöglicht aber die Erklärung der Optimalität einer Strategie für das Entscheidungsmodell (7) mittels der Pareto-Optimalität.

<u>Definition 1:</u> Eine Strategie $e^o \in \Delta(v_t)$ heißt optimal in $(t,N]$, $t<N$, $t, N \in \mathbb{N}$, wenn P_{e^o} - f.s. (bez. v_t)

$$\overline{W}^{*}_{(t,N]}(v_t) := \overline{W}^{e^o}_{(t,N]}(v_t) \in \max{}^{*}\, \overline{W}^{e}_{(t,N]}(v_t) =$$

$$\left\{\overline{W}^{e}_{(t,N]}(v_t) \subset W : \not\exists\ \overline{W}^{e'}_{(t,N]}(v_t) \text{ mit } \overline{W}^{e'}_{(t,N)}(v_t) \succ \overline{W}^{e}_{(t,N]}(v_t)\right\}$$

gilt (W ist die Menge aller vektorwertigen Gewinnfunktionale in $(t,N]$, $\overline{W}^{*}_{(t,N]}(v_t) = \max^{*}_{e\in\Delta(v_t)} \overline{W}^{e}_{(t,N]}(v_t)$ ist das Pareto-Maximum von $\overline{W}^{e}_{(t,N]}(v_t)$).

Für die Herleitung von verallgemeinerten Optimalitätsgleichungen für Ertragsfunktionale (8) wird die Eigenschaft der Maximumvertauschbarkeit (vgl. Lipfert (1980)) modifiziert, und es wird die bekannte von Neumann-Morgenstern Eigenschaft (vNME) benötigt.

Definition 2: Es sei $M^* = \max^* M \subset M \subset R^p$. Dann sagt man M^* hat die (v NME) genau dann, wenn für jedes Element $\varkappa \in M \setminus M^*$ ein Element $\varkappa' \in M^*$ existiert, so daß $\varkappa' \succ \varkappa$ gilt.

Lemma 1: (vgl. Klötzler, Nollau, Lipfert (1985))
Die Mengen $E_t(v_t) \subseteq A \quad t = 1,2,\dots,N-1$, und $\Delta(v_t, a, s), a \in E_t(v_t)$, seien nicht leer. Die $g'_{(t,N]}$, $t = 1,2,\dots,N-1$, seien bez. des dritten Arguments isoton in dem Sinne, daß aus

$$f'^{e}_{(t+1,N]}(v_t) \succ f'^{e'}_{(t+1,N]}(v_t),\ e,\ e' \in \Delta(v_t, a, s),$$

stets

$$g'_{(t,N]}(v_t, a, f'^{e}_{(t+1,N]}(v_t)) \succ g'_{(t,N]}(v_t, a, f'^{e'}_{(t+1,N]}(v_t)) \text{ folgt.}$$

Dann gilt $\overline{W}^*_{(t,N]}(v_t) \subseteq \max\limits_{a \in E_t(v_t)}{}^{*} g'_{(t,N]}(v_t, a, \max\limits_{e \in \Delta(v_t,a,s)}{}^{*} f'^{e}_{(t+1,N]}(v_t))$,

wenn die auftretenden Mengen maximaler Werte nicht leer sind.

Mit Hilfe von Lemma 1 kann ein System von mengenwertigen Gleichungen betrachtet werden, die man in Analogie zum Fall des eindimensionalen Wertebereiches für alle Ertragsfunktionale als verallgemeinerte vektorwertige Optimalitätsgleichungen auffassen kann.

Satz 3: Für das Entscheidungsmodell (7) seien die Voraussetzungen von Lemma 1 erfüllt. Dann genügt die Folge der maximalen bedingten erwarteten Gewinne $\overline{W}^*_{(t,N]}(v_t)$, $t = 1,2,\dots,N-1$, falls sie sämtlich die (v NME) besitzen, dem System der Mengenrelationen

$$\overline{W}^*_{(t,N]}(v_t) = \max_{a \in E_t(v_t)}{}^{*} g'_{(t,N]}(v_t, a, \max_{e \in \Delta(v_t,a,s)}{}^{*} f'^{e}_{(t+1,N]}(v_t)).$$

Bemerkung 2: Im Falle $p = 1$ erhält man inhaltlich ein System von Gleichungen, wie sie in Satz 1 beschrieben wurden.

LITERATUR

Bellman, R. (1957): Dynamic Programming, Princeton Univ. Press

Blackwell, D. (1965): Discounted Dynamic Programming, Ann. Math. Stat., 36, 226-235

Furukawa, N.; Iwamoto, S. (1976): Dynamic Programming on Recursive Reward Systems, Bull. Math. Stat., 17, 103-126

Groenewegen, L.; Wessels, J. (1979): Conditions for Optimality in Multi-Stage Stochastic Programming Problems, Lect. Notes in Economics and Math. Syst. 179, Proceed. Oberwolfach, Springer-Verlag, Berlin.

Hinderer, K. (1970): Foundations of Non-Stationary Dynamic Programming with Discrete Time Parameter, Lect. Notes in Operations Research and Mathematical Systems, Vol. 33, Springer-Verlag, Berlin.

Klötzler, R. (1978): Multiobjektive Dynamic Programming, Math. Operationsforschung, Statist., Ser. Opt., 9/3, 423-426.

Lipfert, W. (1980): Zur stochastischen dynamischen Optimierung mit allgemeinen Ertragsfunktionalen. Diss. B, TH Leuna Merseburg.

(1985): Über ein stochastisches dynamisches Entscheidungsmodell mit allgemeinen Ertragsfunktionalen. Optimization 16, 2, 313-328.

Müller, P.H.; Nollau, V. (1984): Steuerung stochastischer Prozesse. Akademie-Verlag Berlin.

Nollau, V. (1985): A Stochastic Decision Model with Vektor-valued Reward. Optimization, 16, 5, 733-742.

Schäl, M. (1981): Utility Functions and Optimal Policies in Sequential Decision Problems. Game Theory and Math. Economics, 357-365.

Pädagogische Hochschule Dresden
Sektion Mathematik
DDR 8060 Dresden
Wigardstr. 17

ON SOME SUBCLASSES OF OPERATOR SEMI-STABLE AND OPERATOR LÉVY'S MEASURES ON $\mathbb{R}^N$

Andrzej Łuczak

Łódź

Key words: operator semi-stable and operator Lévy's measures, characteristic function of an infinitely divisible measure

ABSTRACT

In the paper two classes of probability measures on $\mathbb{R}^N$ are characterized: the intersection of operator semi-stable measures and Lévy's measures, and the intersection of operator Lévy's measures and semi-stable ones. The description is given in terms of the characteristic functions and generalizes, in a natural way, the known result for measures on $\mathbb{R}$.

INTRODUCTION

In the set of infinitely divisible laws on the real line two subclasses of particular interest have been known for a long time: the class of Lévy's measures and that of stable measures. A new interesting class containing, like Lévy's measures, all stable ones, was introduced and studied by Kruglov in [3]; the members of this class were given the name of semi-stable measures. Now, as the stable laws remained the smallest group among all mentioned above, it looked interesting to describe their "closest relatives", namely, the intersection of the set of Lévy's measures with the set of semi-stable ones. It was done in [1] in terms of the characteristic functions and making use of the extreme-point method, i.e. the integral representation theorem of Choquet for convex compact sets.

The aim of this paper is to carry over such an investigation to the multidimensional case. However, in this situation a far more general setup is possible and let us begin with its brief description.

Let $\{\mu_n\}$ be a sequence of probability measures and $\{A_n\}$ a sequence of non-singular linear operators on $\mathbb{R}^N$. μ is called an operator Lévy's measure if it is the weak limit of the sequence

$$A_n(\mu_1 * \dots * \mu_n) * \delta(h_n),$$

where $h_n \in \mathbb{R}^N$ and $(A\nu)(E) = \nu(A^{-1}(E))$ for a measure ν, a linear operator A and a Borel set E.

μ is called an operator semi-stable measure if it is the weak limit of the sequence

$$A_n \nu^{k_n} * \delta(h_n),$$

where A_n, h_n are as above, the sequence $\{k_n\}$ of positive integers satisfies $k_{n+1}/k_n \to r \geq 1$, and the power ν^{k_n} is taken in the sense of convolution for a probability measure ν.

If the operators A_n are multiplies of the identity, we come to the ordinary notions of Lévy's and semi-stable measures on $\mathbb{R}^N$.

Now, there are two obvious possible generalizations of the above-mentioned result from [1]. The minor one: to characterize the intersection of the set of Lévy's and that of semi-stable measures on $\mathbb{R}^N$ and the major: to do the same for the operator Lévy's and the operator semi-stable mesures.

In this paper we have succeeded in two intermediate cases. Namely, the following classes are characterized:

1. full operator Lévy's measures that are semi-stable
2. full operator semi-stable measures that are Lévy's,

where by "full" we mean a measure which is not concentrated on any $(N-1)$-dimensional hyperplane of $\mathbb{R}^N$ (a customary assumption while considering measures on $\mathbb{R}^N$).

It is interesting to observe that in both the cases the proofs are considerably simpler than that in [1], moreover, neither of them involves the extreme-point method.

1. PRELIMINARIES AND NOTATION

Let μ be an infinitely divisible measure on $\mathbb{R}^N$ and let φ be its characteristic function. We say that φ is represented in the *Lévy form* if

$$(1)\qquad \varphi(z) = \exp\{i(m,z) - \frac{1}{2}(Dz,z) + \int_{\mathbb{R}^N-\{0\}} [e^{i(x,z)} - 1 - \frac{i(x,z)}{1+\|x\|^2}] M(dx)\}$$

where $m \in \mathbb{R}^N$, D is a non-negative linear operator on $\mathbb{R}^N$ and M is the *Lévy spectral measure* of μ, i.e. a Borel measure finite outside each neighbourhood of zero and such that $\int_{\|x\|\leq 1} \|x\|^2 M(dx) < \infty$. For this representation we shall use in the sequel the notation $\mu = [m,D,M]$.

φ is said to be represented in the *Lévy-Khintchine form* if

$$(2)\qquad \varphi(z) = \exp\{i(m,z) - \frac{1}{2}(Dz,z) + \int_{\mathbb{R}^N-\{0\}} [e^{i(x,z)} - 1 - \frac{i(x,z)}{1+\|x\|^2}] \frac{1+\|x\|^2}{\|x\|^2} \gamma(dx)\}$$

where m and D are as above, and γ is a finite Borel measure on $\mathbb{R}^N-\{0\}$ called the *Lévy-Khintchine spectral measure* of μ.

Now we recall some basic facts concerning operator Lévy's measures and operator semi-stable measures. For a more comprehensive treatment the reader is referred to [2], [5], [7].

Let μ be a full probability measure on $\mathbb{R}^N$. μ is an operator Lévy's measure if and only if it is infinitely divisible and

$$(3)\qquad \mu = e^{tQ}\mu * \mu_t, \qquad t > 0,$$

where Q is a non-singular linear operator on $\mathbb{R}^N$ such that all its eigenvalues have negative real parts and μ_t is a probability measure. Moreover, according to [7, Theorem 7.1], the formula

$$(4)\qquad \int_{\mathbb{R}^N-\{0\}} g(x)\gamma(dx) = \int_{\mathbb{R}^N-\{0\}} \int_0^\infty g(e^{tQ}x) \frac{\|e^{tQ}x\|^2}{1+\|e^{tQ}x\|^2} dt \frac{\nu(dx)}{\log(1+\|x\|^2)}$$

for every continuous bounded function g on $\mathbb{R}^N-\{0\}$, establishes a one-to-one correspondence between the Lévy-Khintchine spectral measure γ of μ and a finite Borel measure ν on $\mathbb{R}^N-\{0\}$. Rewriting (4) for the Lévy spectral measure M we obtain

$$\int_{\mathbb{R}^N-\{0\}} f(x)M(dx) = \int_{\mathbb{R}^N-\{0\}} \int_0^\infty f(e^{tQ}x)dt \frac{\nu(dx)}{\log(1+\|x\|^2)}$$

for every continuous function f on $\mathbb{R}^N-\{0\}$ such that $|f(x)| \leq$ $\leq C \frac{\|x\|^2}{1+\|x\|^2}$. Put $\lambda(dx) = \frac{\nu(dx)}{\log(1+\|x\|^2)}$. By virtue of the finiteness of ν, λ is a Lévy spectral measure such that

$$\int_{\mathbb{R}^N-\{0\}} \log(1+\|x\|^2)\lambda(dx) < \infty$$

and we come to the following

PROPOSITION 1. The formula

$$(5) \qquad \int_{\mathbb{R}^N-\{0\}} f(x)M(dx) = \int_{\mathbb{R}^N-\{0\}} \int_0^\infty f(e^{tQ}x)dt\lambda(dx)$$

for every continuous function f on $\mathbb{R}^N-\{0\}$ such that $|f(x)| \leq$ $\leq C \frac{\|x\|^2}{1+\|x\|^2}$, establishes a one-to-one correspondence between the Lévy spectral measure M of a full operator Lévy's measure on $\mathbb{R}^N$ and a Borel measure λ on $\mathbb{R}^N-\{0\}$ having the property $\int_{\mathbb{R}^N-\{0\}} \log(1+\|x\|^2)\lambda(dx) < \infty$. In particular, it follows that λ is also a Lévy spectral measure.

A full probability measure μ on $\mathbb{R}^N$ is operator semi-stable if and only if it is infinitely divisible and

$$(6) \qquad \mu^a = A_\mu * \delta(h)$$

for some $0 < a < 1$, $h \in \mathbb{R}^N$ and a non-singular linear operator A on $\mathbb{R}^N$, $\|A\| < 1$. There are decompositions

$$\mu = \mu_X * \mu_Y, \qquad \mathbb{R}^N = X \oplus Y$$

such that X and Y are A-invariant subspaces of $\mathbb{R}^N$, $\mathrm{sp}\, A|X \subset$

$\{|\alpha|^2 < a\}$, μ_X is a purely Poissonian measure concentrated on X and μ_Y is a Gaussian measure concentrated on Y. Equality (6) is equivalent to

(7) $\quad aM = AM, \qquad aD = ADA^*$

for $\mu = [m,D,M]$.

If, instead of operaotr, we consider ordinary multidimensional Lévy's and semi-stable measures, then equalities (3), (4), (6), (7) become

(3') $\quad \mu = T_t\mu * \mu_t, \qquad 0 < t < 1,$

(4') $$\int_{\mathbb{R}^N-\{0\}} g(x)\gamma(dx) = \int_{\mathbb{R}^N-\{0\}} \int_0^1 g(tx) \frac{t\|x\|^2}{1+t^2\|x\|^2} dt \frac{\nu(dx)}{\log(1+\|x\|^2)},$$

(6') $\quad \mu^a = T_b\mu * \delta(h),$

(7') $\quad aM = T_bM$

for some $0 < |b| < 1$, and $T_r x = rx$, $x \in \mathbb{R}^N$, $r \in \mathbb{R}$. The counterpart of Proposition 1 is now

PROPOSITION 1'. The formula

(5') $$\int_{\mathbb{R}^N-\{0\}} f(x)M(dx) = \int_{\mathbb{R}^N-\{0\}} \int_0^1 f(tx) \frac{dt}{t} \lambda(dx)$$

for every continuous function f on $\mathbb{R}^N-\{0\}$ such that $|f(x)| \leq C \frac{\|x\|^2}{1+\|x\|^2}$, establishes a one-to-one correspondence between the Lévy spectral measure M of a full Levy's measure on $\mathbb{R}^N$ and a Borel measure λ on $\mathbb{R}^N-\{0\}$ having the property $\int_{\mathbb{R}^N-\{0\}} \log(1+\|x\|^2)\lambda(dx) < \infty$. In particular, it follows that λ is also a Lévy spectral measure.

Moreover, if μ is semi-stable, then either μ is Gaussian or μ is purely Poissonian (cf. [4]).

2. SEMI-STABLE OPERATOR LÉVY'S MEASURES

In this section we characterize operator Lévy's measures that are semi-stable.

THEOREM 2. Let μ be a full probability measure on $\mathbb{R}^N$ and let φ be its characteristic function. μ is semi-stable operator Lévy's measure if and only if either

$$\varphi(z) = \exp\{i(m,z) - \tfrac{1}{2}(Dz,z)\} \tag{8}$$

where $m \in \mathbb{R}^N$ and D is a non-negative linear operator on $\mathbb{R}^N$, or

$$\varphi(z) = \exp\{i(m,z) + \sum_{n=-\infty}^{\infty} a^{-n} \int_{|b|<\|x\|\leq 1} \int_0^{\infty} [e^{ib^n(e^{tQ}x,z)} - 1 + \\ - \frac{ib^n(e^{tQ}x,z)}{1+b^{2n}\|e^{tQ}x\|^2}]\, dt\nu(dx)\} \tag{9}$$

where $m \in \mathbb{R}^N$, $0 < a < 1$, $0 < |b| < 1$, Q is a non-singular linear operator on $\mathbb{R}^N$ such that all its eigenvalues have negative real parts and ν is a finite Borel measure on $\{x : |b| < \|x\| \leq 1\}$.

P r o o f. Assume that μ is a full semi-stable operator Lévy's measure. If μ is Gaussian, then (8) follows; otherwise μ is purely Poissonian, $\mu = [m,0,M]$, and M satisfies (7'). μ^a as well as $T_b\mu$ are full operator Lévy's measures having aM and T_bM, respectively, as their spectral Lévy measures, hence, according to (5), we have

$$\int_{\mathbb{R}^N-\{0\}} f(x)(aM)(dx) = a \int_{\mathbb{R}^N-\{0\}} f(x)M(dx) = \\ = a \int_{\mathbb{R}^N-\{0\}} \int_0^{\infty} f(e^{tQ}x)dt\lambda(dx) = \int_{\mathbb{R}^N-\{0\}} \int_0^{\infty} f(e^{tQ}x)dt(a\lambda)(dx)$$

and

$$\int_{\mathbb{R}^N-\{0\}} f(x)(T_bM)(dx) = \int_{\mathbb{R}^N-\{0\}} f(bx)\, M(dx) = \\ = \int_{\mathbb{R}^N-\{0\}} \int_0^{\infty} f(be^{tQ}x)dt\lambda(dx) = \int_{\mathbb{R}^N-\{0\}} \int_0^{\infty} f(e^{tQ}x)dt(T_b\lambda)(dx).$$

The one-to-one correspondence between the measures M and λ in representation (5) and equality (7') yield the relation

$$a\lambda = T_b\lambda. \tag{10}$$

By virtue of [5, Theorem 1.2] λ has the form

$$(11)\qquad \lambda(E) = \sum_{n=-\infty}^{\infty} a^n \nu(b^n E \cap Z), \qquad E \in \mathcal{B}(\mathbb{R}^N - \{0\}),$$

where $Z = \{x : |b| < \|x\| \leq 1\}$ and ν is a finite Borel measure on Z, moreover $\nu = \lambda|Z$.

Thus, for every λ-integrable function g we have

$$\int_{\mathbb{R}^N - \{0\}} g(x)\lambda(dx) = \sum_{n=-\infty}^{\infty} a^{-n} \int_{|b|<\|x\|\leq 1} g(b^n x)\nu(dx)$$

which, together with (5), gives the equality

$$(12)\qquad \int_{\mathbb{R}^N - \{0\}} f(x)M(dx) = \sum_{n=-\infty}^{\infty} a^{-n} \int_{|b|<\|x\|\leq 1} \int_0^{\infty} f(b^n e^{tQ}x)dt\ \nu(dx),$$

for every continuous function f such that $|f(x)| \leq C \frac{\|x\|^2}{1 + \|x\|^2}$.

Now, taking into account representation (1) and applying (12) to the function $f(x) = e^{i(x,z)} - 1 - \frac{i(x,z)}{1 + \|x\|^2}$, we get formula (9).

Let now φ be of form (9). Define a measure λ by (11). Then λ satisfies (10) and

$$\varphi(z) = \exp \{i(m,z) + \int_{\mathbb{R}^N - \{0\}} \int_0^{\infty} [e^{i(e^{tQ}x,z)} - 1 + - \frac{i(e^{tQ}x,z)}{1 + \|e^{tQ}x\|^2}]\, dt\lambda(dx)\}.$$

From [5, Lemma 3.2] it follows that there exists an $\alpha > 0$ such that $\int_{\|x\|\geq 1} \|x\|^{\alpha}\lambda(dx) < \infty$, thus $\int_{\|x\|\geq 1} \log(1 + \|x\|^2)\lambda(dx) < \infty$.

Since λ is a Lévy spectral measure (this being a consequence of definition (11) and [5, Theorem 1.2]), we have $\int_{\|x\|\leq 1} \log(1 + \|x\|^2)\lambda(dx) < \infty$ and, consequently, $\int_{\mathbb{R}^N - \{0\}} \log(1 + \|x\|^2)\lambda(dx) < \infty$.

By virtue of Proposition 1, we obtain that

$$\varphi(z) = \exp \{i(m,z) + \int_{\mathbb{R}^N - \{0\}} [e^{i(x,z)} - 1 - \frac{i(x,z)}{1 + \|x\|^2}]\, M(dx)\},$$

where M is a Lévy spectral measure defined by (5). Consequently, φ is the characteristic function of a full operator Lévy's measure.

Since λ satisfies (10), the reasoning analogous to that in the first part of the proof yields that for M (7') holds, which shows that the measure corresponding to φ is semi-stable.

Obviously, if φ is of the form (8), then the assertion follows and the proof of the theorem is complete. □

3. OPERATOR SEMI-STABLE LÉVY'S MEASURES

These measures are characterized by the following theorem.

THEOREM 3. Let μ be a full probability measure on $\mathbb{R}^N$ and let φ be its characteristic function. μ is operator semi-stable Lévy's measure if and only if

$$(13)\qquad \varphi(z) = \exp\{i(m,z) - \frac{1}{2}(Dz,z) + \sum_{n=-\infty}^{\infty} a^{-n} \int_{Z_A\cap X} \int_0^1 [e^{it(A^n x,z)} - 1 - \frac{it(A^n x,z)}{1 + t^2\|A^n x\|^2}] \frac{dt}{t}\nu(dx)\}$$

where $m \in \mathbb{R}^N$, $0 < a < 1$, A is a non-singular linear operator in $\mathbb{R}^N$, $\|A\| < 1$, X is an A-invariant subspace of $\mathbb{R}^N$ such that $\mathrm{sp}A|X \subset \{|\alpha|^2 < a\}$, D is a non-negative linear operator in $\mathbb{R}^N$ satisfying $aD = ADA^*$ and ν is a finite Borel measure on $Z_A \cap X$; Z_A being defined by

$$Z_A = \{x : \|x\| \le 1\} \cap \{x : \|A^{-1}x\| > 1\}.$$

P r o o f. Assume that μ is operator semi-stable Lévy's measure. The measure μ_t in decomposition (3') is infinitely divisible (cf. [6]), thus, denoting $\mu = [m,D,M]$, $\mu_t = [m_t,D_t,M_t]$, we have

$$(14)\qquad M = T_t M + M_t, \qquad 0 < t < 1.$$

The measure μ is decomposed into the convolution

$$\mu = \mu_X * \mu_Y$$

where $\mu_X = [m_X,0,M]$, $\mu_Y = [m_Y,D,0]$ and the measures μ_X,μ_Y are full on subspaces X and Y, respectively; $X \oplus Y = \mathbb{R}^N$. Equality (14) yields

$$\mu_X = T_t\mu_X * [m_t,0,M_t], \qquad 0 < t < 1,$$

showing that μ_X is a Lévy's measure. Let us regard μ_X as a measure on X. It is easily seen that μ_X^a and $A\mu_X$ are full Lévy's measures on X having aM and AM, respectively, as their spectral Lévy measures. The argument analogous to that in the proof of Theorem 2 shows that

(15) $a\lambda = A\lambda$

in representation (5'), thus λ has the form

(16) $$\lambda(E) = \sum_{n=-\infty}^{\infty} a^n \nu(A^n E \cap Z_A), \qquad E \in \mathcal{B}(X - \{0\})$$

where ν is a finite Borel measure on Z_A (cf. [5, Theorem 1.2]). Consequently,

(17) $$\int_{X-\{0\}} f(x)M(dx) = \sum_{n=-\infty}^{\infty} a^{-n} \int_{Z_A \cap X} \int_0^1 f(tA^n x)\,\frac{dt}{t}\,\nu(dx)$$

for every continuous function f such that $|f(x)| \le C \dfrac{\|x\|^2}{1+\|x\|^2}$. Since M is concentrated on X, formula (17) applied to the function $f(x) = e^{i(x,z)} - 1 - \dfrac{i(x,z)}{1+\|x\|^2}$ gives (13).

Now, let us assume that φ is of form (13). Define a measure λ by (16). Then λ satisfies (15) and the property $\operatorname{sp} A|X \subset \{|\alpha|^2 < a\}$ implies that λ is a Lévy spectral measure (cf. [5, Theorem 1.2]). The rest of the proof is *mutatis mutandis* the same as that of Theorem 2.

REFERENCES

[1] Hensz E., Jajte R. (1978): On a class of limit laws. Teor. Verojatnost. i Primen. 23, No 1, 215-221.

[2] Jajte R. (1977): Semi-stable probability measures on $\mathbb{R}^N$. Studia Math. 61, No 1, 29-39.

[3] Kruglov V.M. (1972a): Ob odnom rasshirenii klassa ustoichivykh raspredelenii. Teor. Verojatnost. i Primen. 17, No 4, 723-732.

[4] ———— (1972b): Ob odnom klasse predel'nykh raspredelenii v gilbertovom prostranstve. Litovsk. Mat. Sb. 12, No 3, 85-88.

[5] Łuczak A. (1981): Operator semi-stable probability measures on $\mathbb{R}^N$. Colloq. Math. 45, No 2, 287-300.

[6] Urbanik K. (1969): Self-decomposable probability measures on $\mathbb{R}^N$. Applications Math. 10, No 1, 91-97.

[7] ———— (1972): Lévy's probability measures on Euclidean spaces. Studia Math. 44, No 1, 119-148.

Institute of Mathematics
Łódź University
ul. Banacha 22
90-238, Łódź, POLAND

ASYMPTOTIC ALMOST EQUIDISTRIBUTION ON A BANACH SPACE

Harald Luschgy

Münster

Key words: Shift group, Stam group, reproducing kernel Hilbert space

ABSTRACT

We prove some properties of the shift group of sequences of probability measures on a Banach space. Further, given a probability measure P on a l^p-space, $1 \leq p < \infty$, or a Hilbert space, conditions are derived to ensure that P is concentrated on some shift group.

INTRODUCTION

Suppose that B is a real separable Banach space. Let $M_+^1(B)$ be the set of all probability measures on the Borel σ-algebra $\mathcal{B}(B)$ of B. For any sequence $(\mu_n)_n$ in $M_+^1(B)$, we introduce the set

$$S((\mu_n)_n) = \{x \in B : \lim_{n\to\infty} \| \mu_n * \varepsilon_x - \mu_n \| = 0\},$$

where $\|\cdot\|$ denotes the total variation norm. One easily verifies that $S((\mu_n)_n)$ is a Borel measurable subgroup of B; it is called the shift group of the sequence $(\mu_n)_n$. Given $P \in M_+^1(B)$, we ask for the existence of an asymptotic P-almost equidistribution on B by which is meant a sequence $(\mu_n)_n$ in $M_+^1(B)$ such that $P(S((\mu_n)_n)) = 1$ holds. Our interest in this question has its origin in Hunt-Stein type theorems for invariant statistical decision problems when the group involved is a Banach space.

In the sequel, I have collected some properties of the shift group and some conditions which ensure the existence of an asymptotic P-almost equidistribution on l^p-spaces, $1 \leq p < \infty$, and Hilbert spaces for second order probability measures P.

SOME RESULTS

Let $(\mu_n)_n$ be a sequence in $M_+^1(B)$. First, we shall give upper bounds for the shift group $S((\mu_n)_n)$. A vector $x \in B$ is a partially admissible shift of $\mu \in M_+^1(B)$ if $\mu * \varepsilon_x$ contains a component absolutely continuous with respect to μ. We denote the set of all partially admissible shifts of μ by $C(\mu)$, so that

$$C(\mu) = \{x \in B : \| \mu * \varepsilon_x - \mu \| < 2\}.$$

Obviously, we have:

$$S((\mu_n)_n) \subset \bigcup_{n=1}^{\infty} C(\mu_n). \tag{1}$$

The Stam group $S(\mu)$ of $\mu \in M_+^1(B)$ is defined as the shift group of the sequence $(\mu^n)_n$ of nth convolution powers of μ.

$$S(\mu) = \bigcup_{n=1}^{\infty} C(\mu^n). \tag{2}$$

This is known for locally compact abelian groups. The proof given by Heyer (1977), Theorem 2.6.10 for this case works for Banach spaces. Notice that in general, the Stam group is no subspace of B. In fact, let $B = \mathbb{R}$ and $\mu = (\varepsilon_1 + \varepsilon_2)/2$. Then by (2), $S(\mu) = \mathbb{Z}$. It is well known that if B is a finite dimensional Banach space, there exists even a Stam group which is all B. However, if B is infinite dimensional, $C(\mu)$ is of the first category in B for every $\mu \in M_+^1(B)$, see Takahashi (1979), Corollary 4.4. Combining this result and (1), we obtain:

(3) Proposition. Suppose that B is infinite dimensional. Then $S((\mu_n)_n)$ is of the first category in B. In particular, $S((\mu_n)_n)$ cannot coincide with B.

For $\mu \in M_+^1(B)$, we let τ_μ be the topology of convergence in μ-probability on the topological dual B' of B. The kernel K_μ of μ is the topological dual of (B', τ_μ).

(4) (Borell (1976)) $K_\mu = \bigcap \{F \in \mathcal{B}(B) : \mu(F) = 1$, F is a subspace of B and $F = \bigcup_{n=1}^{\infty} C_n$ with C_n convex and compact$\}$.

(5) Lemma. (a) $S((\mu_n)_n) \subset \bigcup_{n=1}^{\infty} K_{\mu_n}$.

(b) $S(\mu) \subset K_\mu$.

Proof. According to Theorem 1 of Dudley (1966), $C(\nu) \subset K_\nu$ for every $\nu \in M_+^1(B)$. In view of (1), this implies assertion (a). Further, if F is a Borel measurable subspace of B and $n \in \mathbb{N}$, it is not difficult to see that $\mu(F) = 1$ if and only if $\mu^n(F) = 1$. By (4), this yields $K_\mu = K_{\mu^n}$ for every $n \in \mathbb{N}$, so that assertion (b) follows from (2).

If $\mu \in M_+^1(B)$ is of weak second order (i.e. $\int_B \langle y,x\rangle^2 \, d\mu(x) < \infty$ for every $y \in B'$), the linear operator $U_\mu : B' \longrightarrow B$ defined by

$U_\mu y = \int_B \langle y,x\rangle x \, d\mu(x) - \langle y,m_\mu\rangle m_\mu$ is its covariance operator, where $m_\mu = \int_B x \, d\mu(x)$ is the mean of μ and the integrals are Pettis integrals, see Weron (1976). Then U_μ is positive and symmetric, that is, $\langle y,U_\mu y\rangle \geqslant 0$ and $\langle y,U_\mu z\rangle = \langle z,U_\mu y\rangle$ for every $y,z \in B'$. Let H_μ be the reproducing kernel Hilbert space associated with U_μ; $H_\mu \subset B$ and the inclusion map $H_\mu \longrightarrow B$ is continuous, see e.g. Vakhania (1981).

(6) Proposition. $S((\mu_n)_n) \subset K_\mu \subset H_\mu$ for some symmetric $\mu \in M_+^1(B)$ of strong second order (i.e. $\int_B \|x\|^2 \, d\mu(x) < \infty$).

Proof. We define $\bar{\mu}_n(A) = \mu_n(-A)$, $A \in \mathcal{B}(B)$. Then $S((\mu_n)_n) \subset S((\mu_n * \bar{\mu}_n)_n)$, since for any $\nu, \lambda \in M_+^1(B)$, $x \in B$

$$\|\nu * \lambda * \varepsilon_x - \nu * \lambda\| \leqslant \|\nu * \varepsilon_x - \nu\| .$$

Choose compact symmetric sets $C_n \subset B$ such that $\mu_n * \bar{\mu}_n(C_n) \geqslant 1 - 1/n$ and define

$$\nu_n(A) = \frac{\mu_n * \bar{\mu}_n(A \cap C_n)}{\mu_n * \bar{\mu}_n(C_n)}, \quad A \in \mathcal{B}(B), \ n \in \mathbb{N}.$$

Clearly ν_n is a symmetric probability measure of strong second order. Moreover

$$\|\nu_n * \varepsilon_x - \nu_n\| \leqslant n(n-1)^{-1} \|\mu_n * \bar{\mu}_n * \varepsilon_x - \mu_n * \bar{\mu}_n\| + 4/(n-1)$$

for every $x \in B$, $n \geqslant 2$ which yields $S((\mu_n * \bar{\mu}_n)_n) \subset S((\nu_n)_n)$. If $\sup_{n \in \mathbb{N}} \int_B \|x\|^2 \, d\nu_n(x) < \infty$, let $\mu = \sum_{n=1}^{\infty} 2^{-n} \nu_n$. Then μ is a symmetric probability measure of strong second order. Also, τ_μ is stronger than τ_{ν_n} for every $n \in \mathbb{N}$, so that by (5a), $S((\nu_n)_n) \subset K_\mu$.

If $\sup_{n\in\mathbb{N}} \int_B \|x\|^2 \, d\nu_n(x) = \infty$, choose a subsequence $(\nu_k)_k$ of $(\nu_n)_n$ such that $\inf_{n\in\mathbb{N}} \int_B \|x\|^2 \, d\nu_k(x) > 0$. Clearly $S((\nu_n)_n) \subset S((\nu_k)_k)$. Let $a_k = \int_B \|x\|^2 \, d\nu_k(x)$, $a = \sum_{k=1}^{\infty} 2^{-k} a_k^{-1}$ and $\mu = \sum_{k=1}^{\infty} 2^{-k}(aa_k)^{-1}\nu_k$. Then μ is a symmetric probability measure of strong second order and as above, we obtain $S((\nu_k)_k) \subset K_\mu$. Finally, since H_μ coincides with the topological dual of B' equipped with the seminorm $y \longmapsto \langle y, U_\mu y\rangle^{1/2}$ (Vakhania (1981), p.102), $K_\mu \subset H_\mu$ holds. This completes the proof.

A Gaussian measure is a probability measure on $\mathscr{L}(B)$ such that the law of each $y \in B'$ is Gaussian. Kuz'min (1977) showed that if B is a separable Hilbert space, $S((\mu_n)_n) \subset H_\mu$ for some symmetric Gaussian measure μ. We can immediately extend this result to spaces of type 2. According to Theorem 4.1 of Hoffmann-Jørgensen (1977), spaces of type 2 are characterized by the fact that for any probability measure of strong second order with mean 0 one can find a symmetric Gaussian measure with the same covariance operator.

(7) <u>Corollary.</u> If B is of type 2, then $S((\mu_n)_n) \subset H_\mu$ for some symmetric Gaussian measure μ.

It would be of interest to know if spaces of type 2 are characterized by this property.

For symmetric Gaussian measures μ it is well known that $C(\mu) = K_\mu = H_\mu$. Therefore, from (2) and (5b) follows:

(8) <u>Corollary.</u> If μ is a symmetric Gaussian measure, then $S(\mu) = H_\mu$.

We now come to the existence of asymptotic almost

equidistributions.

(9) Proposition. Let $B = l^p$, $1 \leqslant p < \infty$, and $P \in M_+^1(B)$ of weak second order. Let $(e_n)_n$ be the sequence of coordinate functionals with respect to the natural basis of B. Suppose that

$$\sum_{n=1}^{\infty} \langle e_n, U_P e_n\rangle^{1/2} < \infty \quad \text{and } m_P \in l^1, \text{ if } p \geqslant 2,$$
$$\sum_{n=1}^{\infty} \langle e_n, U_P e_n\rangle^{p/4} < \infty \quad \text{and } m_P \in l^{p/2}, \text{ if } 1 \leqslant p < 2.$$

Then there exists an asymptotic P-almost equidistribution consisting of convolution powers of some symmetric Gaussian measure.

Proof. An easy calculation shows that the above conditions are equivalent to the following conditions:

$$\sum_{n=1}^{\infty} \left(\int_B \langle e_n, x\rangle^2 \, dP(x)\right)^{1/2} < \infty \text{ , if } p \geqslant 2,$$
$$\sum_{n=1}^{\infty} \left(\int_B \langle e_n, x\rangle^2 \, dP(x)\right)^{p/4} < \infty \text{ , if } 1 \leqslant p < 2.$$

Let $\beta_n = \left(\int_B \langle e_n, x\rangle^2 \, dP(x)\right)^{1/2}$. We may assume that $\beta_n > 0$ for every $n \in \mathbb{N}$. (In case $I = \{n \in \mathbb{N} : \beta_n = 0\} \neq \emptyset$, apply the following arguments to the closed subspace $\bigcap_{n \in I} \{x \in B : \langle e_n, x\rangle = 0\}$ of B.) Let

$$H = \{x \in \mathbb{R}^{\mathbb{N}} : \sum_{n=1}^{\infty} x_n^2/\beta_n < \infty\}.$$

Then H is a Hilbert space with scalar product $\langle x,z\rangle_H = \sum_{n=1}^{\infty} x_n z_n/\beta_n$. From Hölder's inequality follows $H \subset B$. Moreover

$$\|x\| \leqslant \left(\sum_{n=1}^{\infty} \beta_n^{\,p}\right)^{1/2p} \|x\|_H$$

for every $x \in H$, so that the inclusion map $j : H \longrightarrow B$ is continuous. For its adjoint $j' : B' \longrightarrow H$ we obtain $j'y = (\beta_n y_n)_n$.

Since $\sum_{n=1}^{\infty} \beta_n{}^{p/2} < \infty$, $j \circ j'$ is the covariance operator of some symmetric Gaussian measure μ, see Vakhania (1981), Theorem 4.4.8. Then $H = H_\mu$ and from

$$\int_B \sum_{n=1}^{\infty} \langle e_n, x\rangle^2/\beta_n \, dP(x) = \sum_{n=1}^{\infty} \beta_n < \infty$$

follows $P(H) = 1$. In view of (8), this yields the assertion.

Notice that in case $2 \leqslant p < \infty$, by (7), the conclusion of the preceding proposition is equivalent to the existence of an arbitrary asymptotic P-almost equidistribution.

(10) <u>Proposition.</u> Let B be a separable Hilbert space with scalar product $\langle \cdot , \cdot \rangle$.

(a) Let $P \in M_+^1(B)$ be of strong second order. If $U_P{}^{1/2}$ is of trace class and $m_P \in U_P{}^{1/4}(B)$, then there exists an asymptotic P-almost equidistribution.

(b) Let P be a Gaussian measure. Then there exists an asymptotic P-almost equidistribution if and only if there is a positive symmetric Hilbert-Schmidt operator S and a positive symmetric trace class operator T on B such that $U_P = STS$ and $m_P \in S(B)$.

Proof. (a) Let $(e_n)_n$ be an orthonormal basis of B consisting of eigenvectors of U_P. Then by assumption, $\sum_{n=1}^{\infty} \langle e_n, U_P e_n\rangle^{1/2} < \infty$ and for $m_P = U_P{}^{1/4}x$, $x \in B$, one obtains

$$\sum_{n=1}^{\infty} |\langle e_n, m_P\rangle| = \sum_{n=1}^{\infty} |\langle U_P{}^{1/4} e_n, x\rangle|$$

$$= \sum_{n=1}^{\infty} \langle e_n, U_P e_n\rangle^{1/4} |\langle e_n, x\rangle|$$

$$\leqslant \Big(\sum_{n=1}^{\infty} \langle e_n, U_P e_n\rangle^{1/2}\Big)^{1/2} \|x\|^{1/2} < \infty .$$

Thus the assertion follows from (9).

(b) It is well known that a subset of B is the reproducing kernel Hilbert space of some symmetric Gaussian measure if and only if it is the range of some positive symmetric Hilbert-Schmidt operator on B. Thus the assertion follows from (8) and a zero-one law of Baker (1973), Theorem 2.

Using (6) one can show that for every $P \in M_{+}^{1}(B)$ there exists an asymptotic P-almost equidistribution (if and) only if B is finite dimensional.

REFERENCES

Baker C.R. (1973): Zero-one laws for Gaussian measures on Banach spaces. Trans. Amer. Math. Soc. 186, 291 - 308.

Borell C. (1976): Random linear functionals and subspaces of probability one. Ark. Mat. 14, 79 - 92.

Dudley R.M. (1966): Singularity of measures on linear spaces. Z. Wahrsch. Verw. Gebiete 6, 129 - 132.

Heyer H. (1977): Probability Measures on Locally Compact Groups. Springer, Berlin-Heidelberg-New York.

Hoffmann-Jørgensen J. (1977): Probability in Banach space. In: Ecole d'Eté de Probabilités de Saint-Flour VI, Lecture Notes in Math. 598, Springer, Berlin-Heidelberg-New York, 2 - 186.

Kuz'min A.D. (1977): Minimax property of Pitman estimate in Hilbert space. Moscow Univ. Computational Math. and Cybernetics, 48 - 55.

Takahashi Y. (1979): Partially admissible shifts on linear spaces. Hokkaido Math. J. 8, 150 - 166.

Vakhania N.N. (1981): Probability Distributions on Linear Spaces. North Holland, Amsterdam.

Weron A. (1976): On weak second order and Gaussian random elements. In: Probability in Banach spaces, Lecture Notes in Math. 526, Springer, Berlin-Heidelberg-New York, 263 - 272.

Universität Münster
Institut für Mathematische Statistik
Einsteinstr. 62
D-4400 Münster
Federal Republic of Germany

RANDOM BRANCHING PROCESS WITH DETERMINISTIC MOTION AND UNLOCAL BRANCHING LAW

Penka Maĭster

Sofia

Key words: branching process, point-process, expectation semigroup

ABSTRACT

The motion of a particle whose velocity undergoes jumps of random size at random times constitutes the prototype of a transport process. We consider the system in which there are many particles moving by the trajectories of the transport process and reproducing themselves at random times. The number of the particles and their positions in space $R^n \times R^n$ are described by the Markov branching process. Trajectory of the ancestral line of any particle is prescribed by some stochastic differential equation. Stability of the ancestral line and the asymptotic behaviour of the expectation semigroup are considered.

MARKOV BRANCHING PROCESS

As is well known, Markov branching process may be completely specified by the initial distribution, the migration process, the branching rate and the branching offspring distribution. Intuitive description is combined with the stochastic measure $\mu_{zt}(A)$, which explains the number of particles located in A at time t, if the process starts with one particle located at z. It is convenient to use the construction of Marlov branching process, given in [1].

Let $(\mathcal{G}, \mathcal{U})$ be a measurable space, let $\mathcal{G}^{(n)}$ be the symmetrization of the direct product $\mathcal{G}^n$ of n copies of $\mathcal{G}$, $\mathcal{G}^{(0)} = \{\delta\}$, δ is some extra point, $\hat{\mathcal{G}} := \bigcup_{n=0}^{\infty} \mathcal{G}^{(n)}$ and $\hat{\mathcal{U}}$ is the σ-algebra on $\hat{\mathcal{G}}$ induced by $\mathcal{U}$. $(\mathcal{G}, \mathcal{U})$ is called the type space and $(\hat{\mathcal{G}}, \hat{\mathcal{U}})$ -the population space.

Definition 1. The Markov process $(\hat{Z}_t, P^{\hat{z}})$ is a $(\hat{\mathcal{G}}, \hat{\mathcal{U}})$-valued Markov process with a branching transition function $P_t(\hat{z}; \cdot)$ expressing the property of independent branching without immigration: for all $t \in R_+$, $P(\mathcal{G}^{(0)}; \mathcal{G}^{(0)}) = 1$, $\forall A_j \in \mathcal{U} : \bigcup_{j=1}^{m} A_j = \mathcal{G}$,

$$P_t(\langle z_1, \dots z_k \rangle ; \{\hat{Z}_t[A_j] = n_j, j=1,\dots m\}) =$$

$$\sum_{j=1}^{m} \sum_{n_{j_1} + \dots n_{j_k} = n_j} \prod_{\nu=1}^{k} P_t(\langle z_\nu \rangle ; \{\hat{Z}_t[A_j] = n_{j\nu}, j=1,\dots m\}).$$

Then, obviously $\mu_{zt}(A) = 0$, if $\hat{Z}_t = \delta$, and $\mu_{zt}(A) = \sum_{\nu=1}^{m} 1_A(z_\nu)$, if $\hat{Z}_t = \langle z_1, \dots z_k \rangle$. The generating functional of $P_t(\hat{z}; \cdot)$ is given by $F_t(\hat{z}; s) := \int_{\hat{\mathcal{G}}} \tilde{s}(\hat{y}) P_t(\hat{z}; d\hat{y})$ where $\tilde{s}(\hat{y}) := 1$, if $\hat{y} = \delta$, and $\tilde{s}(\hat{y}) := \prod_{\nu=1}^{m} s(y_\nu)$, if $\hat{y} = \langle y_1, \dots y_m \rangle$. The function $s(\cdot)$ is bounded, complex-valued, $\mathcal{U}$-measurable on $\mathcal{G}$, $\|s\| = \sup_{z \in \mathcal{G}} |s(z)|$ and $s \in S$, $S = \{s : \|s\| \leq 1\}$. The mapping

$$F_t : \bar{S} \to \bar{S}, \quad F_t[s](z) := F_t(\langle z \rangle ; s) = E \exp \int_{\mathcal{G}} \log s(y) \mu_{zt}(dy)$$

is called a generating semigroup as the Chapman-Kolmogorov relation for $P_t(\hat{z}, \cdot)$ is equivalent to the semigroup relation:

$$F_{t_1+t_2}[s] = F_{t_1}[F_{t_2}[s]], \quad t_1, t_2 \geq 0.$$

It will be assumed that at time $t=0$ there exists a single particle located at the point $z=(x,y)\in G$, $G=D\times V$, where D is a bounded closed convex domain in the Euclidien space R^n with a smooth boundary ∂D, and $V\subset R^n$ is the set of velocities. ∂G stands for the set of all points $z=(x,y)$ in G such that $x\in\partial D$ and y is a direction exiting D, i.e. $(n_x,y)\geqslant 0$, where n_x is the direction of outernormal at x, $(\cdot;\cdot)$ is a inner product.

During its lifetime, each particle migrates independently, according to the same law. The migration process $\zeta(t)$ is the deterministic motion prescribed by the determinate system

$$\frac{d\zeta}{dt}=a(\zeta(t)),\quad \zeta(0)=z. \tag{1}$$

It is convenient to use the notation $\zeta=(\xi,\eta)$, where

$$\zeta=(\zeta^1,\dots\zeta^n,\zeta^{n+1},\dots\zeta^{2n})=(\xi^1,\dots\xi^n,\eta^1,\dots\eta^n)$$

to distinguish the variables ξ showing the position of the particles in $D\subset R^n$ and η showing the velocity, of them, and $a(\zeta)=$

$$(a^1(\zeta),\dots a^n(\zeta),a^{n+1}(\zeta),\dots a^{2n}(\zeta))=(b^1(\zeta),\dots b^n(\zeta),c^1(\zeta),\dots c^n(\zeta))$$

to notice by $b(\cdot)$ the local group velocity and by $c(\cdot)$ an external force field. Then, the system (1) is replaced by

$$\begin{cases}\dfrac{d\xi}{dt}=b(\xi,\eta)\\[2mm] \dfrac{d\eta}{dt}=c(\xi,\eta),\quad (\xi(0),\eta(0))=(x,y)\end{cases} \tag{2}$$

Assume that this system is perturbed by stochastic impulses acting on the acceleration. More precisely, we suppose that the impulses arise at the moments $\tau_1,\tau_2,\dots$ and have the amplitudes $\alpha_1,\alpha_2,\dots$, i.e. the impulses constitute some multivariate point process (also called marked point process) — (τ_n,α_n) determined by the strictly increasing sequence τ_n of positive random variables defined on a measurable space $(\Omega,\mathcal{F})$ and the sequence of random variables α_n, such that the α_n is associated to τ_n, for each n. The variables α_n take their values in a measurable space (V, $\mathcal{V}$) (the set of "marks"). A multivariate point process (τ_n, α_n) is completely characterized by the following discrete random measure on $]0,\infty] \times V$:

(3) $\nu(\omega; dt, d\theta) = \sum_{n\geq 1} \varepsilon_{(\tau_n(\omega), \alpha_n(\omega))}(dt, d\theta)\, 1_{\{\tau_n < \infty\}}$,

where ε_a denotes the Dirac measure located at point a.

In addition, we suppose that these stochastic impulses cause branching of the particles moving by the trajectory of $\zeta(t)$. In this way, the branching rate is connected with the sequence $(\tau_n)_{n\geq 1}$. Let $\tau_n, n\geq 1$, be i.i.d. random variables with exponential distribution :

(4) $$P\{\tau_1 > t\} = \exp[-\int_0^t R(\xi(s), \eta(s))ds],$$

i.e. if $\zeta(t) = (x,y)$ than the probability a particle branches in the time interval ($t, t+\Delta t$) is $R(x,y)\Delta t + o(\Delta t)$. $R(x,y)$ is called the branching rate (killing rule). It does not depend on t, it depend only on $\zeta(t)$, and it is a continuous function, with $0 \leq R(\cdot) \leq M < \infty$. Under this condition, with probability 1, only finitely many impulses (branchings) occur in every finite time interval and $(\tau_n)_{n\geq 1}$ are totally inaccessible stopping times [2].

The branching offspring distribution is connected with the amplitudes of the impulses ($\alpha_n)_{n\geq 1}$. Given that a particle branches at a point $z = \zeta(\tau_-) = (x,y)$, it produces m offsprings with probability $p_m(z)$, $m = 1, 2, \ldots$. To avoid trivialities, assume $p_1(z) \neq 1$ for all z. The branching law is not local: the m offsprings emerge at the points $z_1, \ldots z_m \neq z$. In this connection location of the offsprings in domain $\mathcal{D} \subset R^n$ is the same of the original particle $\xi_j(\tau) = \xi(\tau_-)$, $j = 1, \ldots m$, but every one offspring receives a different initial velocity $\eta_j^i(\tau) = \eta^i(\tau_-)[1 + \alpha^i]$, $i = 1, \ldots n$, $j = 1, \ldots m$, where $\alpha = (\alpha^1, \ldots \alpha^n) \in V$, $\alpha \in (\alpha_n)_{n\geq 1}$ corresponds to $\tau \in (\tau_n)_{n\geq 1}$.

ANCESTRAL LINE OF A PARTICLE

It will be assumed that $(\alpha_k)_{k\geq 1}$ are i.i.d.random variables with probability density $P\{\alpha_k \in d\theta\} = g(\theta)d\theta$. It is easy to notice and verify the following propositions:

* The random measure $\nu(\omega; dt, d\theta)$ is a Poisson measure with independent values and mathematical expectation

$$E\,\nu(\omega; dt, d\theta) = P\{\tau_k \in dt, \alpha_k \in d\theta\} = \varkappa(x,y)g(\theta)dt\,d\theta \quad \text{if } \zeta(t_-) = (x,y).$$

* The ancestral line of any particle is a Markov process with independent increments satisfying the following stochastic differential equation in R^{2n}:

$$\xi^i(t) = x^i + \int_0^t b^i(\xi, \eta(s))ds$$

$$\eta^i(t) = y^i + \int_0^t c^i(\xi(s-), \eta(s-))ds + \int_0^t\int_V \eta^i(s-)\,\theta^i\,\nu(\omega; ds\,d\theta). \tag{5}$$

The sample paths of the ancestral line are continuous by the components $\xi^1, \dots \xi^n$ and undergoes jumps by the components $\eta^1, \dots \eta^n$, directly proportional to the velocity.

* The infinitesimal operator of the semigroup $T_t f(x,y) = E_{(x,y)} f(\zeta(\omega)) = v(t,x,y) : R_+ \times G \xrightarrow{v} R$ is the following

$$\mathcal{L}v = \sum_{i=1}^n b^i(x,y)\frac{\partial v}{\partial x^i} + \sum_{i=1}^n c^i(x,y)\frac{\partial v}{\partial y^i} + \int_V \varkappa(x,y)g(\theta)\left[v(x, y(1+\theta)) - v(x,y)\right]d\theta \tag{6}$$

Now, we shall consider the problem of stability of the ancestral line with respect to the initial conditions $\zeta(0) = (x,y)$.

<u>Definition 2.</u> The trivial solution $\zeta(t) \equiv 0$ of the equation (5) is said to be stable in probability P, if $\lim_{(x,y)\to 0} P\{\sup_{t\geq 0} |\zeta(t)| > \varepsilon\} = 0$, where $\zeta(0) = (x,y)$ [3].

We shall use an analogy of the Liapunov second method provided that $b(0,0) = 0$, $c(0,0) = 0$ and b, c, g satisfy the sufficient conditions for existence and uniqueness of solution of (5), i.e. $\|a(z)\|^2 \leq C_1(1+|z|^2)$ and $|a(z_1)-a(z_2)| \leq C_2|z_1-z_2|$.

Theorem 1. If there exists a continuously differentiable Liapunov function $V(t,x,y)$ such that $V(0,0,0)=0$, and $\forall t>0$ $V(t,x,y) > W(x,y) > 0$ for $(x,y) \neq (0,0)$, and

$$\frac{\partial V}{\partial t} + \mathcal{L}V \leq 0,$$

then the trivial solution of (5) is stable in probability.

Proof. By the regularity of the process $\zeta_t(\omega)$ and by the given assumptions about $V(t,x,y)$ we have

$$P_{(x,y)}\{\sup_{s\leq t}|\zeta_s|>\varepsilon\} \leq E_{(x,y)}V(t,\zeta_t) \Big/ \inf_{|z|\geq\varepsilon} V(t,z)$$

and

$$E_{(x,y)}V(t,\zeta_t) = V(0,x,y) + E_{(x,y)}\int_0^t \left(\frac{\partial}{\partial s}+\mathcal{L}\right)V(s,\zeta_s)\,ds \leq V(0,x,y)$$

Using the continuity of $V(t,x,y)$ we complete the proof.

THE GENERATING FUNCTIONAL OF $P_t(\hat{z}, \cdot)$

Let us denote by $\pi(z,A)$, $z\in\mathcal{G}$, $A\in\mathcal{U}$, the conditional stochastic measure describing the offsprings, provided that branching occurs at the point z and by $H[s](z)$ its generating functional, i.e. $H[s](z) := E\exp\int_{\mathcal{G}} \log s(\theta)\,\pi(z;d\theta)$, $\|s(\cdot)\| \leq 1$.

When the branching law is local, the stochastic measure $\pi(z,A) = \pi(z,\langle z\rangle)$, $\forall A\in\mathcal{U}$, and $H[s]$ is an operator of superposition:

$$H[s](z) = \sum_{n=0}^{\infty} [s(z)]^n P\{\pi(z,\langle z\rangle)=n\} = H(s(z)),$$

but in our model, it is an integral operator. We have, $z=(x,y)$

$$P\{\pi(z,A_\nu)=n_\nu,\ \nu=1,\dots j;\ n=\sum_{\nu=1}^{j} n_\nu\} = p_n(z)\,n!\sum_{\nu=1}^{j}\frac{1}{n_\nu!}\Big(\int_{A_\nu\cap V} y(1+\theta)g(\theta)\,d\theta\Big)^{n_\nu}$$

where $\{A_\nu\}_{0\leq\nu\leq j}$ varies over the measurable decompositions of $\mathcal{G}$.

The mathematical expectation of the offsprings is

$$\rho(z,A)=E\pi(z,A)=\sum_n n\,p_n(z)\int_{A\cap V} y(1+\theta)g(\theta)d\theta\,,\quad z=(x,y).$$

It can be studied by the linear operator $\mathcal{A}[s](z):=\int_{\mathcal{G}} s(\theta)\rho(z,d\theta)$.

Further, the following notations are introduced: $C(\mathcal{G})$ the set of all continuous functions on $\mathcal{G}$ with sup-norm $\|s\|$ and $u(t,x,y)=F_t[s](x,y)$.

Theorem 2. If $s\in C(\mathcal{G})$, then $u(t,x,y)$ satisfies the following quasi-linear hyperbolic equation

$$\left|\begin{array}{l}\dfrac{\partial u}{\partial t}=\sum_{i=1}^{n} b^i(z)\dfrac{\partial u}{\partial x^i}+\sum_{i=1}^{n} c^i(z)\dfrac{\partial u}{\partial y^i}+r(x,y)\left[H[u](x,y)-u(t,x,y)\right],\\ u(0,x,y)=s(x,y),\quad u(t,x,y)\big|_{(x,y)\in\partial\mathcal{G}}=1.\end{array}\right.$$

The existence and uniqueness of the nontrivial solution of this equation is obtained from the general theory of the hyperbolic systems [6].

Theorem 3. If $s\in C(\mathcal{G})$, then $m(t,x,y)=M_t[s](x,y)=$
$=\int_{\mathcal{G}} s(\theta)E\mu_{zt}(d\theta)$ satisfies

$$\frac{\partial m}{\partial t}=\sum_{i=1}^{n} b^i(z)\frac{\partial u}{\partial x^i}+\sum_{i=1}^{n} c^i(z)\frac{\partial u}{\partial y^i}+r(x,y)\left[\mathcal{A}[m](x,y)-m(t,x,y)\right],\ m(0,z)=s(z).$$

The asymptotic behaviour of the $M_t[s]$ when $t\to\infty$ depends on the spectral properties of the operator $\mathcal{L}$, compactness and positivity of the $M_t[s]$. The representation

$$M_t[s](x,y)=e^{\lambda_0 t}\left[\varphi_0(x)\varphi_0^*(s)+o(e^{\lambda t})\right],\ |\lambda|<\lambda_0,$$

plays an important role in the proof of the limit theorems [4,9]. The monoenergetic and isotropic transport process on bounded domain in R^3 is completely studied in [7]. A particular case, when $V=\{1,2,\dots N\}$ and $C(z)\equiv 0$, is similar to the model considered in [9].

REFERENCES

1. Asmussen S. and Hering H. (1983): Branching processes. Birkhauser, Boston. Basel. Stuttgart.

2. Gihman I. and Skorohod A. (1982): Stochastic differential equations and their applications. "Naukova Dumka", Kiev.

3. Has'minskii R.(Khasminskii) (1969): Stability of the systems of differential equations under random perturbations of their parameters. "Nauka", Moscow.

4. Maister P. (1981): Branching diffusion processes in an unbouded domain. Math. USSR Sbornik. 114(156), No.3, 406-424, English transl.,(1982), 42, No.3, 357-373.

(1986a): Stability of the solutions of stochastic equation with integrals on martingales and random measures. In: Proc. Third Conf. Differ. Equat. Appl. (Rousse'85), Bulg. Acad. Sci., Sofia, 1986.

6. Mizohata S. (1977):Theory of partial differential equations. "Mir", Moscow, Russian transl. from Japanese.

7. Mori T., Watanabe Sh. and Yamada T. (1971): On neutron branching processes. Publ. RIMS, Kyoto Univ., 7, 153-179.

8. Papanicolaou G. (1975): Asymptotic analysis of transport processes. Bull.Amer.Math.Sos.,81, No.2, 330-392.

9. Volkova E. (1985): A limit theorems for critical branching processes with motion of the particles. Teor. Verojatnost. i Primenen., 30, No. 2, 372-377, (Russian).

Inst. for Foreign Students
ul. Assen Velchev 27,
1111 Sofia, Boulgaria.

INFORMATION ASSEMBLY VALUE
IN DECISION-MAKING

Piotr Marczyński

Cracow

Key words: information value, game theory, automata theory

ABSTRACT

The concept of information assembly value is introduced. It is applied in goal oriented decisive systems: two-person zero-sum game and the collective automata game.

1. GENERAL DESCRIPTION OF THE DECISIVE-INFORMATION SYSTEM

The decisive-information system is constitued by the following elements:

$D = \{d_1, d_2, \ldots, d_\mu\}$ - finite set of decision-maker's decisions;

Data item $q = (l_i, z_j)$, where

l_i - data name;

z_j - data value;

$L = \{l_1, l_2, \ldots, l_N\}$ - information assembly;

L is the set of names of data items which are received by the decision-maker.

B - the set of information transfers $b_r = \{q_{1j_1}, q_{2j_2}, \ldots, q_{Nj_N}\}$;

The b_r is the set of data items which are received by the decision-maker - the q_{ij_i} is the data item which name is l_i.

$W = \{w_1, w_2, \ldots, w_\delta\}$ - the set of payments;

$J : B \rightarrow D$ - the strategy which - taking under the consideration the b_r - chooses the decision d ;

$U : J \rightarrow J'$ - the procedure of strategy modification;

The process of decision-making is performed as following: in the moment t the decision-maker receives the information transfer $b(t)$. Then the decision $d(t) \in D$ is chosen using the strategy $J(t)$. After the payment $w(t) \in W$ is given to the decision-maker the strategy modification is performed:

$$J(t + 1) = U[J(t), d(t), w(t)] .$$

2. I-VARIABLE STRUCTURE STOCHASTIC AUTOMATA

2.1. Variable Structure Stochastic Automata as a decision-maker in stochastic environment and in two-person zero-sum game.

Variable structure stochastic automaton is defined as the 6-tuple [Narendra, 1974] :

$$SAVS = \langle W, \Phi, D, p(t), U, G \rangle , \text{ where}$$

W - the set of input signals;

$\Phi = \{\varphi_1, \varphi_2, \ldots, \varphi_s\}$ - the set of internal states;

$D = \{d_1, d_2, \ldots, d_\mu\}$ - the set of output signals;

$p(t)$ - the vector of the probability distribution;

$p_i(t)$ - is the probability of the state φ_i choice;

U - the reinforcement scheme which counts $p(t+1)$ from $p(t)$ and $w(t)$.

G - the output function. It is assumed that $G : \Phi \rightarrow D$ is deterministic and one to one.

In the moment t the automaton chooses /according to the distribution $p(t)$/ its internal state $\varphi(t)$, it generates the adequate output signal $d(t)$ and receives from the environment the input signal $w(t)$ which is the payment. Then the new vector of probability distribution is counted: $p(t+1) = U[p(t), \Phi(t), w(t)]$. For the P-model the set $W = \{0, 1\}$.

If $d(t) = d_i$ then $w(t) = 0$ /penalty/ with probability c_i and $w(t) = 1$ /penalty/ with probability $1 - c_i$.

The environment in this case is described by the vector

$$C = \{c_1, c_2, \ldots, c_\mu\} .$$

For the Q-model $W = \{w_1, w_2, \ldots, w_\delta\} \in [0,1]$.

In [Narendra 1974] the following reinforcement scheme was given:

If after the signal d_i had been generated the automaton received the payment $w(t)$ then the elements of the vector p are changed:

$$p_i(t+1) = p_i(t) + a(t) \cdot w(t) \cdot [1-p_i(t)] ,$$

$$p_j(t+1) = p_j(t) - a(t) \cdot w(t) \cdot p_j(t) , \quad j \neq i ,$$

$a(t)$ is the element of the sequence: $\lim_{t \to \infty} a(t) = 0$.

It was proved that this scheme ensures ε-optimality, it is that expected value of the average payment is infinitesimally close to $c^* = \max_i \{c_i\}$.

It was also verified [Viswanathan 1974] that the described VSSA may be used in the two-person zero-sum game. The simulation of two VSSAs acting in this game results the oscillation of the average payment around the optimal value of game /game without saddle point/ or the convergence of the average payment to the value of game /game with saddle point/.

2.2. I-Variable Structure Stochastic Automaton as an extension of VSSA

I-variable structure stochastic automaton is defined as

I-VSSA = $\langle B,W,D,P,U \rangle$, where

B - the set of information transfers;

D - the set of decisions;

P - the probabilistic matrix $k \times \mu$, where $k = \overline{\overline{B}}$, $\mu = \overline{\overline{D}}$;

W - the set of payments;

U - ε-optimal reinforcement scheme.

In the moment t the automaton receives the information transfer $b(t) = b_r$ and the choice of the decision $d(t)$ according to the r-th row of P /the probability distribution/ is performed. After the payment $w(t)$ is received the r-th row of P is modified using the scheme U.

2.3. Two person zero-sum game with additional information

Two-person zero-sum game described by the matrix X $\mu \times m$

is considered.

The player I chooses the pure strategy $s_1 \in S^1 = \{1,2,\ldots,\mu\}$ and the player II chooses the pure strategy $s_2 \in S^2 = \{1,2,\ldots,m\}$. It is assumed that the player I receives additional information: the subset $S' \subset S^2$ where the pure strategy chosen by the player II in the current trial belongs to.

Generally - the player I has the information assembly

$L = \{1S_1^2, 1S_2^2, \ldots, 1S_N^2\}$, where

$1S_i^2$ is a data name: "the attachment of the player's II pure strategy to the subset S_i^2".
The player I can construct the information assembly having on disposal 2^m subsets of S^2.

Of course if $L = \emptyset$ then the player I is applying the optimal mixed strategy and receives the payment which expected value is $E(V|L) = E(V|\emptyset)$.

The transformation $L \leftarrow L \cup 1S_i^2$ constitutes the new situation, in which the player I has two mixed strategies: p^1 and p^2. The strategy p^1 is applied when $q = (1S_i^2, 0)$, it is when the player's II pure strategy does not belong to S_i^2 and the strategy p^2 is applied when $q = (1S_i^2, 1)$, it is when the player's II pure strategy belongs to S_i^2. Then the expected value of the payment for the player I is $E(V|L) = E(V|\{1S_i^2\})$.

It is evident that $E(V|\{1S_i^2\}) \geqslant E(V|\emptyset)$.

In the general case it can be shown that the transformation $L \leftarrow L \cup L'$ results that the expected value of the payment does not change or increases.

So the concept of the value of the information assembly L' /relativised to L/ may be introduced. Its measure is the increase of the expected value of payment:

$$Q(L'|L) = E(V|L \cup L') - E(V|L) .$$

This concept seems to be similar to the definition of the value of information introduced in [Bartoszyński, 1964].

It is possible to use I-VSSA in the examination of the influence of information assembly change to the effectivity of decision-making concerning to the conflict described as a two-person zero-sum game.

2.4. Game of I-Variable Structure Stochastic Automata.

The set of acting automata constitutes the game if the payment of every automaton depends on its past action and on the action of unless one of other players.

In the certain game of I-VSSAs the influence of the change of information assemblies had been examined.

2.4.1. Description of game.

Three players take part in the game. Each of them is acting in the proper environment. The environment of player I is described by the set of states H^1 and 4 transition matrices Y^{1k} /k =1,2,3,4/. The y_{ij}^{1k} is the probability of the fact, that the environment being in the state h_i^1 changes it to h_j^1. In the same time the player I is choosing the decision $d^1(t) \in D^1$ and receives the payment $w^1(t) = w^1[h^1(t),d^1(t)]$. In the instant $t + \tau$ the player's II environment is activied. It is described by the set of states H^2 and the set of conditional probability distributions Y^2: $y^2[h^2(t+\tau)|d^1(t),h^1(t)]$ is the probability of the choice of $h^2(t+\tau)$ if $h^1(t)$ and $d^1(t)$ had been chosen.

In the moment $t + \tau$ the player II is choosing the decision $d^2(t+\tau) \in D^2$ and receives the payment $w^2(t+\tau) = w^2[h^2(t+\tau), d^2(t+\tau)]$.

In the moment $t + 2\tau$ the player III and its environment are activied. This environment is influenced by the player II and its environment.

In the moment $t + 3\tau = t + 1$ the next cycle of the game is beginning.

In the considered game $w_1,w_2,w_3 \in \{0,1\}$.

The choice of matrix Y^{1k} depends on the results received in the previous cycle:

$$Y^{1k}(t+1) = \begin{cases} Y^{11} & \text{if } w^2(t+\tau) = 1;\ w^3(t+2\tau) = 1 \\ Y^{12} & \text{if } w^2(t+\tau) = 0;\ w^3(t+2\tau) = 1 \\ Y^{13} & \text{if } w^2(t+\tau) = 1;\ w^3(t+2\tau) = 0 \\ Y^{14} & \text{if } w^2(t+\tau) = 0;\ w^3(t+2\tau) = 0. \end{cases}$$

Information assemblies were constructed using data names:

$d^i(t + (i-1)\tau - k]$, $w^i[t + (i-1)\tau - k]$ - where
$d^i[t + (i-1)\tau - k]$ is "the decision chosen by the i-th player in k-th cycle before the current cycle",
$w^i[t + (i-1)\tau - k]$ is "the payment obtained by the i-th player in k-th cycle before the current cycle".
G_1, G_2, G_3 denotes average payments of the players.

2.4.2. Results of simulation studies.

If information assemblies of players /automata/ are empty, it is $L^1 = \emptyset$, $L^2 = \emptyset$, $L^3 = \emptyset$, the following results were obtained: $G_1 = 0.249$, $G_2 = 0.251$, $G_3 = 0.248$.

Some other results are shown in the Table 1.

TABLE 1

Nr	L^1	L^2	L^3	G_1	G_2	G_3
1	$d^1(t-1)$	$d^2(t+\tau-1)$	$d^3(t+2\tau-1)$	0.316	0.294	0.291
2	$w^1(t-1)$	$w^2(t+\tau-1)$	$w^3(t+2\tau-1)$	0.499	0.412	0.366
3	$d^1(t-1), w^1(t-1)$	$d^2(t+\tau-1)$, $w^2(t+\tau-1)$	$d^3(t+2\tau-1)$, $w^3(t+2\tau-1)$	0.613	0.456	0.401
4	$d^1(t-1), w^1(t-1)$, $w^2(t+\tau-1)$, $w^3(t+2\tau-1)$	$d^1(t)$	$d^2(t+\tau)$	0.731	0.605	0.512
5	- " -	$d^1(t), w^1(t)$	$d^1(t)$, $d^2(t+\tau)$	0.784	0.727	0.6
6	- " -	- " -	$d^1(t)$, $d^2(t+\tau)$, $w^1(t)$, $w^2(t+\tau)$	0.791	0.726	0.731

Results obtained by automata strongly depend on their information assemblies. One can observe that the change the i-th player's information assembly influences not only his payment but - sometimes - results obtained by other players /compare Nr.4, Nr.5,

Nr.6/.

Such simulation studies can be performed in order to find the best information flows connecting decision makers.

The value of the whole information assembly can be determined comparing the player's average payment with the result obtained if his information assembly is empty.

REFERENCES

1. Bartoszyński R. 1964 : On a certain concept of value of information in games. In: Transactions of the 3rd Prague Conference on Information Theory, Statistical Functions, Random Processes, Prague 1964.

2. Narendra Kumpati S., Thathachar M.A.L. 1974 : Learning automata - a survey. In: IEEE Transactions on System, Man and Cybernetics, Vol.SMC-4, Nr. 4, July 1974.

3. Viswanathan R., Narendra Kumpati S. 1974 : Games of stochastic automata. In: IEEE Transactions on System, Man and Cybernetics, Vol.SMC-4, Nr. 4, January 1974.

University of Mining
and Metallurgy
Institute of Control
Al. Mickiewicza 30
31-059 Cracow/Poland

DECISION MAKING UNDER COMPARATIVE EVALUATION OF RESULTS

Milan Mareš

Prague

Key words: decision making, criterion, global criterion

ABSTRACT

Rational methods of reconstruction of the global criterion in multicriterial decision making are considered and briefly discussed. It is shown how to choose "experimental" decisions which should be Pareto optimal and which should bring more information about the unknown global evaluation of the consequences.

BASIC NOTIONS

In the whole paper we denote by R the set of all real numbers.

Let us consider a multicriterial decision making problem $(C, n, \{u_i\}_{i=1}^n)$ where C is a non-empty set of consequences (one--to-one connected with decisions), $n \geq 2$ is the number of criteria, and $u_i: C \to R$, $i = 1,\dots,n$, are utility functions of the particular criteria.

Let us denote

$$(1.1) \qquad \mathcal{U} = \{(x_i)_{i=1}^n \in R^n: \exists c \in C, \forall i=1,\dots,n, x_i=u_i(c)\}$$

the set of achievable utility vectors, and $P_{\mathcal{U}}$ the Pareto optimum of $\mathcal{U}$, i. e. the set

$$(1.2) \qquad P_{\mathcal{U}} = \{x=(x_i)_{i=1}^n \in \mathcal{U}: \ \forall y=(y_i)_{i=1}^n \in \mathcal{U}, \ y \neq x \Rightarrow \exists i \in \{1,\dots,n\}, \ x_i > y_i\}.$$

A consequence $c \in C$ is called Pareto optimal iff $(u_1(c),\dots,u_n(c)) \in P_{\mathcal{U}}$.

Let us suppose that, besides the particular criteria represented by the utility functions $u_1,\dots,u_n$, there exists a global criterion represented by global utility function $u: C \to R$ such that for all $c \in C$

$$(1.3) \qquad u(c) = w_1u_1(c) + \dots + w_nu_n(c)$$

where $w_i \in R$, $i = 1,\dots,n$ are weights,

$$(1.4) \qquad w_i \geq 0, \ i = 1,\dots,n, \ w_1 + \dots + w_n = 1.$$

PROBLEM FORMULATION

In this contribution, as well as in Mareš (1986) we shall suppose that the decision maker knows the particular utility functions $u_1,\dots,u_n$ but he does not know the weights $w_1,\dots,w_n$ and, consequently, he does not know the global utility function u. Only when choosing some decision and corresponding consequence $c \in C$ the decision maker obtains some information about the value $u(c)$. By means of rational choice of such "experimental" consequences the decision maker can estimate the possible values of $w_1,\dots,w_n$. Different types of information about the obtained values of the global utility function and the corresponding methods of estimation were discussed in Mareš (1986). Here we shall be interested in a specific problem connected with.

We shall suppose that after choosing consequences c_1, $c_2 \in C$ the decision maker obtains the information wheather

$$(2.1) \qquad u(c_1) \geq u(c_2) \quad \text{or} \quad u(c_1) \leq u(c_2)$$

i. e. which one of c_1 and c_2 is globally preferable. It is obvious and was shown in Mareš (1986) that choosing proper consequences $c_1, c_2,\dots$ it is possible to derive a system of inequalities limiting the possible values of $w_1,\dots,w_n$. It is desirable to choose the "experimental" consequences which not only lead to effective inequalities for $w_1,\ldots,w_n$ but which are also rational, i. e. whose utility vectors belong to the Pareto optimum $P_{\mathcal{U}}$ of $\mathcal{U}$. Our goal is to show some conditions under which such choice is possible.

IMPROVEMENT OF INEQUALITIES

Let us consider the decision making problem $(C, n, \{u_i\}_{i=1}^n)$, let $i \in \{1,\ldots,n\}$, and let $c_1 \in C$, $c_2 \in C$ be such that $u_i(c_1) \neq u_i(c_2)$. It is easy to see that

$$(3.1) \qquad 1 \geq w_i \geq \frac{\sum_{j=1,j\neq i}^n w_j \cdot (u_j(c_2)-u_j(c_1))}{u_i(c_1) - u_i(c_2)}$$

if $u(c_1) \geq u(c_2)$, i. e. if c_1 is globally preferred to c_2. On the other hand,

$$(3.2) \qquad 0 \leq w_i \leq \frac{\sum_{j=1,j\neq i}^n w_j (u_j(c_2)-u_j(c_1))}{u_i(c_1) - u_i(c_2)}$$

if $u(c_1) \leq u(c_2)$.

If $u(c_1) \geq u(c_2)$, then it is useful to choose $c_3 \in C$ which increases the right hand side of (3.1). It means such that

$$(3.3) \quad \frac{\sum_{j=1,j\neq i}^n w_j (u_j(c_3)-u_j(c_1))}{u_i(c_1) - u_i(c_3)} > \frac{\sum_{j=1,j\neq i}^n w_j (u_j(c_2)-u_j(c_1))}{u_i(c_1) - u_i(c_2)} .$$

If $u(c_1) \geq u(c_3)$ then the inequality (3.1) is improved. If $u(c_1) <$ $< u(c_3)$ then (3.2) with c_3 instead of c_2 can be applied and the value of w_j is limited from both sides: from below by $u(c_1) \geq u(c_2)$ and (3.1), from above by $u(c_1) \leq u(c_3)$ and (3.2). If all the consequences, c_1, c_2, c_3 are to be Pareto optimal then none of the relations

$$u_i(c_r) \geq u_i(c_s) \quad \text{for} \quad i = 1,\ldots,n$$

can be fulfilled for $r, s \in \{1,2,3\}$, $r \neq s$.

It is not difficult to derive some general conditions under which (3.3) holds. For example, if $u_i(c_1) > u_i(c_2)$, $u_j(c_1) < u_j(c_2)$ for all $j \in \{1,\ldots,n\}$, $j \neq i$, if $\varepsilon > 0$, $\eta > 0$, then $c_3 \in C$ such that

$$(3.4) \qquad u_i(c_3) = u_i(c_2) - \eta, \; u_j(c_3) = u_j(c_2) + \varepsilon, \; j=1,\ldots,n, \; j \neq i,$$

leads to (3.3) iff ε and η are such that

$$(3.5) \qquad \frac{\varepsilon}{\eta} > \frac{\sum_{j=1,j\neq i}^{n} w_j(u_j(c_2)-u_j(c_1))}{(u_i(c_1)-u_i(c_2)(1-w_i)}.$$

The existence of such ε, η for which c_3 is also Pareto optimal depends on the actual shape of the set $\mathcal{U}$, but it is not excluded by (3.4) and (3.5).

Condition (3.5) is applicable only if the values of w_j and w_i on its right-hand-side are at least approximately known from the previous steps of the described procedure. However, for small number of criteria, i. e. for $n = 2$ and $n = 3$, conditions analogous to (3.5) and independent on the values of weights w_j, $j = 1,\dots,n$, can be derived.

Let $n = 2$, $c_1, c_2 \in C$, let $u_1(c_2) < u_1(c_1)$, $u_2(c_2) > u_2(c_1)$, and let $u(c_1) \geq u(c_2)$. Then

$$(3.6) \qquad w_1 \geq \frac{u_2(c_2) - u_2(c_1)}{u_2(c_2) - u_2(c_1) + u_1(c_1) - u_1(c_2)},$$

as follows from $u(c_1) \geq u(c_2)$ and from (1.3), (1.4).
Then it can be easily seen that for $\varepsilon > 0$, $\eta > 0$ and $c_3 \in C$ such that $u_1(c_3) = u_1(c_2) - \eta$, $u_2(c_3) = u_2(c_2) + \varepsilon$ the r-h-s of (3.6) increases iff

$$\frac{\varepsilon}{\eta} > \frac{u_2(c_2) - u_2(c_1)}{u_1(c_1) - u_1(c_2)}.$$

On the other hand, if $0 < \varepsilon < u_2(c_2) - u_2(c_1)$, $0 < \eta < u_1(c_1) - u_1(c_2)$, then the r-h-s of (3.6) increases iff $u_1(c_3)=u_1(c_2)+\eta$, $u_2(c_3)=u_2(c_2)-\varepsilon$, and

$$\frac{\varepsilon}{\eta} < \frac{u_2(c_2) - u_2(c_1)}{u_1(c_1) - u_1(c_2)}.$$

If $n = 3$, $c_1, c_2 \in C$, $u_1(c_1) > u_1(c_2)$, $u_3(c_1) < u_3(c_2)$, and if $u(c_1) \geq u(c_2)$ then, using (1.3) and (1.4),

$$(3.7) \qquad w_1 \geq \frac{w_2[(u_2(c_2)-u_2(c_1))-(u_3(c_2)-u_3(c_2))]}{(u_3(c_2)-u_3(c_1))-(u_1(c_2)-u_1(c_1))}.$$

There exist many possible strategies of choice of $c_3 \in C$ increasing the r-h-s of (3.7). Let us mention two of them.

If $\varepsilon > 0$, $0 < \eta < (u_3(c_2)-u_3(c_1)) - (u_1(c_2)-u_1(c_1))$ and $u_1(c_3) = u_1(c_2)$, $u_2(c_3) = u_2(c_2) + \varepsilon$, $u_3(c_3) = u_3(c_2) - \eta$ then the r.h.s of (3.7) increases iff

$$\frac{\varepsilon+\eta}{\eta} > \frac{u_3(c_2)-u_3(c_2)-u_2(c_2)+u_2(c_1)}{u_3(c_2)-u_3(c_1)-u_1(c_2)+u_1(c_1)}.$$

If $\varepsilon > 0$, $0 < \eta < u_3(c_2)-u_3(c_1)-u_1(c_2)+u_1(c_1)$, and $u_1(c_3) = u_1(c_2) - \eta$, $u_2(c_3) = u_2(c_2) + \varepsilon$, then the r-h-s of (3.7) increases iff

$$\frac{\varepsilon}{\eta} > \frac{u_2(c_2) - u_2(c_1) - u_3(c_2) + u_3(c_1)}{u_3(c_2) - u_3(c_1) - u_1(c_2) + u_1(c_1)} .$$

TESTING OF GIVEN VALUES OF WEIGHTS

It is also possible to proceed in another way. Considering a weight w_j, $j \in \{1,\dots,n\}$ and a real value α, $0 < \alpha < 1$, we can find such c_1, $c_2 \in C$ whose global comparison $u(c_1) \gtreqless u(c_2)$ implies the proper one of the inequalities $w_j \gtreqless \alpha$. This method was considered in Mareš (1986), and it was shown that for c_1, $c_2 \in C$ such that

$$u_i(c_2) \neq u_i(c_1) \quad \text{for some } i \neq j,\ i \in \{1,\dots,n\},$$

(4.1) $$u_k(c_2) = u_k(c_1)-u_i(c_1)+u_i(c_2) \quad \text{for } k=1,\dots,n,\ i\neq k\neq j$$

$$u_j(c_2) = u_j(c_1) + \frac{1-\alpha}{\alpha}(u_i(c_1) - u_i(c_2))$$

the inequalities $u(c_1) \gtreqless u(c_2)$ lead to $w \gtreqless \alpha$. The set of the utility vectors $(u_1(c_2),\dots,u_n(c_2))$ fulfilling conditions (4.1) forms a line in R^n, and $u_i(c_2) > u_i(c_1)$ iff $u_j(c_2) < u_j(c_1)$. It means that the existence of a Pareto optimal $c_2 \in C$ fulfilling (4.1) is not generally excluded, and it depends on the actual shape of the set $\mathcal{U}$. The course of the line mentioned above depends on the value α and can be rather regulated by it, as follows from (4.1).

REFERENCES

Mareš M. (1986): Problem of ambicious decision-maker. Accepted for publication in Ekonomicko-matematický obzor, 1986.

Czechoslovak Academy of Sciences
Institute of Information Theory and Automation
182 08 Prague 8
Pod vodárenskou věží 4
Czechoslovakia

SUR LA CONVERGENCE PONCTUELLE DE T^nf/a_n, DANS L_p

Radko Mesiar

Bratislava

Key words: linear operators, a.e. convergence

ABSTRACT

Let T be a linear operator in L_p, $1 \leq p \leq \infty$, and let $\{a_n\}$ be a sequence of positive reals. We study the almost everywhere convergence of T^nf/a_n for $f \in L_p$.

RÉSUMÉ

Soit $\{a_n\}$ une suite des positifs réels, T un opérateur linéaire dans L_p, $1 \leq p \leq \infty$. On étudie la convergence presque sûre de T^nf/a_n pour f appartenant à L_p.

Soit (X, $\mathcal{S}$, m) un espace mesuré (m σ-finie) et T un opérateur linéaire de L_p(X, $\mathcal{S}$, m) dans L_p(X, $\mathcal{S}$, m), $1 \leq p \leq \infty$. Soit $\{a_n\}$ une suite des positifs réels. Nous étudions la convergence ponctuelle de T^nf/a_n comme la propriété de la suite $\{a_n\}$ (on obtient sans difficulté la convergence forte) .

Nous supposons que le radius spectral de l'opérateur T

$$\lambda_T = \inf \sqrt[n]{\|T^n\|_p} = \lim \sqrt[n]{\|T^n\|_p} = 1 .$$

Si non, nous étudierons son opérateur normalisé $T' = T/\lambda_T$, alors $\lambda_{T'} = 1$.

PROPOSITION 1. Soit $\lim_n a_n = \infty$.

i) Si $T = T^*$ (T opérateur auto-adjoint) est une contraction de L_p ($p > 1$), alors pour chaque $f \in L_p$ on a

$$T^nf/a_n \xrightarrow[n]{} 0, \text{ p.s.}$$

ii) Si T est a puissances bornées dans L_∞ , on obtient pour chaque $f \in L_\infty$

$$T^n f/a_n \xrightarrow[n]{} 0, \text{ p.s.}$$

Démonstration. D'aprés Stein (1961) on a $\sup_n |T^n f| \in L_p$ et donc $(\sup_n T^n |f|)/a_n \xrightarrow[n]{} 0$, p.s., qu'implique le résultat i).
Le résultat ii) est immédiat.

PROPOSITION 2. Soit $1 \leq p < \infty$.

i) Si $\{1/a_n\} \in l_p$ (i.e. $\sum 1/a_n^p < \infty$), alors pour chaque opérateur linéaire T a puissances bornées dans L_p et chaque $f \in L_p$ on a

$$T^n f/a_n \xrightarrow[n]{} 0, \text{ p.s.}$$

ii) Si $\{1/a_n\} \notin l_p$, il existe (sur un espace mesuré convenable X) un opérateur linéaire positif T à puissances bornées dans L_p et $f \in L_p$ tels que

$$m(\limsup T^n f/a_n = \infty) > 0 .$$

iii) Si $\{1/a_n\} \notin l_p$, et il existe une constante C telle que $a_{i.j} \leq C a_i a_j$, $i,j = 1, 2, \ldots$, alors il existe une contraction positive P de $L_p(\langle 0 ; 1 \rangle)$ et $f \in L_p(\langle 0 ; 1 \rangle)$ telles que

$$\limsup P^n f/a_n = \infty \text{ , p.s.}$$

Démonstration. Soit T un opérateur linéaire dans L_p, $1 \leq p < \infty$. Alors il existe un opérateur linéaire T_1 unique dans L_1 tel que pour $g \in L_1^+$ on a $T_1 g = |Tg^{1/p}|^p$. On a

$$\|T_1\|_1 = \|T\|_p^p \text{ , } T_1^n g = |T^n g^{1/p}|^p \text{ , } g \in L_1^+ .$$

Au contraire, soit T_1 un opérateur linéaire dans L_1. Alors il existe un opérateur linéaire T unique dans L_p, $1 \leq p < \infty$, tel que pour $f \in L_p^+$ on a $Tf = |T_1 f^p|^{1/p}$. On a

$$\|T\|_p = \|T_1\|_1^{1/p} \text{ , } T^n f = |T^n f^p|^{1/p} \text{ , } f \in L_p^+ .$$

i) Soit $\{1/a_n\} \in l_p$ et soit T un opérateur linéaire a puissances bornées dans L_p, $\sup_n \|T^n\|_p = M < \infty$. Nous posons

$U = \sum_n \frac{T_1^n}{a_n^p}$. On a $\|U\|_1 \leq M^p \sum_n \frac{1}{a_n^p} < \infty$, alors l'opérateur linéaire U est bien défini dans L_1. Soit $f \in L_p$. Ensuit $|f|^p \in L_1$, $U|f|^p \in L_1$, alors

$$\frac{T_1^n|f|^p}{a_n^p} = \frac{|T^n f|^p}{a_n^p} \xrightarrow[n]{} 0, \text{ p.s.} ,$$

qu'implique le résultat.

ii) Soit $\{1/a_n\} \in l_p$. D'aprés Assani, Mesiar (1985) il existe un opérateur linéaire T_1 dans $L_1(\langle 0 ; 2))$ et une fonction positive $g \in L_1^+(\langle 0 ; 2))$ tels que

$$m(\limsup T_1^n g/a_n^p = \infty) = 1$$

(la construction d' Assani, Mesiar (1985) est convenable même si la suite $\{a_n\}$ n'est pas croissante) .

Alors pour $f = g^{1/p}$, $Tf = (T_1 g)^{1/p}$ on a

$$\limsup T_1^n g/a_n^p = \infty \Longleftrightarrow \limsup T^n f/a_n = \infty \quad ,$$

qu'implique le résultat.

iii) L'idée principale est dans la modification da la construction de Chacón (1964). Pour le cas p = 1 elle est faite en détail dans Mesiar (1986b). La contraction P sera induite par une transformation affine t. Si $t: \langle a ; b) \longrightarrow \langle c ; d)$, alors pour $a \leq x < b$ on a

$$tx = c + \frac{d-c}{b-a} . (x-a) .$$

Pour $y \in \langle c ; d)$, $f \in L_p(\langle a ; b))$ on pose

$$Pf(y) = (\frac{b-a}{d-c})^{1/p} . f(t^{-1}y) .$$

Alors P est une contraction de $L_p(\langle a ; b))$ dans $L_p(\langle c ; d))$.

Si nous utilisons la construction de Mesiar (1986b) qui est faite par étapes, et en passant à (n+1)-ème étape nous nous servirons des constantes c_n, K_n telles que

$$1/(2^n c_n) \leq n a_{N_n}^p \quad , \quad 0 < 1 - (1/2^n + c_n . \sum_{i=1}^{K_n - 2} (1/a_{i+1}^p)) < 1/2^n ,$$

(telles constantes existent toujours parceque $\{1/a_n\} \notin l_p$), on obtient le résultat cherché.

CORROLAIRE. Si $\inf_n \|T^n\|_p < 1$, alors pour chaque suite $\{a_n\}$ pour laquelle $\liminf \sqrt[n]{|a_n|} \geq 1$, on obtient pour chaque $f \in L_p$

$$T^n f/a_n \xrightarrow[n]{} 0, \text{ p.s.}$$

Démonstration. Car $\inf_n \|T^n\|_p < 1$, on a $\lambda_T < 1$. Soit $B = \frac{1+\lambda_T}{2}$, $b_n = |a_n|/B^n$. On a $\limsup \sqrt[n]{1/b_n} = B/(\liminf \sqrt[n]{|a_n|}) \leq B < 1$, alors la série $\sum_n 1/b_n^p$ est convergente. L'opérateur R = T/B est à puissances bornées dans L_p ($1 \leq p < \infty$) . La proposition 2, part i) implique le résultat. Le cas de $T^n f/n^r \xrightarrow[n]{} 0$, p.s. , pour chaque réel r, $f \in L_p$, montré dans Assani, Mesiar (1985) , est le conséquence de $\liminf \sqrt[n]{n^r} = 1$.

REMARQUE. D'après Mesiar (1986a), il est possible de généraliser les parts i) et ii) de Proposition 2 pour les opérateurs linéaires T tels que $\|T^n\|_p \lesssim n^k$, $k \in \{0, 1, 2, \dots\}$ (ici $\|T^n\|_p \lesssim n^k$ est equivalent de $\limsup_n \|T^n\|_p/n^k < \infty$) . Si $\{n^k/a_n\} \notin l_p$, alors pour n'importe quels T, $\|T^n\|_p \lesssim n^k$, et $f \in L_p$, on a $T^n f/a_n \xrightarrow[n]{} 0$, p.s. . Au contraire, si $\{n^k/a_n\} \notin l_p$, on peut construire un opérateur linéaire T (dans un convenable $L_p(X, \mathcal{S} , m)$) tel que $\|T^n\|_p \lesssim n^k$, et une fonction $f \in L_p$, tele que la convergence $T^n f/a_n \xrightarrow[n]{} 0$, p.s. , n'a pas lieu, plus précise que $m(\limsup T^n f/a_n = \infty) > 0$.

REFERENCES

Assani I. , Mesiar R,

(1985) : Sur la convergence ponctuelle de $T^n f/n^a$, dans L_p. Ann. Sc. de L'Univ. de Clermont-Ferrand II, 1985, fasc. 3, 21-29.

(1986) : On the a.e. convergence of $T^n f/a_n$ in L_1-space. A paraitre in: Trans. of the 13-th Winter School, Srní, 1985.

Chacon R. V. (1964) : A class of linear transformation. P AMS 15, 560-564 .

Mesiar R. (1986a) : On the limit properties of $T^n f/a_n$. A paraître in: Trans. of the conf. "Ergodic theory and related topics II", Georgenthal, 1986.

(1986b) : O kontrakciách v L_1 (On the contractions in L_1) . A paraître.

Stein E. M. (1961) : On the maximal ergodic theorem. Proc. Nat. Acad. Sci. 47, 1894-1897.

Katedra matematiky a desk. g.
Stavebná fakulta SVŠT
Radlinského 11
813 68 Bratislava
ČSSR

INFORMATION GEOMETRY

OF NONIDEAL GASES

Ryszard Mrugała

Toruń

Key words: Fisher information matrix, Riemannian metric, thermodynamic phase space, contact structure

ABSTRACT

A matrix closely related to the Fisher information matrix is calculated for the Boguslavski distribution of statistical physics. This matrix is assumed to be a Riemannian metric tensor in the space of thermodynamical parameters. The resulting scalar curvature of this space is strictly zero for ideal gases while for more realistic models it may be positive, negative and zero. Negative curvature, however, is not allowed by the laws of thermodynamic. Thus, any model should be considered physically inadequate in the region of parameters where the curvature defined here becomes negative.

INTRODUCTION

In recent years we observe an increasing interest in the use of differential geometry to thermodynamics - equilibrium and nonequilibrium. Any thermodynamic problem may be approached either on phenomenological or statistical way.

In phenomenological thermodynamics prevailed common belief that the space of thermodynamic parameters lacks an intrinsic geometrical structure, in particular a metric structure. It was shown, however, by Mrugala (1978) that it is possible to construct a new thermodynamic space, the so-called thermodynamic phase space M, which has a natural contact structure generated either by the energy or by entropy differential 1 - forms θ . States of the system are then represented by integral manifolds $\mathcal{S}$ of the Pfaff equation $\theta=0$, $\mathcal{S}\subset M$. A new question arises, however, if another geometric structure, possibly independent of that induced by θ, can be given to $\mathcal{S}$.

Weinhold (1975) showed that to some thermodynamic space (which is in fact isomorphic with $\mathcal{S}$) an abstract metric structure may be given which is fully consistent with empirical laws of thermodynamics. Ruppeiner (1979) included the theory of fluctuations into axioms of thermodynamics and showed that this leads to a reasonable Riemannian metric on the manifold of equilibrium states. Theory of fluctuations, which is usually connected with statistical physics, may be included into phenomenological thermodynamics due to the Einstein formula for the probability density function of fluctuations,

$$W(x,x') = B\exp(\Delta S/k),$$

where $\Delta S=S(x')-S(x)\leqslant 0$ is the difference of the instantaneous entropy $S(x')$ at the fluctuation state labeled by x' and the equilibrium entropy $S(x)$ at the equilibrium state x, k is the Boltzmann constant and B is a normalization factor.
Mrugała (1984) showed that metric structures may be linked by a contact transformation and that they are conformally equivalent.

On the other hand one can try to geometrize thermodynamics trough statistical physics. In statistical physics we have many possibilities because we may start from different probability distribution functions, e.g., canonical, grand canonical, Boguslavski distribution functions and so on. Differential geometry enters when one considers varying the probability distribution either directly or trough changing parameters on which the distribution function depends. In the paper we will use the Boguslavski distribution and we will change two parameters: pressure P and temperature T. Janyszek and Mrugała (1986) investigated also geometry of the grand canonical distribution. Relations between differential geometry and information theory , statistics and thermodynamics were investigated extensively by Amari (1985), Campbell (1985), Gilmore (1984), Ingarden et al (1979) and many others.

GEOMETRY OF GENERALIZED GIBBS DISTRIBUTIONS

Let us consider a physical system with (mechanical) phase space Γ of generalized coordinates q and generalized momenta p. Let us assume that the system is in thermodynamic equilibrium and that we measured certain independent quantities $m_1,\ldots,m_n$ which correspond to random variables $F_1(q,p),\ldots,F_n(q,p)$. From statistical point of view $m_1,\ldots,m_n$ are equal to mean values of $F_1(q,p),\ldots,F_n(q,p)$ calculated with some distribution function $f(q,p)$.
The principle of maximum of information (entropy) subject to a set of constraints

(1) $$\int_\Gamma f(q,p)\, F_i(q,p)\, dq\, dp = m_i, \quad i=1,\ldots,n,$$

leads to the generalized Gibbs distribution function

(2) $$f(q,p;\beta) = Z^{-1}(\beta)\, e^{-\beta^i F_i(q,p)} .$$

We indicated here explicitly the dependence of f on $\beta=(\beta^1,\ldots \beta^n)$ and the sumation convention is assumed over repeated indices. $Z^{-1}(\beta)$ stands for the normalization factor and the Lagrange multipliers β^i are called (statistical) temperatures.

Let N denote the manifold of parameters β and let us assume that N is a differentiable manifold. The Riemannian metric on N is introduced by means of the formula

(3) $$dl^2 := \langle (d \ln f)^2 \rangle = \left\langle \frac{\partial \ln f}{\partial \beta^i} \frac{\partial \ln f}{\partial \beta^k} \right\rangle d\beta^i\, d\beta^k ,$$

where differentiation has been performed with respect to β's and $\langle A \rangle$ denotes the mean value of A.
For

(4) $$g_{ik}(\beta) := \left\langle \frac{\partial \ln f}{\partial \beta^i} \frac{\partial \ln f}{\partial \beta^k} \right\rangle$$

dl^2 may be written in the standard form

(5) $$dl^2 = g_{ik}(\beta)\, d\beta^i\, d\beta^k ,$$

with g_{ik} playing the role of a metric tensor.

One can prove that the same result may be reached from the dispersion of relative entropy, c.f. Amari (1985).
The matrix $G=(g_{ik})$ is just the Fisher information matrix.

GEOMETRY OF THE BOGUSLAVSKI DISTRIBUTION

The meaning of β's may be revealed only after refering to phenomenological thermodynamics. If we choose $F_1(q,p)=H(q,p)$ - Hamiltonian and $F_2(q,p)=V$ - volume of the system , then it must be: $\beta^1=T^{-1}$, $\beta^2=P\,T^{-1}$, and consequently

$$(7)\quad f(q,p;T,P) = Z^{-1}(T,P)\, e^{-\frac{H(q,p)}{T} - \frac{P}{T}V} .$$

The distribution function (7) is called the Boguslavski or P-T distribution. Physically it means that energy and volume may freely fluctuate whereas pressure P and temperature T are parameters fixed by external conditions (environment). Changing P and T one obtains two parameter family of distribution functions (7).

The matrix G in this case has rather complicated and inconvenient form. It is much easier and better to calculate it in a transformed representation. Let us notice that if we write formally $Z^{-1}(\beta)=\exp(-Q)$ then from (2) we have

$$(8)\qquad \theta := \langle - d \ln f \rangle = dQ + m_i\, d\beta^i .$$

If one treats Q as a new coordinate independent of m's and β's then θ is a contact form on (2n+1)-dimensional space M with local coordinates (Q, m_i, β^i). One of solutions of the Pfaff equation $\theta=0$ may be given as

$$(9)\quad Q=Q(\beta^1,\dots,\beta^n)\ , \qquad m_i=-\frac{\partial Q}{\partial \beta^i}\quad ,\quad i=1,\dots,n.$$

These n+1 equations define n-dimensional Legendre manifold $\mathcal{S}$ in M and as a matter of fact they give an imbedding of N into M, i.e., N is diffeomorphic with $\mathcal{S}$.

For the Boguslavski distribution (7) we have

$$(10)\quad \theta = dQ + U\, d(T^{-1}) + V\, d(P\,T^{-1}) ,$$

where Q has physical meaning of a thermodynamic potential, namely it is the so-called Planck function. For $\theta=0$ it becomes a function of T and P. If one notices that on

$$(11)\quad g_{ik}(\beta)= \frac{\partial^2 \ln Z}{\partial\beta^i\partial\beta^k} = \frac{\partial^2 Q}{\partial\beta^i\partial\beta^k}\ (\equiv Q_{,ik}) = -\frac{\partial m_i}{\partial\beta^k}$$

then it is obvious that for calculating g_{ik} we may use the apparatus of phenomenological thermodynamics. First of all , by an appropriate contact transformation $\lambda : M \to M$ we transform θ into a new contact form

(12) $\theta' = dU - T\,dS + P\,dV$,

where U and S denote the internal energy and the entropy of the system, respectively, and we used the phenomenological relation: $Q = S - T^{-1}U - PT^{-1}V$. Because $\theta = 0$ and $\theta' = 0$ define the same surface $\mathcal{S}$ we may calculate the metric tensor of $\mathcal{S}$ in the transformed representation (12). Adapting (11) for (12) we have

$$(13) \quad G = \begin{pmatrix} \frac{\partial T}{\partial S} & \frac{\partial T}{\partial V} \\ -\frac{\partial P}{\partial S} & -\frac{\partial P}{\partial V} \end{pmatrix} = \begin{pmatrix} \frac{T}{C_V} & -\frac{T\alpha_P}{k_T C_V} \\ -\frac{T\alpha_P}{k_T C_V} & \frac{C_P}{V k_T C_V} \end{pmatrix},$$

where

$C_P = T(\partial S/\partial T)_P$ - the heat capacity by constant pressure,

$C_V = T(\partial S/\partial T)_V$ - the heat capacity by constant volume,

$\alpha_P = V^{-1}(\partial V/\partial T)_P$ - the coefficient of thermal expansion,

$k_T = -V^{-1}(\partial V/\partial P)_T$ - the isothermal compressibility.

One can see that components of G are expressed trough directly measurable thermodynamic parameters (T,V) and quantities (C_P, C_V, α_P, k_T).

CURVATURE OF THE SURFACE $\theta = 0$ FOR IDEAL AND REAL GASES

According to (11) the Christoffel symbols can be reduced to the form

$$(14) \quad \Gamma^i_{jk} = \frac{1}{2} g^{il}\, Q_{,ljk} ,$$

where (g^{il}) is the inverse of the matrix (g_{il}).

After simple calculations the Riemannian curvature tensor and the scalar curvature can be expressed by the following formulas involving the second and the third derivatives of Q only:

$$(15) \quad R_{ijkl} = \frac{1}{4} g^{sr} (Q_{,sjk}\, Q_{,ril} - Q_{,sjl}\, Q_{,rik}) ,$$

$$(16) \quad R = \frac{1}{4} (g^{jl} g^{ik} - g^{jk} g^{li})\, g^{sr}\, Q_{,sjk}\, Q_{,ril} .$$

Obviously for n=2 we have

$$R = 2\, R_{1212}\, (\det g)^{-1} \quad . \tag{17}$$

It is easy to prove that for the ideal gas R is always equal to 0. However, this is not the case for real gases. As a model of a real gas we will choose the well-known van der Waals model for which the equation of state is given in the form (for 1 mole)

$$(P + a\, V^{-2})\, (V - b) = T \ , \tag{18}$$

where a and b are some positive constants.
The entropy of this gas resulting from (18) has the form

$$S = \frac{3}{2} \ln \left[\frac{4}{3h^2} \pi\, m\, (U + \frac{a}{V}) \right] + \ln (V - b) + \frac{5}{2} \ , \tag{19}$$

where h is the Planck constant and m is the molecule mass.
The metric tensor in the representation:

$$\theta = - dS + T^{-1}\, dU + P\, T^{-1}\, dV \ , \tag{20}$$

has the following components (cf. 11):

$$g_{VV} = - \frac{\partial^2 S}{\partial V^2} = (V - b)^2 - 2\, T^{-1}\, a\, V^{-3} + \frac{2}{3} a^2\, T^{-2}\, V^{-4} \quad ,$$

$$g_{UU} = - \frac{\partial^2 S}{\partial U^2} = \frac{2}{3} T^{-2} > 0 \ , \tag{21}$$

$$g_{UV} = - \frac{\partial^2 S}{\partial U \partial V} = - \frac{2}{3} a\, T^{-2}\, V^{-2} < 0 \ .$$

The sign of g_{VV} is not fixed and depends on values of parameters T and V. As a result neither the sign of det g nor of R are fixed. For det g tending to 0, R tends to plus or minus infinity. For these values of parameters for which $R = +\infty$ we observe that system is not stable any longer and phase transitions occur. Negativity of R violates the empirical laws of thermodynamics. Therefore in these regions of parameters for which $R < 0$ the model is completely inacceptable and has to be replaced by a better one.

The author would like to express his gratitude to Dr. H. Janyszek for many discussions and help during preparation of this manuscript.

REFERENCES

Amari S. (1985): Differential - Geometrical Methods in Statistics. Lecture Notes in Statistics. 28, Springer, Berlin

Campbell L.L. (1985): The relation between information theory and the differential geometry approach to statistics. Information Science 35.

Gilmore R. (1984): Length and curvature in the geometry of thermodynamics. Phys.Rev.A. 30, No. 4, 1994-1997.

Ingarden R.S., Sato Y., Sugawa K., and Kawaguchi M. (1979): Information thermodynamics and differential geometry. Tensor, N.S. 33, 347-353.

Janyszek H., and Mrugała R. (1986): Geometrical analysis of generalized Gibbs distributions, to be published in Rep.Math.Phys.

Mrugała R. (1978): Geometrical formulation of equilibrium phenomenological thermodynamics. Rep.Math.Phys. 14., No. 3, 419-427.

(1984): On equivalence of two metrics in classical thermodynamics. Physica 125A, 631-639.

Ruppeiner G. (1979): Thermodynamics: A Riemannian geometric model. Phys.Rev.A.20, No. 4, 1608-1613 .

Weinhold F. (1975): Metric geometry of equilibrium thermodynamics. J.Chem.Phys. 63, No. 6, 2479-2483.

Institute of Physics
Nicholas Copernicus University
ul. Grudziądzka 5
87-100 Toruń, Poland

О СТЕПЕНЯХ НЕДЕТЕРМИНИЗМА И ВЕРОЯТНОСТНЫХ ВЫЧИСЛЕНИЯХ НА АВТОМАТАХ С МАГАЗИННОЙ ПАМЯТЬЮ

Насыров Ильдар Р.

Казань

Ключевые слова: вероятностные автоматы, характеристическая функция вероятностных языков

АННОТАЦИЯ

Исследованы некоторые статистические свойства характеристической функции вероятностных КС языков.

Понятие вероятностных автоматов широко используется для различных приложений, связанных с проблемами построения быстрых алгоритмов вычислений, моделированием и анализом сложных объектов Бухараев (1985). Особое значение в задачах распознавания образов и изучения статистических свойств языков программирования имеют вероятностные КС языки, распознаваемые на вероятностных автоматах с магазинной памятью Фу (1977), Wetherell (1980). В данной работе исследуются некоторые статистические свойства вероятностных автоматов с магазинной памятью, для чего применяются методы, разработанные для получения оценок сложности недетерминированных вычислений Насыров (1985а) и (1985в).

Вероятностный односторонний автомат с магазинной памятью (ВМПА) $B=\langle Q,\Sigma,\Gamma,\delta,q_0,z_0,F,p\rangle$ определяется исходя из обычного недетерминированного автомата с магазинной памятью (НМПА) $A=\langle Q,\Sigma,\Gamma,\delta,q_0,z_0,F,p\rangle$ Ахо, Ульман (1978) недетерминированным тактоам которого сопоставляется вероятность $p((q',y)/(q,a,z))$ для $(q',y)\in\delta(q,a,z)$, где $q,q'\in Q$ состояния автомата, $a\in\Sigma\cup\{e\}$ входной символ, $z\in\Gamma$, $y\in\Gamma^*$, Γ - алфавит магазина. Причем $0\leq p((q',y)/(q,a,z))\leq 1$ и $\sum\limits_{(q',y)\in Q\times\Gamma^*} p((q',y)/(q,a,z))=1$.

Для простоты нормализации предполагаем, что после каждого σ-такта чтения входного символа выполняется не менее одного e-такта без чтения символов входной ленты, что не влияет на класс языков, распознаваемых в ВМПА.

Автомат B может выполнить несколько различных вычислений на одном входном слове. l-му проходу автомата B по слову $\omega\in\Sigma^*$ сопоставляется вероятность

$$pr_l(\omega)=pr_l((q_0,\omega,z_0)\vdash_l^*(q_l,e,y_l))$$

равная произведению вероятностей тактов, выполненных автоматом.

Язык, распознаваемый ВМПА B с точкой сечения λ $0\leq\lambda\leq 1$ есть множество слов для которых суммарная вероятность попадания автомата в допускающее состояние более λ

$L(B,\lambda)=\{\omega \mid \chi(\omega)=\sum pr_l(\omega)\geq\lambda$, сумма по всем l: $q_l\in F\}$

В отличии от конечного автомата Бухараев (1985) ВМПА не имеет удобной интерпретации посредством матриц, поэтому вопросы вычисления суммарной вероятности попадания ВМПА в допускающие состояния представляют специальный интерес.

Для изучения функции $\chi(\omega)$ введем специальную модель. Пусть $\Delta=\{p_j \mid$ существуют q,a,z,q',y такие, что $p_j=p((q',y)/(q,a,z))\}$. Множество Δ конечно. Определим моделирующий НМПА

$C=\langle Q,\Sigma',\Gamma,\delta',q_0,z_0,F\rangle$,

$\Sigma'=(\Sigma\cup\{e\})\times\Delta$,

δ': $\delta'(q,(a,p_j),z)=(q',y)$, если $p((q',y)/(q,a,z))=p_j$.

Очевидно, что если определить гомоморфизм $h: \Sigma' \to \Sigma$, где $h((\sigma,p)) = \sigma$ $(\forall \sigma \in \Sigma)$ и $h((e,p)) = e$ (e – пустое слово), то каждому проходу автомата B по слову ω соответствует слово ω', распознаваемое автоматом C такое, что $\omega' = (a_1, p_{\ell_1})(a_2, p_{\ell_2}) \ldots (a_k, p_{\ell_k})$, $h(\omega') = \omega$ и $pr_\ell(\omega) = p_{\ell_1} \cdot p_{\ell_2} \cdot \ldots \cdot p_{\ell_k}$.

Путь мощность множества Δ равна m и символы из Δ некоторым образом перенумерованы. Определим $\Psi(\omega') = (n_1, n_2, \ldots, n_m)$, где n_j –количество букв σ_j' в слове ω'.

По теореме Парика язык, распознаваемый НМПА C
$L(C) = \{\omega' \mid (q_0, \omega', z_0) \vdash_C^* (q_\ell, e, y_\ell),\ q_\ell \in F\}$
определяет полулинейное множество Гинзбург (1970), которое можно представить в виде объединения конечного числа m-мерных линейных множеств $\Psi(L(C)) = \bigcup_i S_i$, $i = \overline{1,g}$,
$S_i = \{\bar{v} \mid \bar{v} = v_0(i) + c_1 \cdot \bar{v}_1(i) + \ldots + c_{k_i} \bar{v}_{k_i}(i),\ c_\ell \in N,\ \ell = \overline{1,k_i}\}$.

Заметим, что по любому ВМПА B эффективно строится НМПА C и для определния множеств $\{\bar{v}_0(i), \ldots, \bar{v}_{k_\ell}(i)\}$, $i = \overline{1,g}$ можно воспользоваться алгоритмом перехода от НМПА к КС грамматике G Ахо,Ульман (1978), порождающей язык $L(G) = L(C)$, затем к грамматике G применить непосредственно теорему Парика, которая конструктивно описывает множества $\{\bar{v}_0(i), \ldots, \bar{v}_{k_i}(i)\}$, $i = \overline{1,g}$.

Обозначим $T_{ih} = \{p_j \mid \sigma' = (a, p_j),\ n_j(i) \neq 0, \bar{v}_h(i) = (n_1(i), \ldots, n_m(i))\}$ и определим $T = \{T_{ih} \mid i = \overline{1,g},\ h = \overline{1,k_i}\}$, которое будем в дальнейшем называть базисом вероятности автомата B.

Т е о р е м а I. Если базис вероятности T ВМПА $B = \langle Q, \Sigma, \Gamma, \delta, q_0, z_0, F, p \rangle$ содержит одноэлементное множество $T_{ih} = \{1\}$, то существует константа c такая, что для бесконечного множества слов $\omega \in \Sigma^*$ существуют ℓ для которых $pr_\ell((q_0, \omega, z_0) \vdash_\ell^* (q_\ell, e, y_\ell)) \geq c$, $q_\ell \in F$.

Действительно, пусть $T_{ih} = \{1\}$, тогда для $\bar{v}_h(i) = (n_1(i), \ldots, n_m(i))$ из $n_j(i) > 0$ следует $p_j = 1$ для $\sigma_j' = (a, p_j)$. Причем $a \neq e$, т.к. если $a = e$ и $p_j = 1$, то это означает, что с вероятностью I ВМПА B зацикливается на некотором e-такте. Тогда согласно определению множества T для любого $f = 1, 2, \ldots$ существует слово $\omega_f' \in L(C)$ такое, что $\Psi(\omega_f') = \bar{v}_0(i) + f \cdot \bar{v}_h(i)$ и $\{h(\omega_f') \mid f = 1, 2, \ldots\}$ будет искомым множеством. Определяем

$c = \prod p_j$, где произведение берется по всем j таким, что j-й элемент вектора $\overline{v}_0(i)$ не равен нулю.

Таким образом, если выполняются условия теоремы I, то выбрав точку сечения λ меньше c получаем бесконечный язык $L(B, \lambda)$

Т е о р е м а 2. Если базис вероятности T ВМПА $B = \langle Q, \Sigma, \Gamma, \delta, q_0, z_0, F, p \rangle$ не содержит множеств $T_{ih} = \{1\}$, то существуют константы $d_1 < 1$ и d_2 такие, что для любого $\omega \in \Sigma^*$ и любого прохода $\ell : q_\ell \in F$ $pr_\ell((q_0, \omega, z_0) \vdash_\ell^* (q_\ell, e, y_\ell)) \le d_1^{d_2 \cdot |\omega|}$

Здесь $|\omega|$ – длина слова ω.

Для любого прохода по слову ω, переводящего автомат B в допускающее состояние имеем $pr_\ell(\omega) = p_{\ell_1} \cdot \ldots \cdot p_{\ell_K}$, где для слова $\omega' = (a_{\ell_1}, p_{\ell_1}) \ldots (a_{\ell_K}, p_{\ell_K})$ $\omega' \in L(C)$ $h(\omega') = \omega$ и $\Psi(\omega') \in S_i$, а именно $\Psi(\omega') = \overline{v}_0(i) + c_1 \overline{v}_1(i) + \ldots + c_{K_i} \overline{v}_{K_i}(i)$ для некоторого $i \in \{1, \ldots, g\}$.

Для векторов $\overline{v}_h(i) = (n_1^h(i), \ldots, n_m^h(i))$ определим $\varkappa_{hj} = 1$, если $n_j^h(i) = 0$ и $\varkappa_{hj} = p_j \uparrow n_j^h(i)$, где $j : \delta_j' = (a, p_j)$, $n_j^h(i) \ne 0$. Тогда согласно условиям теоремы для любого $h \in \{1, \ldots, m\}$ существует $\varkappa_{hj} < 1$. Вероятность ℓ-го прохода можно представить

$$pr_\ell(\omega) = \varkappa_{01} \cdot \varkappa_{02} \cdot \ldots \cdot \varkappa_{0m} \cdot \prod_{h=1}^{K_\ell} (\varkappa_{h1} \cdot \ldots \cdot \varkappa_{hm})^{c_h}$$

где $c_h \ge 0$ и $\sum_h c_h (n_1^i(h) + \ldots + n_m^i(h)) \ge |\omega| - m$. Достаточно взять

$d_1 = max\{p_j < 1 : p_j \in \Delta\}$ и

$d_2 = 1/(m + max\{\sum_j n_j^h(i) \mid \overline{v}_h^i = (n_1^h(i), \ldots, n_m^h(i)),\ i = \overline{1,g},\ h = \overline{1,K_i}\})$.

Заметим, что условия теорем I и 2 проверяются эффективно для любого заданного ВМПА.

Для изучения предельных свойств функции $\chi(\omega)$ можно использовать меры сложности степень недетерминизма и значность недетерминированного автомата.

Пусть A НМПА. Для слова ω обозначим $D_A(\omega)$ – число попарно различных вычислений автомата A на ω. Определим $D_A(n) = max\{D_A(\omega) \mid \omega : |\omega| \le n\}$, которую будем называть степенью недетерминизма. Аналогично определяется функция $K_A(n)$ – значность НМПА A, но про определении $K_A(n)$ учитываются только проходы автомата по слову, переводящие его в допускающее состояние.

В работах Насыров (1985а) и (1985в) показано, что существуют только три семейства НМПА у которых функция $D_A(n)$ ограничена сверху константой, полиномом от длины входного слова или экспонентой.

Любому ВМПА $B = \langle Q, \Sigma, \Gamma, \delta, q_0, z_0, F, p \rangle$ можно сопоставить НМПА $A = \langle Q, \Sigma, \Gamma, \delta, q_0, z_0, F \rangle$, который будем называть каноническим для B.

Следствие 1. Если для НМПА A канонического для ВМПА B степень недетерминизма ограничена полиномом и базис вероятности автомата B не содержит элементов $T_{ih} = \{1\}$, то для любой точки сечения λ $0 < \lambda \leq 1$ язык $L(B, \lambda)$ конечен.

Если условия следствия 1 выполнены для ВМПА B, то это приводит к тому, что для всех $\omega \in L(A)$ $\chi(\omega) \to 0$ при $|\omega| \to \infty$, т.к. $\chi(\omega)$ содержит не более полинома слагаемых $pr_\ell(\omega)$, которые стремятся к нулю как экспонента (см. теорему 2).

Следствие 2. Если для НМПА A канонического для ВМПА B не существует констант c_1 и c_2 таких, что значность автомата $K_A(n) = \Omega(c_1^{c_2 \cdot n})$ и базис вероятности автомата B не содержит $T_{ih} = \{1\}$, то для любой точки сечения λ $0 < \lambda \leq 1$ язык $L(B, \lambda)$ конечен.

Доказательство аналогично следствию 1.

Следствие 3. Если для НМПА A канонического для ВМПА B существуют константы d_1 и d_2 такие, что значность автомата $K_A(n) = \Theta(d_1^{d_2 n})$ и базис вероятности автомата B не содержит $T_{ih} = \{1\}$, то из $d_2 \log d_1 < \beta \log(1/\alpha)$ следует конечность языка $L(B, \lambda)$ для любого λ $0 < \lambda \leq 1$.

Здесь $\alpha = \max\{p_j \mid p_j \in \Delta, p_j < 1\}$, $\beta =$
$1/(m + \max\{\sum_j n_j^h(i) \mid S_i = \{\bar{v}_0(i) + c_1 \cdot \bar{v}_1(i) + \ldots + c_{k_i} \bar{v}_{k_i}(i) \mid c_h \in \mathbb{N}, h = \overline{1, k_i}\}, i = \overline{1, g}\}$.

Действительно, пусть $pr_\ell(\omega) = x_{01} \cdot \ldots \cdot x_{0m} \cdot \prod_h (x_{h1} \cdot \ldots \cdot x_{hm})^{c_h}$ для некоторого $i \in \{1, \ldots, g\}$, где $k_i - 1$ – число периодов множества S_i. Тогда, т.к. $\alpha < 1$, имеем $pr_\ell(\omega) \leq \alpha^{\beta n}$ и следовательно $\chi(\omega) \leq d_0 \cdot d_1^{d_2 n} \cdot \alpha^{\beta n} = d_0 \cdot \exp(n(d_2 \log d_1 - \beta \log(1/\alpha)))$ где $n = |\omega|$ стремится к бесконечности.

ЛИТЕРАТУРА

Ахо А., Ульман Дж. (1978): Теория синтаксического анализа перевода и компиляции, т.I, 1978, Мир.

Бухараев Р.Г. (1985) : Основы теории вероятностных автоматов, 1985, Наука.

Гинзбург С. (1970) : Математическая теория контекстно-свободных языков, 1970, Мир.

Насыров И.Р. (1985а) : О сложности распознавания контекстно-свободных языков. В: Тезисы докладов, т.I, Седьмая Всесоюзная конференция по проблемам теоретической кибернетики, Иркутск, 1985, 148-149.

Насыров И.Р. (1985в) : Классификация недетерминированных автоматов с магазинной памятью. В: Теория управляющих систем и программирование, Казань, 1985, деп. в ВИНИТИ, № 892-В86, 79-81.

Wetherell S. (1980) : Probabilistic languages: A review and some open questions. Computing Surveys, v. 12, N4, 1980, 361-379.

Kazan university
departament of theoretical
cybernetics
ul. Lenina, 18
420008, Kazan,
USSR

A CONSTRUCTION OF PROTOCOL SEQUENCES FOR COLLISION CHANNELS WITHOUT FEEDBACK

Nguyen Quang A

Hanoi, Budapest

Key words: Coding of multiple-access channels, collision channels

ABSTRACT

Protocol set is introduced for collision channels. Method of accessing a collision channel without feedback via protocol sequences is described. Some constructions of protocol sequences for multiple-access collision channels without feedback are presented based on the cyclic sets and binary cyclic constant weight codes. Procedures for active user identification , for frame-sychronization as well as for decoding messages of active users are given.

1. INTRODUCTION

The collision channel without feedback was first investigated by Massey (1982), and then by Massey and Mathys (1985), Bassalygo and Pinsker (1983), Tsybakov and Likhanov (1983), Hui (1984), A and Zeisel (1985). The technical problem that leads to such investigations can be stated briefly as follows: suppose each active user of a time-slotted frame-asynchronous multiple-access collision channel controls his packet transmission via a **periodically repeated** binary protocol sequence of length N whose 1's correspond to slots where a packet is sent. There are T potential users in the network and it is assumed that at any instant the number of active users is at most M; $2 \leq M \leq T$. How can one organize the channel access, via appropriate protocol sequences, such that based on the received patterns of idles, successes and collisions at the channel output one can determine the active set of users as well as decode their messages. There is no coordination (frame-synchronization) between the active users. The network is completely given if a protocol sequence (a binary vector of length N) is assigned to each user. Thus the problem reduces to choose a set of T such protocol sequences such that if at most M of the T users are active then each active user achieves at least s successful transmissions during every N slots.

It turns out that such protocol sequences can be constructed from certain cyclic constant weight codes, and cyclic sets. Constructions of such constant weight codes and their properties are given in Nguyen Quang, Gyorfi and Massey (1985).

In the next section the procedure for accessing of multiple-access collision channel without feedback is presented. Algorithms for active user identification, for synchronization as well as for decoding the messages of active users will be given in detail.

2. MULTIPLE-ACCES COLLISION CHANNEL WITHOUT FEEDBACK

In this section, we first briefly recall the definition of the multiple access collision channel then the concept of protocol set is introduced. The following notations are used throughtout:

T the number of all users (terminals) in the networks,

M the maximum number of active users,

N the length of protocol sequences,

w the weight of the protocol sequences,

s is the minimum number of successful slots during a period of protocol sequence.

For simplicity, suppose that a packet of a user takes values from some finite field GF(Q). The **T-user multiple access collision channel** is defined as a channel with T inputs and one output. Each input can take Q+1 possible values, namely a packet can take Q possible values and no packet (idle) takes value Ø. If all users are silent then the output symbol is idle Ø; if all but one user are silent then the output symbol can be one of the Q elements of GF(Q); if two or more users send packets at the same instant then the packets collide and are assumed to be completely destroyed, the output symbol is * indicating that a collision has occurred. Thus, the output of the channel can take Q+2 possible values. Formally if $x_1, x_2, \ldots, x_T$ stand for the inputs then the output y is given by

$$y = \begin{cases} \emptyset & \text{if } x_j = \emptyset \text{ for all } j \\ x_i & \text{if } x_j = \emptyset \text{ for all } j \neq i \\ * & \text{otherwise} \end{cases} \qquad (1)$$

The channel access is completely specified if a binary vector of length N is assigned to each user in the network. We call such a binary vector the **protocol sequence.** An **P(M,T,s,N) protocol set** is a set of T such protocol sequences such that, when at most M of the T users are active, each active user achieves **at least** s successful transmissions every N slots and the received patterns of idles, successes and collisions in N successive slots uniquely determine the active set of users as well as the subsets of slots where each active user has sent packets, for every possible relative alignment of the protocol sequences in the case of frame-asynchronous network.

Based on the received pattern of idles, successes and collision in N successive slots at the channel output, the receiver has the tasks to

i. identify the set of active users, (2)

ii. decode their messages. (3)

Since there are s successful packets transmissions by every N slots and there are at most M active users at any instant, the throughput, the sum-rate, is defined to be

$$R_s = \frac{M s}{N} \tag{4}$$

3. ACCESSING OF THE FRAME-ASYNCHRONOUS COLLISION CHANNELS

The channel access is given by the following procedure. Since we want to have s source packets by every N slots and during every N slots we can transmit w channel packets we consider a source packet as a symbol in GF(Q) and use a Reed-Solomon code of parameters [w,s] to produce w channel packets (a code-word of length w) from s information packets. We send these w channel packets during the w 1's of one period of the protocol sequence. It is assumed that the network is time-slotted, otherwise there is no coordination between active users. Let x_i denote the protocol sequence of the i-th user with period of N. Because of the assumed asynchronism, it will appear to the receiver that the user i has the protocol sequences $S^j x_i$ for some integer j, where any j such that $0 \leq j < N$ is possible.

Thus, we assume that $x_1, x_2, \ldots, x_T$ are cyclically distinct. The idle (0), success (1), collision (2) can be written as

$$r = S^{j_1} x_{i_1} \oplus S^{j_2} x_{i_2} \oplus \ldots \oplus S^{j_m} x_{i_m} \tag{5}$$

$m \leq M$, where $(i_1, i_2, \ldots i_m)$ is the set of indices of the active users, $0 \leq j_i < N$, and O denotes the component-wise integer summation with limiting at 2. For example

$$[1,0,1,1,0] + [0,0,0,1,1] + [0,0,1,1,0] = [1,0,2,2,1]$$

A set **C** of binary vectors with length N and weight w is called a **cyclic set** C(N,c,w) if it is closed under cyclic shift operation and inner product of distinct pair of elements is at most c. If each element of **C** has N distinct shifts then **C** is said to be of **maximal period**, a subset **P** of **C** is a representative set if its elements are not cyclic shifts of each other.

Let **C** be the cyclic set consisting of $\mathbf{x}_1, \mathbf{x}_2, \ldots \mathbf{x}_T$ and all their distinct cyclic shifts. Suppose that this is a cyclic set C(N,c,w) of maximal period, i.e.,

$$\mathbf{P} = (\mathbf{x}_1, \mathbf{x}_2, \ldots, \mathbf{x}_T)$$

is a representative set of **C**.

For $\mathbf{x}_1, \mathbf{x}_2, \ldots, \mathbf{x}_T$ to constitute an **P**(M,T,s) protocol set **r** in (5) we must first identify uniquely $i_1, i_2, \ldots i_m$ and $j_1, j_2, \ldots j_m$ whenever $m < M$. This must be the case if every $S^j\mathbf{x}_i$ is such that not all its 1's can be covered by 1's in any M of other N-tuples in **C**. But, since $S^j\mathbf{x}_i$ belongs to **C**, $S^j\mathbf{x}_i$ has weight at least w and no other N-tuple in **C** can cover more than c of its 1's, this must be the case whenever

$$w > Mc \qquad (6)$$

The other condition for an **P**(M,T,s) protocol set is that each of the M or fewer active users obtains at least s successes in slots 1,2,...N. But, as each other user can cause at most c collisions with the packets of a given user, this condition must hold whenever

$$w \geq (M-1)c+s \qquad (7)$$

Combining (6) and (7) we have the following sufficient conditions for a set to be protocol set.

Theorem 1. If **P** is a representative set of a cyclic set of maximal period **C**(N,c,w) then it is a protocol set **P**(M,s,N) up to some integer M satisfying

$$M \geq \min[\lfloor (w-1)/c \rfloor, \lfloor (w-s)/c \rfloor + 1] \qquad (8)$$

Remark

It has been mentioned that (5) gives the possibility to solve tasks (2) and (3), i.e., to determine uniquely $i_1,...,i_m$ and $j_1,...,j_m$. However, it tells us nothing about the way how to solve these problems (of course the exhaustive searching always works). Since each protocol sequence has N distinct cyclic shifts the following simple detection rule works: let **y** be a binary sequence of length 2N formed by the received pattern of idles, successes and collisions such that idle corresponds to zero and success as well as collision corresponds to one. We slide the protocol sequence $\mathbf{x}_i$ on **y**. If there is a segment of length N, say starting at slot j, $0 < j < N$, of **y** that covers $\mathbf{x}_i$, then we decide that the i-th user is active and $j_i = j$, otherwise we decide that the i-th user is not active. This relatively simple identification and synchronization algorithm works for both constant and nonconstant weight protocols. Having identified the active users and knowing the beginning of their frames $j_1,...,j_m$, the receiver can pick out from the received sequence w (partially distroyed) packets of each active user and decode them to get s source packets.

4. A CONSTRUCTION OF PROTOCOL SET

Massey and Mathys (1985) as well as Tsybakov and Likhanov (1983) have determined the capacity region of collision channel without feedback constraint to using protocol sequences for accesing the channel. They also gave constructions for protocols that achieve the boundary of the capacity region. For the case of equal rate users the achievable sum-rate is given by

$$C_M = R_s = MR_i = (1 - 1/M)^{M-1} \qquad (9)$$

A pure theoretican would properly point out that the coding of collision channel without feedback, thus, has been completely solved, and there would be no reason for further investigations. From a more practical point of view we see that the goal of precluding long delay, is not achieved by these optimal protocols. Since the period of the optimal protocol sequences is given by

$$N = M^T \qquad (10)$$

for the case of equal rate users.

Although the optimal protocol achieves the channel capacity, the delay determined partially by (10) is too large for a reasonably large population. This gives us the feeling that there must be some trade-off between sum-rate and delay. Bassalygo and Pinsker (1983) have shown, by random coding argument, how this trade-off looks like by giving lower and upper bound on the length of protocol set.

Theorem 2. [Bassalygo, Pinsker]
There exists a protocol **P**(M,T,s) of period N satisfying

$$N_l \lesssim N \lesssim N_u \qquad (11)$$

where

$$N_l = \max(sM/C_M \; ; \; 0.5\, sM + 0.5\, M \log T) \qquad (12)$$

$$N_u = M/C_M [\, s + M \log T\,]\, (1 + \alpha + (\alpha^2 + 2\alpha)^{1/2}) \qquad (13)$$

$$\alpha = [(M - C_M) \ln T]/[s + (M-1)\log T] \leq 1.5 \ln 2 \qquad (14)$$

and C_M is given by (9).

The optimal protocol, Tsybakov and Likhanov(1983), achieves the lower bound (12). It is worthwhile to mention that nonoptimal protocol of Tsybakov and Likhanov, i.e., protocol which does not achieve the boundary of the region also has length which is exponential in T for the case of equal rate users. We give a nonoptimal construction with short period, the construction is based on the cyclic sets given in Gyorfi and Massey (1985).

Let β be a primitive element of GF(q) q being a prime, $2 \leq k \leq n = q - 1$ be fixed integer and **A** be k x n matrix defined as follows

$$\mathbf{A} = \begin{bmatrix} 1 & 1 & 1 & \cdots & 1 \\ 1 & \beta & \beta^2 & \cdots & \beta^{n-1} \\ 1 & \beta^2 & \beta^4 & \cdots & \beta^{2(n-1)} \\ \cdot & \cdot & \cdot & \cdots & \cdot \\ \cdot & \cdot & \cdot & \cdots & \cdot \\ \cdot & \cdot & \cdot & \cdots & \cdot \\ 1 & \beta^{k-1} & \beta^{2(k-1)} & \cdots & \beta^{(n-1)(k-1)} \end{bmatrix} \qquad (15)$$

Theorem 3

Let E be the set of q^{k-2} sequences of length n over GF(q) defined as follows

$$E = \{ \; x : x = a\mathbf{A} \; , \; a = (0, 1, a_2, \ldots, a_{k-1}), a_i \in GF(q) \; \} \qquad (16)$$

Let **P** be the set of binary sequences of length nq obtaining from E by representing each element of GF(q) by a binary vector of length q as follows 0 : 10...0; 1 : 010...0; . . . ; q-1 : 00...1. Then **P** is a representative set of a cyclic set C(N,c,w) with $c = k-1$, $w = n = q-1$, $N = q(q-1)$.

By our construction a given cyclic set C(N,c,w) can result in several (T,M,s) protocol sets having different parameters M and s. The sum-rate of the protocol is given by

$$R_s = M.s.N^{-1} \geq s.N^{-1}.\min[(w-1)/c \,;\, (w-s)/c + 1] \qquad (17)$$

We can use this relation to determine the parameters M,s maximizing the sum-rate. The protocol given by Theorem 3 has the following parameters

$$T = q^{k-2}, \qquad N = q^2 - q \qquad (18)$$

$$M \geq \min[(q-2)/(k-1) \,;\, (q-1-s)/(k-1) + 1] \qquad (19)$$

The sum-rate of this protocol set is maximized when $s = (q-1)/2$, and is given by

$$R_s \geq [(q-1)/q]/[4(k-1)] + 1/(2q) \sim 1/(4k-4) \qquad (20)$$

For fixed k, thus we have strictly positive lower bound of achievable sum-rate with length much shorter than T if $k > 4$. For example if $k = 4$ then we are sure that $R_s \geq 8.33\%$ and period of the protocol increases only linearly in population size. If $k = 5$ then $R_s > 6.25\ \%$ and the period of protocol is much smaller than the population size.

Length of our protocols is strictly larger than the lower bound. Now it is interesting to ask if whether the period of these protocols is less than the upper bound and how it is close to the lower bound. For the case of protocol sets of Theorem 3 the ratio of period to the Bassalygo-Pinsker lower bound is given by the following asymptotic formulas

$$N/N_l \sim (k-1)4/e$$
$$N/N_u \sim 4(k-1)C_M/\log q \to 0 \text{ as } q \to \infty. \qquad (21)$$

The construction can be easily extended to the case of q being a prime power, see Nguyen Quang, Györfi and Massey (1985), and the resulted protocols have the same asymptotic properties as given by *(21)*. Thus, for fixed k and large q our protocols are not far from achieving the Bassalygo-Pinsker lower bound but having much shorter period than the optimal protocol achieving the lower bound.

REFERENCES

Nguyen Quang A, T. Zeisel (1985): Packet Communication on a T-user Collision Channel Without Feedback: Cyclotomic Protocol Sequences. In : Proc. COMNET'85, pp. 8-40 - 8-53, 1-4 Oct. 1985, Budapest.

Nguyen Quang A, L. Gyorfi and J. L. Massey (1985): Some Constructions of Cyclic Sets, Constant Weight Codes. Submitted to IEEE Trans. Information Theory, also presented partially at IEEE Int. Symp. on Information Theory, 24-28 Jun. 1985, Brighton, England.

L. A. Bassalygo, M. S. Pinsker (1983): Restricted Multiple-Access of a Nonsynhronous Channel. Problemy Peredachi Informatsii, vol. 19, No-4, pp. 92-96, 1983.

J. Y. N. Hui (1984): Multiple Accessing of the Collision Channel Without Feedback. IEEE J. Selected Areas Comm., SAC-2, pp.575-582, Jul. 1984.

J.L. Massey (1982): The Capacity of the Collision Channel without Feedback. In: Abstracts of Papers, IEEE Int. Symp. on Information Theory, Les Arcs, France, p.101, 1982.

J.L. Massey, P. Mathys (1985): The Collision Channel without Feedback, IEEE Trans. Information Theory, IT-31, pp. 192-204, March 1985.

B. Tsybakov, N.B. Likhanov (1983): Packet Communication on a Channel without Feedback, Problemy Peredachi Informatsii, vol. 19,No-2, pp.69-84, 1983.

Institute for Communication and Electronics
Technical University of Budapest
1111 Budapest, Stoczek u. 2, Hungary
and
Institute of Computer Sciences and Cybernetics
Vietnamese Academy of Sciences, Nghia Do Hanoi

A MINIMAX RESULT

IN A MODEL WITH INFINITELY MANY NUISANCE PARAMETERS

Michael Nussbaum, Silvelyn Zwanzig

Berlin

Key words: asymptotic minimax risk, functional relationship, infinitely many nuisance parameters

ABSTRACT

In a "errors-in-variables" model a local asymptotic minimax risk bound is established. In the case of bounded loss functions it is attained by the maximum likelihood estimator. Furthermore, a certain modified estimator is constructed, which is optimal in the minimax sense for any loss function with polynomial major, possible unbounded.

INTRODUCTION

Consider observations

$$
\begin{aligned}
x_i &= \xi_{in} + \varepsilon_{1i} \\
y_i &= \xi_{in}\beta + \varepsilon_{2i} \quad , \quad i = 1 \ldots n
\end{aligned}
\tag{1}
$$

where ε_{1i} , ε_{2i} are independent random variables with distribution N(0,1) , ß the real parameter of interst and $\xi^{(n)} = (\xi_{in})_{i=1\ldots n}$ is a set of nuisance parameters. We study the problem of asymptotically optimal estimation of ß when the prior information about the incidental parameters $\xi^{(n)}$ consists of some numerical bounds on the expression $\sum_{i=1}^{n} \xi_{in}^2$:

$$\xi^{(n)} \in \mathcal{T}^{(n)}(\alpha)$$
$$\mathcal{T}^{(n)}(\alpha) = \{ \xi^{(n)} / \sum_{i=1}^{n} \xi_{in}^2 \geq \alpha(n) \} \qquad (2)$$

The basic motivation for our study is the fact that the presence of infinitely many nuisance parameters in this model implies that the standard efficency bound of Hajek-Le Cam may not be attained by any estimator.

Furthermore, we consider a generalized asymptotic approach which also includes the case $\alpha(n) \to \infty$ with n remaining fixed. In terms of geometry these models have an increasing spread of points ξ_{in}, i=1,...,n on the line while the number of points may be fixed.

The report contains no proofs. A complete version of the paper will be published elsewhere.

LOWER ASYMPTOTIC RISK BOUNDS

In Nussbaum (1984) an asymptotic minimax risk bound for estimation of ß has been established. The prior information considered there was of the form

$$\xi^{(n)} \in \mathcal{T}^{(n)}$$
$$\mathcal{T}^{(n)} = \{ \xi^{(n)} / c_1 \leq \frac{1}{n} \sum_{i=1}^{n} \xi_{in}^2 \leq c_2 \} \qquad (3)$$

for some constants $c_1, c_2 > 0$. The condition $c_1 > 0$ is essential, since if $(0,\ldots,0) \in \mathcal{T}^{(n)}$ then uniformly consistent (over $\mathcal{T}^{(n)}$) estimators of ß do not exist.

On the other hand it is intuitively clear that large values of $\sum_{i=1}^{n} \xi_{in}^2$ should be advantageous for estimation of β. Thus it appears natural to study sets of the form (2) with $\alpha \to \infty$

Ibragimov and Khasminski (1983) investigated the efficiency bound for estimation of β in the model (1) with prior information (2) for $n \to \infty$. For the convenience of the reader we state the result in question.

Denote L the class of functions $l(x)$, $x \in R$ such that $l(0)=0$, $l(-x)=l(x) > 0$, $l(x_1)=l(x_2)$ for $0 \leq x_1 \leq x_2$ and $\ln l(x)=o(x)$ for $x \to \infty$. Let N be the standard normal distribution.

Theorem 1

(i) If $\liminf_{n\to\infty} \frac{\alpha(n)}{\sqrt{n}} < \infty$ then there are no estimators from observations (1) which are uniformly consistent over $\tilde{\xi}^{(n)}(\alpha)$ for $n \to \infty$.

(ii) If $\liminf_{n\to\infty} \frac{\alpha(n)}{\sqrt{n}} = \infty$ then for estimators β_n from observations (1) and loss functions $l \in L$

$$\liminf_{n\to\infty} \inf_{\beta_n} \sup_{|t-\beta|\leq\delta} \sup_{\xi^{(n)} \in \tilde{\xi}^{(n)}(\alpha)} E_{t,\,\xi^{(n)}}\, l(\,(\beta_n - t)\tau) \geq \int l \, dN$$

with

$$\tau = \frac{\alpha(n)}{(\,n + \alpha(n)\,(1+\beta^2\,)\,)^{\frac{1}{2}}} \qquad \text{for all } \delta > 0.$$

Part (ii) generalizes the risk bound of Nussbaum (1984), which has been established for a parameter restriction (3) and bounded loss functions.

The norming factor τ is also in agreement with the lower bound for the asymptotic variance within a certain class of estimators for β in (1) derived in the paper of Kumon and Amari (1984).

A characteristic feature of the bound in theorem 1 is that in it the supremum is taken over the whole set $\tilde{\xi}^{(n)}(\alpha)$, while typical asymptotic minimax bounds are local in character.

A natural question which arises is that of a possible localization around a given value $\xi_0^{(n)}$.
Let $x, y \in \mathbb{R}^n$

$$\mu(x,y) = \frac{\| x - y \|}{\| x \|}, \quad \| \cdot \| \text{ denoting the Euclidean norm.}$$

In order to include a more general asymptotics with n possible constant, we will assume from now on that both n and $\alpha(n)$ depend on a parameter ε, and all limits are for $\varepsilon \to 0$. We will also assume that (for $n = n_\varepsilon$, $\alpha(n) = \alpha(\varepsilon)$)

$$\lim_{\varepsilon \to 0} \frac{\alpha_\varepsilon}{\sqrt{n}} = \infty \tag{4}$$

Theorem 2

For any $\xi^{(\varepsilon)} \in \mathcal{T}^{(\varepsilon)}(\alpha)$ and any estimator β_ε from observations (1) and any loss function $l \in L$ and all $\delta > 0$

$$\liminf_{\varepsilon \to 0} \inf_{\beta_\varepsilon} \sup_{|t-\beta| \le \delta} \sup_{\mu(\xi^{(\varepsilon)}, \xi_0) \le \delta} E_{t, \xi_0^{(\varepsilon)}} \, l \left((\beta - t) \left(\frac{\sum \xi_{in}^2}{1+\beta^2} \right)^{\frac{1}{2}} \right) \ge \int l \, dN$$

ATTAINMENT OF THE BOUNDS

The bound in Nussbaum (1984) is attained by the maximum likelihood estimator $\hat{\beta}_\varepsilon$ in model (1) calculated without restrictions on $\xi^{(\varepsilon)}$ for $\varepsilon = n^{-1}$.
To show the attainment of the bounds in theorem 1 and 2 it is necessary te extend the convergence results to uniform ones over sets $\mathcal{T}^{(\varepsilon)}(\alpha)$.

Theorem 3

The maximum likelihood estimator $\hat{\beta}_\varepsilon$ from (1) fulfills

$$E_{\beta, \xi^{(\varepsilon)}} \, l \left((\hat{\beta}_\varepsilon - \beta) \frac{\sum \xi_{in}^2}{(n + \sum \xi_{in}^2 (1+\beta^2))^{\frac{1}{2}}} \right) \to \int l \, dN$$

The convergence is uniform over $|\beta| \le c$ and $\xi^{(\varepsilon)} \in \mathcal{T}^{(\varepsilon)}(\alpha)$ for any bounded $l \in L$.

Theorem 3 implies that for bounded loss functions and $\varepsilon = n^{-1}$ the bound of theorem 1 is sharp and in the case

$$\liminf_{\varepsilon \to 0} \frac{\alpha_\varepsilon}{n_\varepsilon} = \infty$$

also the bound of theorem 2 .

It is well known that the maximum likelihood estimator $\hat{\beta}_\varepsilon$ has infinite variance for any sample size so that theorem 3 will not be valid for e.g. quadratic loss.

Let $z_i = (x_i, y_i)'$ and $Q = (q_{ij})$ with

$$Q = \sum_{i=1}^{n} (z_i z_i' - I_2)$$

then $\hat{\beta}_\varepsilon$ is given by

$$\hat{\beta}_\varepsilon = \frac{q_{22} - q_{11} + ((q_{22} - q_{11})^2 + 4 q_{12}^2)^{\frac{1}{2}}}{2 q_{12}}$$

The presence of $q_{12} = \sum x_i y_i$ in the denominator and the fact that the density of q_{12} is bounded away from 0 in a neihbourhood of 0 are the reasons.

To establish the attainment of the bounds for unbounded loss functions we shall consider a certain modified estimator β_ε^*. Unlike the estimator $\hat{\beta}_\varepsilon$ (which is maximum likelihood estimator for unristricted $\xi^{(\varepsilon)}$) this estimator makes use of the prior information (2); it involves a certain truncation at the bound α .

The procedure is as follows.

In Nussbaum (1979) an asymptotically equivalent alternative to $\hat{\beta}_\varepsilon$ was proposed:

$$\bar{\beta}_\varepsilon = \frac{q_{12} (q_{11} + q_{22})}{q_{11}^2 + q_{12}^2} \tag{5}$$

The expected value of $q_{11} = \sum x_i^2 - 1$ equals $\sum \xi_i^2$ which is known to exceed α_ε , so using

$$q_{11}^* = \max (q_{11} , \frac{\alpha_\varepsilon}{2})$$

instead of q_{11} in (5) appears justified.

So we propose the modified estimator

$$\beta_\varepsilon^* = \frac{q_{12}\,(\,q_{11}^* + q_{22}\,)}{q_{11}^{*\,2} + q_{12}^2} \qquad (6)$$

Let L^* be the class of loss functions l from L with polynomial major :

$$l(x) \leq (\,1 + |x|^m\,) \text{ for some m.}$$

Theorem 4

The estimator β_ε^* fulfills

$$E_{\beta,\xi(\varepsilon)}\; l\left(\,(\,\beta_\varepsilon^* - \beta)\,\frac{\sum \xi_{in}^2}{(\,n + (1+\beta^2)\sum \xi_{in}^2\,)^{\frac{1}{2}}}\right) \underset{\varepsilon\to 0}{\longrightarrow} \int l \, dN$$

for any $l \in L^*$.
The convergence is uniform over $|\beta| \leq c$, $\xi^{(\varepsilon)} \in \mathcal{T}^{(\varepsilon)}(\alpha)$.

Theorem 4 implies that β_ε^* attains the bound of theorem 1 for $\varepsilon = n^{-1}$ and the bound of theorem 2 in the case

$$\liminf_{\varepsilon\to 0} \frac{\alpha_\varepsilon}{n_\varepsilon} = \infty$$

for any l possible unbounded with polynomial major.

ACKNOWLEDGEMENT
The authors are greatly indebted to Professor R.Z. Khasminski whose suggestions stimulated the writing of this paper.

REFERENCES

Ibragimov I.A. Khasminski R. Z. (1983): On the efficient estimation in the presence of an infinite dimensional nuisance parameter (in Russian). In: Prooceedings of the USSR - Japan Symposium on Probality theory and Mathematical Statistics, Tiblissi , Springer Lecture notes in statistics, to appear

Kumon M. Amari S. (1984): Estimation of a structural parameter the presence of a large number of nuisance parameters. Biometrika 71, 445-459

Nussbaum . (1979): Asymptotic efficiency of estimators of a multivariate linear functional relation. Math. Operationsforsch. Statist. Ser. Statist. 10, 505-527

(1984): An asymptotic minimax risk bound for the estimation of a linear functional relationship. J. Multivariae Anal. 13, 300-314.

Akademie der Wissenschaften der DDR
Karl Weierstraß Institut für Mathematik
Mohrenstr 39
Berlin
DDR 1086

SPECTRUM DECOMPOSITION FOR STATIONARY WEAKLY ISOTROPIC RANDOM FIELDS IN THE PLANE

Antonín Otáhal

Prague

Key words: stationary weakly isotropic random field, covariance function, spectral density

ABSTRACT

A covariance function and a spectral density of a (second order) stationary weakly isotropic random field in the (discrete) plane are decomposed as countable linear combinations of some "standard" covariance functions or spectral densities and the coefficients of the linear combinations are expressed explicitly by means of the covariance function.

1. INTRODUCTION

In the frame of "multiple time" series approach to stationary random fields, problems of extrapolation and filtration lead to the problem of factorization of spectral density, i. e. how to express the spectral density $f = |h|^2$ for a holomorphic "factor function" h. Rudin (1969) has proved, under some reasonable assumptions, the existence of such a factorization but the result does not give an effective way of obtaining the factor function.

For stationary weakly isotropic random fields a spectrum decomposition we consider in this paper may be used to overcome the difficulty - cf. Otáhal (1984).

In the second part of the paper basic notions are introduced, the results are stated in the third one.

2. STATIONARY WEAKLY ISOTROPIC RANDOM FIELDS

We will study random fields in the discrete plane $Z \times Z$ where Z is the set of all integers.

Stationary random field on $Z \times Z$ is a system X of complex random variables, $X = (X(m,n): m, n \in Z)$, such that every variable from X has zero mean and finite variance and there exists *covariance function* B defined on $Z \times Z$ such that, for every $j, k, m, n \in Z$, it is

$$EX(j,k).X(m,n)^* = B(j-m,k-n)$$

where asterisk denotes the complex conjugate.

Stationary random field on $Z \times Z$ is *weakly isotropic* if, for every $m, n \in Z$ and every choice of signs $\pm$, the equalities

$$B(m,n) = B(\pm m,\pm n) = B(\pm n,\pm m) \tag{1}$$

hold.

We will consider only those stationary random fields whose covariances decay quickly enough when the distance grows. More exactly, we suppose the covariance function B to be *absolutely summable*, that is,

$$\sum_{m=-\infty}^{\infty} \sum_{n=-\infty}^{\infty} |B(m,n)| < \infty. \tag{2}$$

The absolute summability of B provides the existence of the *spectral density* f, that is, the existence of such a function f defined on $(-\frac{1}{2},\frac{1}{2}) \times (-\frac{1}{2},\frac{1}{2})$ that, for every $m, n \in Z$, it holds

$$B(m,n) = \int_{-\frac{1}{2}}^{\frac{1}{2}} \int_{-\frac{1}{2}}^{\frac{1}{2}} e^{2\pi i(mx+ny)} f(x,y)\,dxdy. \tag{3}$$

In fact, the spectral density f is expressed

$$f(x,y) = \sum_{m=-\infty}^{\infty} \sum_{n=-\infty}^{\infty} e^{-2\pi i(mx+ny)} B(m,n); \tag{4}$$

the absolute summability of B ensures the uniform convergence of the series on the right hand side of (4).

The relation (3) is called the *spectral representation* of B, (4) is called its inverse formula.

A stationary random field with the spectral density f is weakly isotropic if and only if the equalities

$$f(x,y) = f(\pm x,\pm y) = f(\pm y,\pm x)$$

hold for almost every $(x,y) \in (-\frac{1}{2},\frac{1}{2}) \times (-\frac{1}{2},\frac{1}{2})$ (cf. Otáhal (1984), Thm. 1.1.).

Specially, the spectral density of a stationary weakly isotropic random field is an even function. Hence the covariance function is a real one.

3. SPECTRUM DECOMPOSITION

We introduce an "infinite triangle" of indices $T = \{(j,k): j = 0,1,2,..., k = j,j+1,j+2,...\}$. Let us fix $\varepsilon > 0$ and define a countable system $b = (b_{jk}: (j,k) \in T)$ where every b_{jk} is a sequence, i. e. a function defined on Z. Viz. to every $m = 0,1,2,...$ we ascribe the corresponding value $b_{jk}(m)$:

$$(5) \quad \begin{cases} b_{00}(m) = \delta_{0m} \\ b_{0j}(m) = \delta_{0m} + \varepsilon\delta_{jm} \\ b_{jj}(m) = \delta_{0m} - \varepsilon\delta_{jm} \\ b_{jk}(m) = \delta_{0m} + \varepsilon(\delta_{jm}+\delta_{km}) \end{cases}$$

where $j = 1,2,3,...$, $k = j+1,j+2,j+3,...$ and δ is the usual Kronecker symbol. We extend, for every $(j,k) \in T$, the definition of b_{jk} onto the whole Z through the condition

$$(6) \qquad b_{jk}(m) = b_{jk}(-m).$$

Further we define, for every $m, n \in Z$

$$(7) \qquad B_{jk}(m,n) = b_{jk}(m) \cdot b_{jk}(n), \qquad (j,k) \in T.$$

Our aim is to express an absolutely summable covariance function B of a stationary weakly isotropic random field in the form

$$(8) \qquad B(m,n) = \sum_{j=0}^{\infty} \sum_{k=j}^{\infty} a_{jk} B_{jk}(m,n)$$

for all $m, n \in Z$. From the definition of B_{jk} it follows that

$B_{jk}(m,n) = B_{jk}(\pm m,\pm n) = B_{jk}(\pm n,\pm m)$ for every $(j,k) \in T$ and every $m, n \in Z$. Hence, with respect to (1), it is sufficient to find coefficients a_{jk}, $(j,k) \in T$, such that (8) is true for every $(m,n) \in T$. In fact, $T \cup \{(n,m): (m,n) \in T\}$ covers the whole "quadrant" $Q = \{(m,n) \in Z \times Z: m \geq 0, n \geq 0\}$ and for every $(m,n) \in Z \times Z$ a proper choice of signs makes $(\pm m,\pm n) \in Q$.

2.1. Theorem. Let B_{jk} be given by means of (5), (6), (7) and B be an absolutely summable covariance function of a stationary weakly isotropic random field. If we put

$$(9)\ \begin{cases} a_{00} = B(0,0) - \frac{1}{\varepsilon^2} \sum_{k=1}^{\infty} B(k,k) + \frac{1}{\varepsilon^2} \sum_{k=1}^{\infty} \sum_{j=1}^{k-1} B(j,k) \\ a_{0m} = \frac{1}{2\varepsilon^2}(B(m,m)+\varepsilon B(0,m)) - \frac{1}{\varepsilon^2}\left(\sum_{j=1}^{m-1} B(j,m) + \sum_{k=m+1}^{\infty} B(m,k)\right) \\ a_{mm} = \frac{1}{2\varepsilon^2}(B(m,m) - \varepsilon B(0,m)) \\ a_{mn} = \frac{1}{\varepsilon^2} B(m,n) \end{cases}$$

for $m = 1,2,\dots$, $n = m+1,m+2,\dots$ then the expression (8) of B is true.

Proof. On the base of the preceding remark we verify (8) just for $(m,n) \in T$.

As B is absolutely summable, all the a_{jk}'s in (9) are well defined. Substituting in (8) for B_{jk}'s according to (5), (6), (7) and for a_{jk}'s according to (9) we get, for every $(m,n) \in T$, a relation which after an easy calculation reduces to $B(m,n) = B(m,n)$ (going separately through each of the "typical" cases $m = n = 0$, $m = n$, $0 < m < n$). ∎

Let us denote, for every $(j,k) \in T$ and every $x \in (0,1)$

$$(10)\qquad \phi_{jk}(x) = \sum_{m=-\infty}^{\infty} e^{-2\pi i m x}\, b_{jk}(m);$$

it is easy to see that every series in (10) is actually a finite sum and

$$(11)\quad \begin{cases} \phi_{00}(x) = 1 \\ \phi_{0j}(x) = 1 + 2\varepsilon \cos 2\pi jx \\ \phi_{jj}(x) = 1 - 2\varepsilon \cos 2\pi jx \\ \phi_{jk}(x) = 1 + 2\varepsilon(\cos 2\pi jx + \cos 2\pi kx) \end{cases}$$

for $j = 1,2,\ldots$, $k = j+1, j+2, \ldots$. Obviously, for every positive $\varepsilon < \frac{1}{4}$ all ϕ_{jk}'s are strictly positive. So from now on we suppose $0 < \varepsilon < \frac{1}{4}$. Then every ϕ_{jk} may be viewed on as a spectral density of a stationary random sequence with covariance function b_{jk} and a function

$$f_{jk}(x,y) = \phi_{jk}(x) \cdot \phi_{jk}(y), \tag{12}$$

defined on $(0,1) \times (0,1)$, may be viewed on as a spectral density of a stationary weakly isotropic random field with the covariance function B_{jk}.

2.2. Theorem. Let f be a spectral density corresponding to an absolutely summable covariance function B of a stationary weakly isotropic random field and let a_{jk}, $(j,k) \in T$, be given by (9). Then the *spectrum decomposition*

$$f = \sum_{j=0}^{\infty} \sum_{k=j}^{\infty} a_{jk} f_{jk} \tag{13}$$

holds and the series on the right hand side of (13) converges uniformly.

Proof. From (9) and from the absolute summability of B

$$\sum_{j=0}^{\infty} \sum_{k=j}^{\infty} |a_{jk}| < \infty$$

follows. As for every $(j,k) \in T$ it is $|f_{jk}| \le (1 + 4\varepsilon)^2$, there exists an integrable majorant for partial sums of the series on the right hand side of (13) and the series converges uniformly. So, for every $m, n \in Z$,

$$\int_0^1\int_0^1 e^{2\pi i(mx+ny)} \sum_{j=0}^{\infty} \sum_{k=j}^{\infty} a_{jk} f_{jk}(x,y)\,dxdy = \sum_{j=0}^{\infty} \sum_{k=j}^{\infty} a_{jk} \int_0^1\int_0^1 e^{2\pi i(mx+ny)} \cdot f_{jk}(x,y)\,dxdy.$$

The last expression equals, with regard to (8) and to the definition of f_{jk}, to $B(m,n)$. ¤

2.3. Remark. Let us consider a stationary weakly isotropic random field with bounded range covariances, that is, there exists $r > 0$ such that $B(m,n) = 0$ whenever $|m| > r$ (or, equivalently, $|n| > r$). In this case obviously the coefficients a_{0m}, a_{mm}, a_{mn} in (9) vanish for $|m| > r$. In other words, for a bounded range covariance function, on the right hand sides of both (8) and (13) are in fact finite sums.

REFERENCES

Otáhal A. (1984): Spectrum decomposition for stationary weakly isotropic random fields with bounded range interactions. Kybernetika 20, No. 6, 438-444.

Rudin W. (1969): Function Theory in Polydiscs. W. A. Benjamin, New York - Amsterdam.

Czechoslovak Academy of Sciences
Institute of Information Theory and Automation
182 08 Prague 8
Pod vodárenskou věží 4
Czechoslovakia

MULTI-PULSE L.P.C. ANALYSIS OF SPEECH AND SEISMIC SIGNALS

Abdeldjalil OUAHABI

Alger

Key words:Multi-Pulse Linear Predictive Coding, speech,seismic traces

ABSTRACT

Classical linear prediction and autoregressive modeling techniques have widely shown their interest in a large range of applications.

Nevertheless, they also exhibit some defects that can be attributed to various reasons: rough hypotheses on the input (white noise) or on the model (autoregressive only), delection of phase information and short time duration events.....

Consequently, born in the domaine of speech processing, the idea came out of a more accurate modeling of the input as a complex train of pulses of convenient location and size.

This paper provides an analytical derivation of a simple non-iterative technique for extracting a multiple impulse excitation model directly from the L.P.C. residual sequence.

While suboptimal with respect to "multipulse" techniques, this method is very applicable for speech enhancement and can solve the deconvolution problem which occurs in seismics where processor capability is limited.

The results suggest an additional "orthogonality" requirement between the excitation sequence and the resulting prediction error, which aids in the intuitive understanding of the method.

Additionally, a method is developped which allows a more accurate recomputation of the autoregressive coefficients based upon incorporation the multipulse excitation.

I.INTRODUCTION

The new concept of L.P.C. excitation "**Multi-Pulse Linear Predictive Coding (MP-LPC)**" is suggested by Atal and Remde (1982).

In speech synthesis this model is developped for producing natural-sounding speech and eliminates the difficult problems of voiced-unvoiced decision and pitch analysis.

This recent method consist to excit the AR filter with an input as multiple pulse.

The optimum pulse sequence -the locations and amplitudes of the pulses-is chosen to minimise a spectrally-weighted mean-squared error between the original and estimate signals.

However, the methods of Atal(1982), Singhal and Atal(1984) are somewhat computationnally complex, and for this reason, there has been recent interest in suboptimal, but computationally simple methods of speech excitation modeling.

This communication derives an alternative to the above methods which is simple and non-iterative.Since it uses the L.P.C. prediction residual to compute the best set of M excitations impulses, it will be suboptimal with respect to the methods of Atal(1982). However, its simplicity may be of advantage in situations in which computaional power is limited.

These techniques can solve an interesting **deconvolution** problem which occurs in many other domains (seismics....)than speech.

II.SUBOPTIMAL MP-LPC ALGORITHM

The multi-pulse model is

$$\hat{s}_M(n) = \sum_{k=1}^{P} a_k . s(n-k) + \sum_{j=1}^{M} G_j \, \delta(n-t_j) \qquad (1)$$

.Autoregressive (L.P.C.) methods estimate the present signal sample s(n) based upon a knowledge of past signal sample, where the a_k (AR coefficients) may be found by standard method Makhoul(1975) and (1978).

.The subscript on $\hat{s}_M(n)$ signifies: M excitations during the frame produces a prediction $\hat{s}_M(n)$

Therefore, the problem becomes that of estimating the coefficients a_k, the impulse strength G_j, and the impulse locations t_j directly from the data.

The prediction residual for the M-impulse excitation is

$$e_M(n) = s(n) - \hat{s}_M(n) \qquad (2)$$

and the error measure is $\varepsilon_M = \sum_{n=1}^{N} e_M^2(n)$ (3)

Under:
.quasi-ergodic hypothesis for correlation and other averages
.quasi-stationnary signals
the sum over n in Eq.(3) is usually carried out a time interval during which the spectral characteristics of the signal are approximately constant.

Minimise (3) with respect to the G_j:

$$\frac{\partial}{\partial G_j}\left[\varepsilon_M(n)\right] = 0 \qquad 1\leqslant j\leqslant M \qquad (4)$$

Substituting (3) into (4) and differentiating, we obtain

$$\sum_{n=1}^{N} e_M(n)\frac{\partial}{\partial G_j} e_M(n) = 0 \qquad 1\leqslant j\leqslant M \qquad (5)$$

With $e_M(n)$ defined by (2) and $\hat{s}_M(n)$ defined by (1), we have for each j

$$\frac{\partial}{\partial G_j}\left[e_M(n)\right] = \frac{\partial}{\partial G_j}(s(n) - \hat{s}_M(n)) = -\frac{\partial}{\partial G_j}\left[\sum_{m=1}^{M} G_m\ \delta(n-t_m)\right] \quad (6)$$

or

$$\frac{\partial}{\partial G_j} = -\delta(n-t_j) \quad (7)$$

A necessary assumption in deriving (6) is that the a_k are not functions of the G_j. This is equivalent to having first found a set of a_k and then holding them fixed while the G_j selection is made.

Next, substituting (7) into (5) produces

$$\sum_{n=1}^{N} e_M(n)\,\delta(n-t_j)=0 \qquad 1\leqslant j\leqslant M \qquad (8)$$

note the resemblance of (8) to the **orthogonality conditions** [Makhoul (1975),Ouahabi (1986b)] for the optimal autoregressive (or all-pole) LPC coefficients:

$E\{e(n)s(n-k)\} = 0 \qquad 1\leqslant k\leqslant P$ under stationnarity and ergodism , these conditions are approximated by

$$\sum_{n=1}^{N} e(n)s(n-k)=0 \quad 1\leqslant k\leqslant P$$

One may interpret (8) as requiring the prediction residual to be orthogonal to the excitation set.

Enforcement of the excitation orthogonality conditions (8) will produce a simple method for excitation determination which requires only the autoregressive residual sequence e(n) over a frame of data. The conditions (8) require the products to sum to zero.However, note that $\delta(n-t_j)=0$, $n\neq t_j$.

Thus, the M equations of (8) are equivalent to the requirements

$$e_M(t_j)=0, \quad 1\leqslant j\leqslant M \qquad (9)$$

Using (2), (1) and (9), we obtain immediatly

$$G_j = e(t_j) \qquad 1 \leq j \leq M$$

where $e(t_j) = s(t_j) - \sum_{k=1}^{P} a_k \, s(t_j - k)$ (10)
is simply the prediction residual at $n=t_j$ incurred from using the autoregressive predictor.

This result is very important for the following reasons

1)No extra computation is needed in order to find the optimal impulse strengths.The optimal strengths are given by the amplitude of specific elements in the autoregressive residual sequence.

2)To find the impulse locations, we simply locate the M largest residual magnitudes and excite the synthesis filter with impulses of strength $e(t_j)$ at these time locations t_j.

This property follows directly from enforcing (4) for the minimisation of the M-pulse summed residual ε_M.
This is easily seen by expanding (3)

$$\varepsilon_M = \sum_{n=1}^{N} \left\{ s(n) - \sum_{k=1}^{P} a_k s(n-k) - \sum_{m=1}^{M} G_m \delta(n-t_m) \right\}^2$$

$$= \sum_{n=1}^{N} \left\{ e(n) - \sum_{m=1}^{M} G_m \delta(n-t_m) \right\}^2$$

Then, using (10), we may derive an expression for the M-pulse residual that is entirely a function of e(n), the autoregressive LPC residual:

$$\varepsilon_M = \sum_{n=1}^{N} \left\{ e(n) - \sum_{m=1}^{M} e(t_m) \delta(n-t_m) \right\}^2 \qquad (11)$$

But $e(t_m)\delta(n-t_m) = e(t_m)$ for $n=t_m$ and vanishes elsewhere.
Using this property in (11) then produces

$$\varepsilon_M = \sum_{\substack{n=1 \\ n \neq t_m}}^{N} e^2(n) \qquad (12)$$

Since the elements of $e^2(n)$ are all positive (or zero), exclusion of the M largest $e^2(n)$ will therefore minimize the summation in (12).Thus, the locations of the M largest e(n) are the desired locations of the G_m impulses.

III.OPTIMAL LPC PARAMETERS

We know that the prediction error is reduced by using the MP-LPC method. Then, it is natural to develop a method which allows a more accurate recomputation of the autoregressive coefficients based upon incorporating knowledge of the estimated pulse locations and amplitudes along with error weighting.

Unfortunatly, this optimal method entails large computational costs. Thus, it is possible to circumvent this problem by deriving satisfactory solution at substantially lower computational cost:

a.The MP-LPC method used differs from Atal (1982) in that pulses are not determined via a search procedure minimizing a mean-square criteria; instead they are found by
-using the suboptimal MP-LPC algorithm
or -thresholding the prediction residual (l_1 norm) which in turn directly reduces the prediction error.

b.In the second etape, modified covariance incorporating the multipulse information produces final LPC parameters.

Finally the multi-pulse detection -covariance (example) analysis sequence can be iterated if desired to converge to a best set of LPC parameters and multi-pulse sequence.

IV.PREMILINARY RESULTS

a. Speech signal

It is well knownthat an important source of error in LPC analysis procedures is introduced by the rectangular window used for computing the squared prediction error.
The rectangular window causes the resultant LPC parameters to become sensitive to the position of the analysis frame in the utterance.
This problem can be avoided by using a non-rectangular window, such as Tuckey window.(Ouahabi(1985a).

As an example of the expected performance of the method, some arabic sentences was analysed.
While suboptimal with respect the original method, the current one demonstrates its use.
Good performances of the analysis are obtained with:
.the frame size was chosen to be 20ms
.the autoregressive predictor order was held constant at p=8
.the estimated multi-pulse sequence was limited to no more than 8 pulses per 10 msec interval.

b.Seismic traces

The goal of the processing of seismic traces is to determine the position and nature of strata in order to know if there are some worthy minerals like oil, natural gas....

The acoustic wave produced by an explosion on the surface, is transmitted and reflected by the different geological strata, and registred at various dephts (Ouahabi (1985 b)).

The received wave can be represented as the convolution of the source wavelet with the multiple pulse constituted by the environnement.

The temporal position t_j of pulses corresponds to the spatial location of interfaces and their amplitude G_j to their reflection coefficients.

The idealized model of seismic propagation is then

$$s_R(t)\text{(Received signal)} = s_E(t)\text{(Transmitted signal)} * \sum_{j=1}^{M} G_j \delta(t-t_j)$$

where * indicates the convolution

Thus our work on developping a new model of seismic propagation aims at determining the locations and amplitudes of the pulses by using a computationnally efficient non-iterative analysis procedure.

The difficult problem of **deconvolution** of seismic traces will be solved.

It is obvious that in presence of noise, the performances of the proposed algorithm deteriorates rapidly.

Fortunatly, it is possible to eliminate or reduce this noise before the analysis, by Wiener filtering with one or several references (Ouahabi (1985 c)).

REFERENCES

Atal B.S., Remde J.R.(1982): A new model of LPC excitation for producing natural-sounding speech at low bit rates. In:IEEE ICASSP, 614-617.

Makhoul J.(1975): Spectral linear prediction: Properties and application. In:IEEE trans. ASSP.

(1978): Stable and efficient lattice method for linear prediction. In:IEEE trans ASSP.

Ouahabi A.(1985 **a**): Use of weighting function in spectral analysis. In:6ème J.T.E.A., Tunis, 32.1-32.4

(1985 b): Auto regressive method applied to seismics. In: Journal Applied Informatics, Acta-Press, 66-69.

(1985 c): On the cascad structure of filtering with several inputs. In:Proc. Colloque GRETSI, Nice

(1986 a): Sur l'algorithme "DTW": Application à la reconnaissance des mots isolés. In: Proc. Congrés AFCET-INSA-IASTED, Toulouse.

Ouahabi et al (1986 b): On the new model LPC analysis. In: IEEE ASSP (à paraitre).

Singhal S.,Atal B.S. (1984): Improving performance of

multi-pulse LPC coders at low bit rates. In: Proc. IEEE ICASSP., San Diego, 1.3.1-1.3.4.

Unite de recherche "Electronique"
Institut d'Electronique
USTHB BP 9 Dar-El-Beida ALGER
Algerie

MINIMAX LINEAR APPROGRESSION ESTIMATION FOR FINITE DIMENSIONAL CLASSES OF REGRESSION FUNCTIONS

Jürgen Pilz

Freiberg

Key words: Inadequate model, minimax linear estimation, Bayes linear estimation, least favourable prior distribution

ABSTRACT

We consider the problem of model robust linear regression estimation where possible deviations from the ideal "parsimonious" setup are modelled by including additive contamination functions from a parametric family. We determine a minimax linear estimator of the response surface, subject to prior knowledge described by a restricted parameter space, which minimizes the maximum expected squared error loss w.r.t. some weight function. The solution is obtained by changing to the corresponding Bayesian estimation problem with respect to a least favourable prior distribution on the parameter space.

MODEL AND PROBLEM

In applied regression problems there is often uncertainty about the specification of the regression setup. Thus it becomes a natural task to approximate the unknown regression function belonging to some class $\mathcal{F}$ on the basis of a regression experiment in some ideal class $\mathcal{G}$ of sufficiently simple structure. We consider the case in which $\mathcal{F}$ and $\mathcal{G}$ are finite - dimensional linear function spaces and $\mathcal{G} \subset \mathcal{F}$. A typical example of this situation con-

cerns polynomial regression with a class $\mathcal{G}$ of low-degree polynomials and possible deviations from this "parsimonious" setup modelled by a class $\mathcal{F} \supset \mathcal{G}$ including higher order terms.
Generally, possible deviations of the regression function from some ideal setup may be modelled by including additive contamination functions from some prespecified family, say C, which can be viewed as a neighbourhood of the constant zero. The case of a parametric family C was first treated by Box, Draper (1959) and has been continued by Karson et al. (1969). A Bayesian solution was given by Kiefer (1973). Petersen (1973) and Läuter (1975) discussed a minimax approach leading to a very intrinsic optimization problem which is not feasible for a solution. The case of a nonparametric contamination family C has been considered by Marcus, Sacks (1977) and Sacks, Ylvisaker (1978). Finally, there are several papers dealing with optimum experimental design to lessen the sensitivity of standard estimators w.r.t. model inadequacy, a good literature survey is given in Gaffke (1984).
In this paper we give an explicit form solution of the minimax problem under parameter restrictions, where we confine attention to linear estimators, as it is the case in all the papers mentioned above.

Suppose the regression function $\eta: X \to R^1$; $\eta(x) = E\, Y(x)$, $x \in X \subset R^k$; to be a member of the class

$$(1) \qquad \mathcal{F} = \{ \alpha^T g + c : \alpha \in A \subset R^p,\ c \in C \}$$

where $g=(g_1,\dots,g_p)^T: X \to R^p$ is a vector of given real-valued functions, $\alpha^T g$ stands for the ideal setup and $c=c(x)$ is an additive contamination function from some specified parametric family

$$(2) \qquad C = \{ \beta^T h : \beta \in B \subset R^q \},$$

where $h=(h_1,\dots,h_q)^T: X \to R^q$ is a further vector of given real-valued functions . Let a seminorm on $\mathcal{F}$ be defined by

$$(3) \qquad \|f\|^2 = \int_X f(x)^2 \, dw(x), \quad f \in \mathcal{F}$$

where w is some nonnegative weight function defined on X, and let be satisfied

Assumption 1: $g_1,\dots,g_p,h_1,\dots,h_q$ are linearly independent.

Further, suppose we are given prior knowledge such that it holds

Assumption 2: (i) A is a compact subset of R^p

(ii) $B = \{\beta\in R^q : \beta^T Q\beta \leq 1\}$, Q positive definite.

Restrictions of the type (ii) occur in a natural way when we have e.g. prior knowledge about η such that with $\mathcal{G} = \{\alpha^T g : \alpha\in A\}$ it holds

(4) $$\inf_{\gamma\in\mathcal{G}} \|\eta - \gamma\|^2 \leq \delta_1$$

for some constant $\delta_1>0$. Also, such a restriction may arise from some smoothness condition such as

(5) $$\int_X [\eta^{(m)}(x)]^2\, dx \leq \delta_2$$

where $\eta^{(m)}$ means the m-th derivative ($1\leq m<p+q$) of η and $\delta_2>0$ is some given constant. For example, if η is assumed to be a third-degree polynomial and we are interested in deviations from linearity (i.e. $\mathcal{G} = \{\alpha_1 + \alpha_2 x : (\alpha_1,\alpha_2)^T\in A\subset R^2\}$ and $C = \{\beta_1 x^2 + \beta_2 x^3 : (\beta_1,\beta_2)^T\in B\subset R^2\}$) then condition (5) with m=2 and X=[0,1] leads to the ellipsoid constraint $\beta_1^2 + 3\beta_1\beta_2 + 3\beta_2^2 \leq \delta_2/4$.

Remark 1: If we have no prior restrictions on the parameter α then this can be modelled by assuming $A = \{\alpha\in R^p : \alpha^T\alpha \leq \delta\}$ and considering the limiting situation as $\delta\to\infty$. A similar proposal was made by Kiefer (1973) who suggested a Bayesian approach with $E\,\alpha\alpha^T = \delta I_p$ and $\delta\to\infty$.

Consider the class of linear estimators

(6) $$D = \{\hat{\eta}(x) = g(x)^T SY : S\in\mathcal{M}_{p\times n}\}$$

of the regression function, where $\mathcal{M}_{p\times n}$ stands for the set of all real matrices of the type $p\times n$ and $Y = (Y(x_1),\dots,Y(x_n))^T$ denotes the n- vector of observations taken at experimental points $x_1,\dots,x_n\in X$. Writing $\theta^T = (\alpha^T,\beta^T)$ and $f(x)^T = (g(x)^T,h(x)^T)$ we have $EY(x_i) = \eta(x_i) = f(x_i)^T\theta$ and thus the observation model reads

(7) $$E\,Y = F\theta \quad , \quad F = (f(x_1,\dots,f(x_n))^T .$$

We wish to find a minimax linear estimator $\hat{\eta}^*$ subject to our prior knowledge, i.e. an estimator which minimizes the maximum expected distance

(8) $$\sup_{\eta \in \mathcal{F}} \int_X E_{Y|\eta} (\eta(x) - g(x)^T SY)^2 \, dw(x) .$$

THE BAYESIAN ESTIMATION PROBLEM

Let be h_1 and h_2 arbitrary vector functions and define

$$W_{h_1 h_2} = \int_X h_1(x) h_2(x)^T \, dw(x) .$$

Observing that

$$E_{Y|\eta} (\eta(x) - g(x)^T SY)^2 = \theta^T f(x) f(x)^T \theta - 2\theta^T f(x) g(x)^T SF\theta + g(x)^T S(\Sigma + F\theta\theta^T F^T) S^T g(x) ,$$

where Σ= Cov Y means the covariance matrix of the observations, the risk

$$R(\eta,\hat{\eta}) = \int_X E_{Y|\eta} (\eta(x) - g(x)^T SY)^2 \, dw(x)$$

of the estimator $\hat{\eta} = g(x)^T SY$ comes out as

(9) $$R(\eta,\hat{\eta}) = \operatorname{tr} W_{gg} S\Sigma S^T + \theta^T G(S)\theta$$

where

$$G(S) = F^T S^T W_{gg} SF - F^T S^T W_{gf} - W_{fg} SF + W_{ff} .$$

The first term in (9) refers to the variance and the second to the squared bias of the estimator $\hat{\eta}$. Denoting W_{gg}^- some generalized inverse of W_{gg}, the matrix G(S) may be rewritten as

(10) $$G(S) = W_{ff} - W_{fg} W_{gg}^- W_{gf} + (SF - W_{gg}^- W_{gf})^T W_{gg} (SF - W_{gg}^- W_{gf}) .$$

Here the matrix

(11) $$W = W_{ff} - W_{fg} W_{gg}^- W_{gf}$$

refers to the minimum bias matrix, the minimum squared bias obtainable by some linear estimator $\hat{\eta} = g(x)^T SY$ is given by

$$\theta^T W \theta = \beta^T (W_{hh} - W_{hg} W_{gg}^- W_{gh}) \beta$$

and thus does not depend on the parameter $\alpha \in A$.

Lemma 1: The matrix W defined by (11) is positive semidefinite.

Proof: Let be $z^T = (z_1^T, z_2^T)$ with arbitrary $z_1 \in R^p$, $z_2 \in R^q$. Then we have

$$z^T W z = z^T (W_{ff} - W_{fg} W_{gg}^- W_{gf}) z = z_2^T (W_{hh} - W_{hg} W_{gg}^- W_{gh}) z_2 .$$

The matrix $\tilde{W} = W_{hh} - W_{hg} W_{gg}^- W_{gh}$ can be written as

$$\tilde{W} = \int_X \left[h(x) - W_{hg} W_{gg}^- g(x)\right] \left[h(x) - W_{hg} W_{gg}^- g(x)\right]^T dw(x) ,$$

which is obviously positive semidefinite, and thus $z^T W z = z_2^T \tilde{W} z_2 \geq 0$ whatever z is chosen.

Using (9), the estimation problem amounts to solving the minimax problem

$$(12) \qquad \sup_{\theta \in \Theta} \left\{ \operatorname{tr} W_{gg} S \Sigma S^T + \theta^T G(S) \theta \right\} \longrightarrow \inf_{S \in \mathcal{M}_{p \times n}}$$

where $\Theta = A \times B$ is a compact subset of R^{p+q}. In this form, however, the problem is not feasible for a solution. We shall solve the problem by changing to a Bayesian estimation problem.
Let Θ^* denote the set of all (prior) probability distributions over the measurable space $[\Theta, \mathcal{B}_\Theta]$ where $\mathcal{B}_\Theta$ is some σ- algebra of subsets of Θ, including all the single-element subsets. For every $P \in \Theta^*$ let

$$(13) \qquad \bar{R}(P, \hat{\eta}) = \int_\Theta R(\eta, \hat{\eta}) \, P(d\theta)$$

denote the Bayes risk of the estimator $\hat{\eta} = g(x)^T SY$ w.r.t. P. Since Θ is compact, we have

$$(14) \qquad \sup_{\eta \in \mathcal{F}} R(\eta, \hat{\eta}) = \sup_{P \in \Theta^*} \bar{R}(P, \hat{\eta})$$

for all estimators $\hat{\eta}$. Letting

$$(15) \qquad M_P = \int_\Theta \theta \theta^T \, P(d\theta)$$

the matrix of second moments associated with the prior $P \in \Theta^*$, it follows from (9) and (13) that the Bayes risk may be written as

(16) $$\bar{R}(P,\hat{\eta}) = \operatorname{tr}\left[W_{gg}S\Sigma S^T + G(S)M_P\right] =: r(M_P,S)$$

i.e. the Bayes risk depends on the prior only through M_P.

Lemma 2: Let be $\mathcal{M} = \{M_P : P \in \Theta^*\}$. Then it holds

$$\inf_{\hat{\eta}\in D}\sup_{\eta\in\mathcal{F}} R(\eta,\hat{\eta}) = \sup_{M_P\in\mathcal{M}}\inf_{S\in\mathcal{M}_{p\times n}} r(M_P,S) .$$

Proof: According to (14) and (16) we have

$$\inf_{\hat{\eta}\in D}\sup_{\eta\in\mathcal{F}} R(\eta,\hat{\eta}) = \inf_{S\in\mathcal{M}_{p\times n}}\sup_{M_P\in\mathcal{M}} r(M_P,S) .$$

From (16) it can be seen that $r(M_P,S)$ is convex w.r.t. $S\in\mathcal{M}_{p\times n}$, linear and thus concave w.r.t. $M_P\in\mathcal{M}$. Furthermore, $r(\cdot,S)$ is continuous on the compact set $\mathcal{M}$ (w.r.t. the usual sum of squares norm for matrices). The result then follows from a well-known minimax theorem (see Sion (1958), Theorem 4.2).

Lemma 3: For every $P\in\Theta^*$ define $S_P = W_{gg}^-W_{gf}M_PF^T(\Sigma + FM_PF^T)^{-1}$. Then it holds

$$\inf_{S\in\mathcal{M}_{p\times n}} r(M_P,S) = r(M_P,S_P) .$$

Proof: First, rearranging terms in $r(M_P,S)$ gives

(17) $$r(M_P,S) = \operatorname{tr} W_{gg}S(\Sigma + FM_PF^T)S^T - 2\operatorname{tr} W_{gf}M_PF^TS^T + \operatorname{tr} W_{ff}M_P .$$

Then, differentiating w.r.t. S yields

$$\frac{1}{2}\frac{\partial}{\partial S} r(M_P,S) = W_{gg}S(\Sigma + FM_PF^T) - W_{gf}M_PF^T$$

and equating to zero gives the required result.

Lemma 3 reveals that, for any prior $P\in\Theta^*$,

$$\hat{\eta}_P(x) = g(x)^TS_PY = g(x)^TW_{gg}^-W_{gf}M_PF^T(\Sigma+FM_PF^T)^{-1}Y$$

yields a Bayes linear approgression estimator. This estimator is already known from Kiefer (1973). It is simply the projection of the usual Bayes linear estimator

$$\tilde{\eta}_P(x) = g(x)^TM_PF^T(\Sigma+FM_PF^T)^{-1}Y$$

of the response surface η in case of an adequate model (where we have f=g) onto the function space spanned by the components of g.

THE MINIMAX ESTIMATOR

Combining Lemma 2 and 3, we can conclude that the Bayes estimator w.r.t. a least favourable prior distribution from Θ^* yields a minimax estimator.

Theorem 1: Let Assumptions 1 and 2 be satisfied. Then

$$\hat{\eta}^*(x) = g(x)^T W_{gg}^- W_{gf} M_o F^T(\Sigma + FM_oF^T)^{-1} Y$$

is a minimax linear estimator for $\eta(x)$ where $M_o = M_{P_o}$ is the matrix of second moments of a least favourable prior distribution P_o within Θ^* determined by

$$r(M_o, S_o) = \sup_{P \in \Theta^*} r(M_P, S_P), \quad S_o = S_{P_o} .$$

The existence of a least favourable prior P_o maximizing $r(M_P, S_P)$ w.r.t. $P \in \Theta^*$ follows from the fact that $r(M_P, S_P)$ is continuous w.r.t. M_P ranging over the compact set $\mathcal{M}$.
We will now demonstrate that the least favourable moment matrix M_o can be obtained as solution of a convex optimization problem. In the sequel let the following assumption be satisfied.

Assumption 3: $F^T\Sigma^{-1}F$ has full rank.

Then, letting $V = (F^T\Sigma^{-1}F)^{-1}$ and $\hat{\theta} = VF^T\Sigma^{-1}Y$ the Gauss-Markov estimator for $\theta = (\alpha^T, \beta^T)^T$, the minimax estimator can be rewritten as

$$\hat{\eta}^*(x) = g(x)^T W_{gg}^- W_{gf} M_o (M_o + V)^{-1} \hat{\theta} . \tag{18}$$

Theorem 2: Under the Assumptions 1,2 and 3 the minimax linear estimator is given by (18), where the least favourable moment matrix M_o minimizes the convex functional

$$\Psi(M_P) = \operatorname{tr} VUV(V + M_P)^{-1} - \operatorname{tr} WM_P , \quad M_P \in \mathcal{M}$$

with $U = W_{fg}W_{gg}^-W_{gf}$ and $W = W_{ff} - U$.

Proof: Using the matrix identity

$$V(V+M_P)^{-1} = (I+M_PF^T\Sigma^{-1}F)^{-1} = I - M_PF^T(\Sigma+FM_PF^T)^{-1}F ,$$

the matrices S_P, $P\in\Theta^*$, defined in Lemma 3 satisfy the equation

$$W_{gg}S_P(\Sigma+FM_PF^T)S_P^T = W_{gf}M_PF^TS_P^T = W_{gf}\left[I-V(V+M_P)^{-1}\right]M_PW_{fg}W_{gg}^-$$

and thus we get from (17)

$$\begin{aligned} r(M_P,S_P) &= \operatorname{tr} W_{ff}M_P - \operatorname{tr} W_{gf}M_PF^TS_P^T \\ &= \operatorname{tr}(W_{ff}-U)M_P + \operatorname{tr} W_{fg}W_{gg}^-W_{gf}V(V+M_P)^{-1}M_P \\ &= \operatorname{tr} WM_P + \operatorname{tr} UV - \operatorname{tr} VUV(V+M_P)^{-1} . \end{aligned}$$

Hence, maximization of $r(M_P,S_P)$ w.r.t. P is equivalent to minimization of $\Psi(M_P)$ w.r.t. $M_P\in\mathcal{M}$ and the result follows from Theorem 1. The convexity of Ψ follows from the fact that both terms, $\operatorname{tr} VUV(V+M_P)^{-1}$ and $\operatorname{tr} WM_P$, are convex w.r.t. M_P.

Interpreting the matrices M_P as information matrices of experimental design measures P over the region Θ, the convexity of the functional Ψ allows the application of convex experimental design theory to derive necessary and sufficient optimality conditions for a matrix M_o to minimize the functional $\Psi(\cdot)$. In particular, we can apply iteration procedures known from experimental design literature (see e.g. Wu (1978), Pilz (1983)) to obtain a least favourable moment matrix M_o.
In case that f=g the functional Ψ reduces to

$$\Psi_o(M_P) = \operatorname{tr} VUV(V+M_P)^{-1} \quad \text{with} \quad U = W_{ff} = \int_X f(x)f(x)^T dw(x)$$

and the minimax estimator for η becomes $\hat{\eta}_M(x) = f(x)^T\hat{\theta}^*$ where

$$\hat{\theta}^* = M^*(M^*+V)^{-1}\hat{\theta} \quad \text{with} \quad \Psi_o(M^*) = \inf_{M\in\mathcal{M}} \Psi_o(M)$$

is the minimax linear estimator for θ in the adequate model $EY(x) = f(x)^T\theta$ with θ constrained to lie in a parameter set $\Theta (=A) \subset R^p$. This estimation problem has been treated in Pilz (1986), where also an iteration procedure for finding M^* and explicit solutions in case of ellipsoid and linear inequality constraints are given. The iteration procedure developed in Pilz (1986, section 5.4) for the special case of an adequate model can easily be extended to the present case of an inadequate model, the only modification to be made is to replace the directional derivative considered there by

$$\lim_{\varepsilon \downarrow 0} \frac{d}{d\varepsilon} \Psi((1-\varepsilon)M_P + \varepsilon\theta\theta^T) .$$

REFERENCES

Box G.E.P., Draper N.R. (1959): A basis for the selection of a response surface design. J. Amer. Statist. Assoc. 54, 622-654

Gaffke N. (1984): Optimal designs for contaminated linear regression. In: Robustness of Statistical Methods and Nonparametric Statistics (D. Rasch and M.L. Tiku, Eds.), Deutscher Verlag der Wissenschaften, Berlin, 37-42

Karson M.J., Manson A.R., Hader R.J. (1969): Minimum bias estimation and experimental design for response surfaces. Technometrics 11, 461-475

Kiefer J. (1973): Optimum designs for fitting biased multiresponse surfaces. In: Multivariate Analysis III (P.R. Krishnaiah,Ed.), Academic Press, New York, 287-297

Läuter,H. (1975): Linear minimax estimation for inadequate models for the I. Petersen approach. Math. Operationsforsch. Statist. 6, 769-774

Marcus M.B., Sacks J. (1977): Robust designs for regression problems. In: Statistical Decision Theory and Related Topics II (S.S. Gupta and D.S. Moore, Eds.), Academic Press, New York, 245-268

Petersen I. (1973): Linear minimax estimation of inadequate models in L^2- metric. Math. Operationsforsch. Statist. 4, 463-471

Pilz J. (1983): Bayesian Estimation and Experimental Design in Linear Regression Models. Teubner- Texte zur Mathematik, Vol. 55, Teubner - Verlag, Leipzig

Pilz J. (1986): Minimax linear regression estimation with symmetric parameter restrictions. J. Statist. Plann. Infer. 13, 297-318

Sacks J., Ylvisaker D. (1978): Linear estimation for approximately linear models. Ann. Statist. 6, 1122-1137

Sion M. (1958): On general minimax theorems. Pacific J. Math. 8, 171-176

Wu C.F. (1978): Some algorithmic aspects of the theory of optimal design. Ann. Statist. 6, 1286-1301

Bergakademie Freiberg
Sektion Mathematik
Bernhard-v.- Cotta-Str.2
9200 Freiberg
DDR

SOME CHARACTERIZATIONS OF NORMAL LAW BY THE INDEPENDENCE OF LINEAR FUNCTIONS IN A TRIANGULAR SYSTEM.

Agnieszka Plucińska

Warsaw

Key words: characterization problem, conditional moment

ABSTRACT

Certain characterizations of normal distribution are obtained by using properties of conditional moments.

Let $\mathbf{X}=(X_1,\ldots,X_n)$ be n-dimensional random vector, $EX_i=0$, $k_{ij}=E(X_iX_j)$, $K^{(n)}=\left[k_{ij}\right]_{i,j=1}^{n}$, $\det K^{(n)}>0$. Let

$$Y_j = c_{1j}X_1 + \ldots + c_{jj}X_j\ ,\quad c_{jj}=1\ ,\quad j=1,\ldots,n.$$

be n linear uncorrelated functions. A number of characterizations of the distribution of $\mathbf{X}$ based on properties of linear statistics have been made; they are presented for example in the book edited by Patil, Kotz, Ord (1975). In the present paper we obtain some characterizations for dependend r.v.'s using known theorem for independent r.v.'s.

The aim of this paper is to prove the following propositions:

Proposition 1. If there exist c, c_{ij} $(i \leq j, j=1,\ldots,n)$ such that $Y_1,\ldots,Y_n$ are independent,

(1) $\quad m_j = E(X_nY_j) \neq 0\ ,\ j=1,\ldots,n$

(2) $\quad E(X_1|X_n) = cX_n$

(2') $\quad Var(X_1|X_n) = const$

then $\mathbf{X}$ has a normal distribution and

$$c_{ij} = K_{ij}^{(j)}(|K^{(j-1)}|)^{-1}$$

where $K_{ij}^{(j)}$ - the cofactor of k_{ij} in $K^{(j)}$, $|K^{(j-1)}| = \det K^{(j-1)}$
The Condition (1) can be expressed by k_{ij} and by properties of conditional moments.Namely let us denote

$$E(X_n|X_1,\dots,X_j) = m_{1j}X_1 + \dots + m_{jj}X_j.$$

Proposition 2.If the assumptions of Proposition 1 hold then

$$(3) \qquad m_j = \begin{vmatrix} k_{11}, \dots, k_{1,j-1}, k_{1n} \\ \dots\dots\dots\dots \\ k_{j1}, \dots, k_{j,j-1}, k_{jn} \end{vmatrix} (|K^{(j-1)}|)^{-1}$$

and the condition (1) is equivalent to

$$(4) \qquad m_{jj} \neq 0 \text{ , } j=1,\dots,n-1.$$

Proposition 3. If $\mathcal{X} = \{X_t, t \in (0,T)\}$ is a stochastic process with independent increments,$E(X_t) = 0$, for all $t_1 < t_2$

$$(5) \qquad E(X_{t_1}^2) < E(X_{t_2}^2) \text{ ,}$$

$$(6) \qquad E(X_{t_1}|X_{t_2}) \text{ is a linear function of } X_{t_2},$$

$$(7) \qquad Var(X_{t_1}|X_{t_2}) \text{ is non-random}$$

then $\mathcal{X}$ is a gaussian process.

A characterization based on independence of $Y_1,\dots,Y_n$ and on Skitovich-Darmois theoremis given in the paper of the author(1987)

We shall use the following Lukacs,Laha theorem(1964)

Theorem 1.Let $Y_1,\dots,Y_n$ be independently distributed r.v.s with second moments,$EY_i = 0$.The linear functions

$$U = a_1Y_1 + \dots + a_nY_n \text{ , } V = b_1Y_1 + \dots + b_nY_n$$

(with $a_ib_i \neq 0$ for $i=1,\dots,n$) satisfy the relations

$$E(U|V) = k V \text{ , } Var(U|V) = \sigma_0^2$$

almost every where iff the following two conditions are satisfied

Each Y_i for which $a_i \neq kb_i$ is normally distributed

$k=(\sum a_i b_i \sigma_i^2)\ (\sum b_i \sigma_i^2)$, $\sigma_0^2 = \sum \sigma_i^2 (a_i - kb_i)^2$, where $\sigma_j^2 = Var(X_j)$.

Proof of Proposition 1. We shall use Theorem 1. Let us put

(8) $U = a_1 Y_1 + \dots + a_n Y_n = dX_1 + X_n$

(9) $V = b_1 Y_1 + \dots + b_n Y_n = X_n$

We are going to show that $a_i b_i \neq 0$, $a_i \neq kb_i$. We take d such that

(10) $d \neq 0$, $d \neq -k_{1n}/k_{11}$, $d \neq -k_{nn}/k_{1n}$.

It is evident that

$$E(X_1 Y_j) = 0 \text{ for } j \neq 1$$

$$E(Y_i Y_j) = \begin{cases} 0 \text{ for } i \neq j \\ |K^{(j)}|\,(|K^{(j-1)}|)^{-1} \quad \text{for } i = j \end{cases}$$

Multiplying (8) by Y_j we get for $j > 1$

(11) $a_j E(Y_j^2) = E(X_n Y_j) = m_j$

Therefore $a_j \neq 0$ for $j > 1$. Now let us multiply (8) by Y_1. Then we get

(12) $a_1 E(Y_1^2) = d\ E(X_1^2) + E(X_1 X_n)$.

Thus in virtue of (10) $a_1 \neq 0$. Analogically

(13) $b_1 = m_1\ (E(Y_1^2))^{-1}$, $b_j = m_j\ (E(Y_j^2))^{-1}$ for $j > 1$

Finally $a_i b_i \neq 0$ for $i=1,\dots,n$. Formulas for conditional moments have the following form

$$E(U|V) = E((d\ X_1 + X_n)|X_n) = (dk_{1n}/k_{nn} + 1)\ X_n$$

$$Var(U|V) = d^2 Var(X_1|X_n) = \text{const.}$$

Thus in virtue of (10) $k = 1 + dk_{1n}/k_{nn} \neq 1$. It follows from (10)-(13) that $a_i \neq kb_i$. All the assumptions of the Theorem 1 are satisfied, then Y_i / $i=1,\dots,n$/ has a normal distribution. Therefore $\mathbf{X}$ has also a normal distribution.

Proof of Proposition 2. Taking into account that

$$m_j = E(X_n Y_j) = \frac{1}{|K^{(j-1)}|} \sum_{i=1}^{j} k_{in} K_{ij}^{(j)}$$

we get formula (3).

In virtue of the independence of $Y_1,\dots,Y_n$ we get

$$0 = E(Y_n) = E(Y_n|Y_1,\dots,Y_j) = E(Y_n|X_1,\dots,X_j) =$$

$$= E(X_n|X_1,\dots,X_j) + \sum_{i=1}^{j} c_{in} X_i \ .$$

The conditional expectation $E(X_n|X_1,\dots,X_j)$ is thus a linear function of $X_1,\dots,X_j$. The coefficients c_{jn} can be expressed in th the following form

$$-c_{jn} = m_{jj} = m_j |K^{(j-1)}| (|K^{(j-1)}|)^{-1}$$

Conditions (1) ,(4) are thus equivalent. Q.E.D.

Proof of Proposition 3. Let $X_i = X_{t_i}$. For the process with dependent increments $Y_i = X_i - X_{i-1}$, $i > 1$. Condition (1) is satisfied as for $t_1 < t_2 < t_3$ in virtue of (5) we have

$$m_{t_2} = E(X_{t_3}(X_{t_2} - X_{t_1})) = E(X_{t_2}^2) - E(X_{t_1}^2) > 0.$$

Conditions (2),(2') follows from (6),(7). Q.E.D.

The independence of $Y_1,\dots,Y_n$ and conditions (1),(2) do not characterize a gaussian distribution. We are going to give some examples.

Example 1. Let $\mathcal{Z} = \{ Z_t, t > 0 \}$ be a Poisson process with parameter λ and $X_t = Z_t - \lambda t$. Then

$$E(X_{t_1}|X_{t_2} = x) = x t_1 (t_2)^{-1}$$

$$\operatorname{Var}(X_{t_1} | X_{t_2} = x) = x t_1 (t_2 - t_1) t_2^{-2}$$

Therefore $\mathcal{X}$ is a process with independent increments, conditions (1),(2) hold and (2') is not satisfied.

Example 2. Let $P(X_1=1, X_2=1) = P(X_1=1, X_2=0) = P(X_1=-1, X_2=0) = P(X_1=-1, X_2=-1) = {}^1/_4$. It is evident that $EX_1 = EX_2 = 0$, $EX_1^2 = 1$, $EX_2^2 = {}^1/_2$, $E(X_1X_2) = {}^1/_2$. We put $Y_1 = X_1$, $Y_2 = X_2 - {}^1/_2 X_1$. Then we have $P(Y_1 = 1, Y_2 = {}^1/_2) = P(Y_1 = 1, Y_2 = -{}^1/_2) = P(Y_1 = -1, Y_2 = {}^1/_2) = P(Y_1 = -1, Y_2 = {}^1/_2)$; thus Y_1, Y_2 are independent. The r.v.'s X_2, Y_i $(i=1,2)$ are correlated :

$m_1 = E(X_2Y_1) = {}^1/_2 \neq 0$,

$m_2 = E(X_2Y_2) = E(X_2(X_2 - {}^1/_2 X_1)) = {}^1/_4 \neq 0$

The conditional moments are given by the following formulas:

$E(X_1 | X_2=1) = 1$, $E(X_1 | X_2=0) = 0$, $E(X_1 | X_2=-1) = -1$,

$\operatorname{Var}(X_1 | X_2=1) = 0$, $\operatorname{Var}(X_1 | X_2=0) = 1$, $\operatorname{Var}(X_1 | X_2=-1) = 0$.

Therefore conditions (1),(2) are satisfied and condition (2') is not satisfied. Condition (2') is then necessary.

Example 3. Let the distribution of (X_1, X_2) be uniform in D where $D = \{(x_1, x_2): -1 < x_1 < 1, x_1 - 1 < 2x_2 < x_1 + 1\}$ Then $EX_1 = EX_2 = 0$, $EX_1^2 = {}^1/_3$, $EX_2^2 = {}^1/_6$, $E(X_1X_2) = {}^1/_6$. We put $Y_1 = X_1$, $Y_2 = X_2 - {}^1/_2 X_1$. It is evident that the distribution of (Y_1, Y_2) is uniform in $D' = \{(y_1, y_2); -1 < y_1 < 1, -1 < 2y_2 < 1\}$ Thus Y_1, Y_2 are independent. The r.v's $X_2, Y_i (i=1,2)$ are correlated:

$m_1 = E(X_2Y_1) = {}^1/_6$, $m_2 = E(X_2Y_2) = {}^1/_{12} \neq 0$

The conditional moments are given by the following formulas:

$E(X_1 | X_2) = X_2$

$$\mathrm{Var}(X_1|X_2=x_2) = {}^1/_3(1 +(1 + x_2)^3) \quad \text{for } x_2 < 0$$

$$\mathrm{Var}\ (X_1|X_2=x_2) = {}^1/_3(1 +(1 + x_2)^3) \quad \text{for } x_2 > 0$$

Conditions (1),(2) are thus satisfied and condition (2') is not satisfied.Condition (2') is necessary.

In the present paper we consider the dependent r.v's. The dependence is characterized by (1) or equivalently by (4). For such random variables we get the following corollary:

Corollary 1. If (1) /or equivalently (4) / holds then $\mathbf{X}$ has a normal distribution iff $Y_1,\dots,Y_n$ are independent and (2),(2') are satisfied.

REFERENCES

G.P. Patil, S. Kotz, J.K. Ord (1975) A Modern Course on Statistical Distributions in Sciantific Work, Vol 3,Characterizations and Applications,Dordrecht-Holland

A. Plucińska (1987) Some characterizations of gaussian vector through the independence of uncorrelated functions

E.Lukacs, R.G. Laha, Applications of characteristic functions, London, 1964

Warsaw Technical University
Mathematical Institute
00-661 Warsaw
Plac Jedności Robotniczej 1

A LIMIT THEOREM FOR MEASURE VALUED DIFFUSION PROCESSES

Gerd Rosenkranz
Frankfurt

Key words: Measure valued diffusions, limit theorem, amarts

ABSTRACT

Let Ξ be a measure valued diffusion process. We establish conditions under which $\Xi(t)/t^b$, $1 \leq b < \infty$, converges to a random measure.

INTRODUCTION AND THE MAIN RESULT

Measure valued processes are mathematical tools to describe random evolution of particles in time and space. Measure valued diffusions are processes having a.s. continuous paths with respect to a suitable topology on the state space. Like R^d-valued diffusion processes, they can often be obtained as weak limits of a sequence of rescaled processes which evolve in discrete steps as for example branching random walks.

To be precise, let S be a compact rectangular subset of d-dimensional Euclidian space and M(S) the space of Borel measures on S equipped with the vague topology. A measure valued diffusion process with state space M(S) can then be characterized as follows: Let L be a nondegenerated second order differential operator with domain D(L) containing all two times continuously differentiable functions on S whose first partial derivatives vanish on the boundary of S. Furthermore, let α and γ be continuous functions on $S \times M(S)$, $\gamma \geq \gamma_0 > 0$, let ϕ, ψ be elements of D(L), ι some Borel measure on S, and suppose that for every $\xi \in M(S)$

(1a) $$\lim_{t\to 0} \frac{1}{t} E_\xi\{\langle\phi,\Xi(t)-\xi\rangle\} = \langle\phi,\iota\rangle + \langle\alpha(\cdot,\xi)\phi+\gamma(\cdot,\xi)L\phi,\xi\rangle$$

(1b) $$\lim_{t\to 0} \frac{1}{t} E_\xi\{\langle\phi,\Xi(t)-\xi\rangle\langle\psi,\Xi(t)-\xi\rangle\} = \langle\gamma\phi\psi,\xi\rangle,$$

where $\langle\phi,\xi\rangle := \int\phi d\xi$ and E_ξ denotes expectation given $\Xi(0)=\xi$. Assuming $\gamma(x,\xi)=\gamma_1(x)\gamma_2(\xi)$ for some suitable functions γ_1,γ_2, it can be shown using martingale arguments developed in Dawson (1978) that a process with this properties exists (Rosenkranz, 1985). In addition, models characterized by (1a,1b) are diffusion limits of certain interacting particle systems on the lattice Z^d which can describe growth and spread of populations. In this context, α can be interpreted as the mean number of offspring per individual, γ as the sum of the birth and death rate and ι as a stationary immigration rate. The operator L determines the spatial propagation of the population mass on S. Since α and γ are assumed to be state dependent, Ξ is a measure valued diffusion process with interactions. We further assume that L has a discrete spectrum of real eigenvalues such that the corresponding eigenfunctions form a basis for the continuous functions on S. By Sturm-Liouville theory, this is for example the case if S is a finite interval. Then we can formulate

THEOREM. Suppose that $\gamma\equiv\gamma_0$, $\alpha(x,\xi)=\beta(x)\|\xi\|^{\delta-1}$ for some continuous function β on S and some $0\leq\delta<1$. If $\beta\equiv$const>0 or if β is strictly positive and $2\gamma<\|\iota\|$, then given ξ there exists a random variable W with $E_\xi\{W\}>0$ and a measure m on the Borel algebra on S such that

$$\lim_{t\to\infty} \frac{\Xi(t,D)}{t^{1/(1-\delta)}} = m(D)\, W \quad \text{a.s.}$$

for every Borel set D.

We do not know at the present time whether the assumption $2\gamma<\|\iota\|$ can be weakened or removed nor whether the theorem holds under less restrictive conditions on α. The measure m can be given explicitly in terms of the coefficients of L in many cases. (See Rosenkranz, 1985, for details.) If α and γ do not depend on ξ, we obtain a result similar to those of Watanabe (1967) for branching Markov processes, namely that $\Xi(t)\exp(-\lambda t)$ converges to a random measure, where λ is the largest eigenvalue of $\alpha+\gamma L$. Whereas in the interaction-free case the limit theorems can be proved with martingale arguments, the main tool in the proof of the theorem stated above is a convergence theorem for amarts (asymptotic martingales). An amart is defined as follows. Let (Ω,F,P) a probability space with filtration (F_t). Let T be the set of all (F_t)-stopping times taking only finitely many values. With the definition $\tau\leq\sigma$ iff $P\{\tau\leq\sigma\}=1$ for $\tau,\sigma\in T$, T is a directed set. Let Y be a (F_t)-adapted process. If $E\{|Y(t)|\}<\infty$ for every t and if $\{E\{Y(\tau)\},\tau\in T\}$ converges then $\{Y(t)\}$ is called an amart relative to (F_t). The amart convergence theorem then states that if Y is separable and $E\{|Y(t)|\}$ is uniformly bounded Y(t) converges P-a.s. (Edgar and Sucheston, 1976). The main idea of the proof of the main result therefore consists in showing that some function of $\Xi(t)$ divided by some power of t is an amart.

PROOF OF THEOREM

The proof of the limit theorem is done in several steps. First note that for every two times continuously differentiable function g on R and every $\phi\in D(L)$

$$g(<\phi,\Xi(t)>)-\int_0^t g'(<\phi,\Xi(s)>)\{<\phi,\iota>+<\alpha\phi+\gamma L\phi,\Xi(s)>\}\,ds$$
$$-\int_0^t g''(<\phi,\Xi(s)>)<\gamma\phi^2,\Xi(s)>\,ds$$

is a martingale relative to the filtration induced by Ξ. This follows from the construction of the process via martingale problems or from Dynkin's formula. A combination of this martingale property and several classical inequalities (see Rosenkranz, 1985) yields

LEMMA 1. For every $\xi\in M(S)$ there exist nonnegative constants K_1, K_2, and K_3 such that

$$E_\xi\{\|\Xi(t)\|\}\leq K_1(1+t)^{1/(1-\delta)},$$
$$E_\xi\{\|\Xi(t)\|^2\}\leq K_2(1+t)^{2/(1-\delta)},$$
$$E_\xi\{|<\phi,\Xi(t)>|\}\leq K_3(1+t)^{(3+\delta)/4(1-\delta)},$$

for every eigenfunction ϕ of L with eigenvalue $\neq 0$.

Since every constant is an eigenfunction of L with eigenvalue zero because of the assumption of reflecting boundaries, the lemma tells us that the integral of the higher eigenfunctions grows at least in the mean at a lower rate than the total variation. Lemma 1 implies further, that for every $0\leq b<1/4$, $(1+t)^{b-1/(1-\delta)}|\langle\phi,\Xi(t)\rangle|$ is an amart converging a.s. to zero for every $\xi\in M(S)$. Since the eigenfunctions of L constitute a basis for the continuous functions on S, every such function ψ can be approximated by finite linear combinations of eigenfunctions. Hence if $V:=\lim \|\Xi(t)\|/t^{1/(1-\delta)}$ exists then

$$(2) \qquad \lim_{t\to\infty} \frac{\langle\psi,\Xi(t)\rangle}{t^{1/(1-\delta)}} = V\cdot l(\psi)$$

for some suitable linear functional l. Approximating the indicator function of some Borel set D by continuous functions would yield the theorem. Hence it is sufficient to show that the limit on the right hand side of (2) exists. If β is constant and positive, this can be done without further assumptions. If β is not identically constant we need the result that for every nonconstant eigenfunction ϕ of L $\langle\phi,\Xi(t)\rangle(1+\|\Xi(t)\|)^{-1}$ converges a.s. to 0. This follows from

LEMMA 2. If $2\gamma<\|\iota\|$, for every $\xi\in M(S)$ there exists a constant K_4 such that

$$E_\xi\{(1+\|\Xi(t)\|)^{-1}\} \leq K_4(1+t)^{-1/(1-\delta)}.$$

Proof: $$E_\xi\{1/(1+\|\Xi(t)\|)\} = 1/(1+\|\xi\|) - \|\iota\|\int_0^t E_\xi\{(1+\|\Xi(s)\|)^{-2}\}ds - \int_0^t E_\xi\{\langle\beta,\Xi(s)\rangle(1+\|\Xi(s)\|)^{\delta-3}\}ds + 2\gamma\int_0^t E_\xi\{\|\Xi(s)\|(1+\|\Xi(s)\|)^{-3}\}ds.$$

Let $u(t):=E_\xi\{1/(1+\|\Xi(t)\|)\}$. Then u is differentiable and

$$u'(t) \leq (\beta_0+2\gamma-\|\iota\|)E_\xi\{(1+\|\Xi(t)\|)^{-2}\}-\beta_0 E_\xi\{(1+\|\Xi(t)\|)^{\delta-2}\}$$

where $\beta_0:=\inf \beta(x)>0$. If $2\gamma<\|\iota\|\leq\beta_0+2\gamma$,

$$u'(t) \leq -(\|\iota\|-2\gamma)E_\xi\{(1+\|\Xi(t)\|)^{\delta-2}\}.$$

If $\beta_0+2\gamma<\|\iota\|$,

$$u'(t) \leq -\beta_0 E_\xi\{(1+\|\Xi(t)\|)^{\delta-2}\}.$$

Hence there exists a constant $a>0$ such that $u'(t)\leq -au(t)^{2-\delta}$.

Let v be a solution of $v(0)=u(0)$ and $v'(t)=-av(t)^{2-\delta}$. Then

$$v(t)=((1-\delta)at+u(0)^{\delta-1})^{-1/(1-\delta)} \leq K_4(1+t)^{-1/(1-\delta)}.$$

Since $u(t)\leq v(t)$ for every t, the lemma is proved.

We now turn to the last step in the proof of the limit theorem.

LEMMA 3. Under the conditions stated in the introduction, $(1+\|\Xi(t)\|)^{1-\delta}/(t+1)$ is a uniformly integrable amart.

Proof: Uniform integrability follows from Lemma 1. To prove the amart property fix $\varepsilon>0$ and note that there is a natural number N and constants b_n, $n=0,\dots,N$, such that

$$\| \beta - \sum_{n=0}^{N} b_n\phi_n \| \leq \varepsilon$$

for suitable eigenfunctions $\phi_o,\dots,\phi_n$ of L, where $\phi_o\equiv$const. If β is constant, $b_o=\beta\phi_o^{-1}$ and $b_n=0$ for $n>0$. Then if $u_N=\beta-\Sigma b_n\phi_n$

$$\begin{aligned} E_\xi\{\tfrac{1}{1+\tau}(1+\|\Xi(t)\|)^{1-\delta}\} &= (1+\|\xi\|)^{1-\delta}E_\xi\{\tfrac{1}{1+\tau}\} \\ &+ (1-\delta)E_\xi\{\tfrac{1}{1+\tau}\int_0^\tau \sum^N b_n\langle\phi_n,\Xi(s)\rangle/(1+\|\Xi(s)\|)ds\} \\ &+ (1-\delta)E_\xi\{\tfrac{1}{1+\tau}\int_0^\tau \langle u_N,\Xi(s)\rangle/(1+\|\Xi(s)\|)ds\} \\ &+ (1-\delta)\|\iota\|E_\xi\{\tfrac{1}{1+\tau}\int_0^\tau (1+\|\Xi(s)\|)^{-\delta}ds\} \\ &+ (1-\delta)\delta\gamma E_\xi\{\tfrac{1}{1+\tau}\int_0^\tau (1+\|\Xi(s)\|)^{-1-\delta}ds\}. \end{aligned}$$

As $\tau\to\infty$, the first and the fifth summand go to zero, the fourth converges to $\|\iota\|$ if $\delta=0$ or to zero otherwise. The third summand is bounded by $(1-\delta)\varepsilon$. Hence

$$\begin{aligned} 0 &\leq \limsup_{\tau\in T} E_\xi\{\tfrac{1}{1+\tau}(1+\|\Xi(\tau)\|)^{1-\delta}\} \\ &- \liminf_{\tau\in T} E_\xi\{\tfrac{1}{1+\tau}(1+\|\Xi(\tau)\|)^{1-\delta}\} \leq 2\varepsilon(1-\delta). \end{aligned}$$

Since ε was arbitrary, the assertion follows.

The amart convergence theorem then yields the a.s. convergence of $\|\Xi(t)\|/t^{1/(1-\delta)}$ to some random variable W. Easy calculations show that $0<E_\xi\{W\}<\infty$ and the limit theorem is proved. The omitted details can be found in Rosenkranz (1985).

FINAL REMARKS

We want to give an example for which the measure m occuring in the formulation of the theorem can be given explicitly. Let S be a real interval and L be given by $L\phi(x) := a(x)\phi'(x) + b(x)\phi''(x)$. Define

$$r(x) := \frac{1}{b(x)} \exp\{\int^x \frac{a(y)}{b(y)}\,dy\}.$$

Then

$$m(D) = \{\int_D r(x)dx\}/\{\int_S r(x)dx\}.$$

A similar formula holds if L is multidimensional but can be written as the sum of onedimensional operators like for example the Laplacian.

Acknowledgements: The result proved in this paper is taken from the author's doctoral thesis written at the University of Heidelberg, Sonderforschungsbereich 123, Stochastische Mathematische Modelle. This work has been supported by Deutsche Forschungsgemeinschaft. The author wants to thank Prof. P. Tautu and Dr. W. Rittgen, German Cancer Research Centre Heidelberg,for many stimulating discussions.

REFERENCES

Dawson, D.A. (1978): Geostochastic calculus. Canad. J. Statistics 6, 143-168.

Edgar, G.A., Sucheston, L. (1976): Amarts. A Class of asymptotic martingales. B. Continuous Parameter. J. Multivariate Analysis 6, 193-221.

Rosenkranz, G. (1985): Existenz und asymptotisches Verhalten einer Klasse von Maßwertigen Diffusionsprozessen. Dissertation, Universität Heidelberg.

Watanabe, S. (1967): Limit theorem for a class of branching processes. In: Markov Processes and Potential Theory, J. Chow (ed.). New York: John Wiley, 205-232.

Hoechst AG
Pharma Forschung Informatik
D-6230 Frankfurt am Main 80

A COMPARISON OF LIFE TIMES BY RANK TESTS

Jerzy Rudnicki
Warsaw

Key words: two-sample problem, locally maximin rank test, combination of tests.

ABSTRACT

A two-sample testing problem with nonnegative observations and alternatives including both location and scale parameters is considered.The characterization of the locally optimal rank tests in this problem is given. A method of deriving the asymptotically optimal rank tests is described.

INTRODUCTION

Assume that $X_1,\ldots,X_m$ and $Y_1,\ldots,Y_n$ are two independent random samples consisting of independent, identically distributed random variables with cumulative distribution functions F for the first sample and G for the second one. Moreover, assume that all random variables are nonnegative. The problem is to decide on the basis of observations whether Y's are greater in a certain sense than X's.

The above problem often occurs in practice when, for instance, the observations are times - to - failure of technical objects or life times of biological items. On the basis of invariance considerations one can propose rank methods as a natural tool in this problem. Therefore consider the problem of constructing an optimal rank test to test the hypothesis $H: F = G$ against the alternative $K: F \geq G, F \neq G$.

Rank tests for the the two-sample problem have been extensively studied. The usual approach is to consider a set of alternatives that depends on a single real parameter. In such a case, under reasonable regularity conditions, the locally most powerful rank test exists and is based on a linear rank statistic /see Theorem II.4.8 in Hájek and Šidák (1967)/. Usually the location - or scale-change model is used. Both of them are justifiable in practice and it is often hard to choose between them. For this reason the problem of constructing the optimal rank test in the model with both location and scale parameters will be considered in this paper.

In the presence of more than one real parameter the locally most powerful rank test usually does not exist and another definition of test optimality must be used. The idea of construction of the locally maximin rank test, formerly applied to testing of independence /see Rudnicki(1984)/, will be utilized. In order to make this notion reasonable it is necessary that unit changes of location and scale parameters cause the same changes of alternatives. Therefore consider the model

$$G(x) = F\left((x-\mu\cdot\beta)/(\theta+1)\right) \quad , \mu = EX_1$$

in which unit changes of parameters cause the same changes of the mean of distribution G. The problem is to test the hypothesis $H: \beta = \theta = 0$ against the alternative $K: \beta,\theta \geq 0, \max(\beta,\theta) > 0$

Assume that F has a differentiable density f, satistying conditions

$$\int |f'(x)|dx < \infty$$
$$\int x|f'(x)|dx < \infty$$

and $F(0) = 0$. Let $\underline{R} = (R_1,\ldots,R_n)$ be a vector of ranks of $Y_1,\ldots,Y_n$ in the pooled sample $X_1,\ldots,X_m, Y_1,\ldots,Y_n$.

Define functions

$$\phi_1(x) = -\mu\frac{f'(x)}{f(x)}$$
$$\phi_2(x) = -1 - x\frac{f'(x)}{f(x)}$$

and introduce the scores

$$a_k(i) = N\binom{N-1}{i-1} \int \phi_k(x)[F(x)]^{i-1}[1-F(x)]^{N-1} dF(x), k=1,2.$$

where N = m+ n . Define the statistics

$$S_k(\underline{R}) = \sum_{i=1}^{n} a_k(R_i) \qquad k=1,2.$$

Let ψ be a critical function of a test. Define numbers

$$A_\psi^k = \sum_{\underline{R}} \psi(\underline{R})\, S_k(\underline{R}).$$

THE LOCALLY MAXIMIN RANK TEST

Recall the definition of the locally maximin rank test.

Definition 1.

The level α test ψ^* is the locally maximin rank (LMR) test if for any level α rank test ψ there exists $\varepsilon_o > 0$ such that for every $0 < \varepsilon < \varepsilon_o$

$$\inf_{\beta^2+\theta^2=\varepsilon^2} M_{\psi^*}(\beta,\theta) \geq \inf_{\beta^2+\theta^2=\varepsilon^2} M_\psi(\beta,\theta)$$

where $M_{\psi^*}(\beta,\theta)$ is the power of a test ψ under the alternative (β,θ).

The LMR test is characterized by the following theorem.

Theorem 1.

Let

$$W_{\underline{R}}(\beta,\theta) = \int\ldots\int_{\underline{R}} \prod_{i=1}^{m} f(x_i)dx_i \prod_{j=1}^{n} (\theta+1)^{-1} f((y_j-\mu\beta)/(\theta+1))dy_j$$

Assume that $W_{\underline{R}}$ is twice differentiable and, moreover, its second partial derivatives are uniformly bounded for (β,θ) in a certain neighbourhood of (0,0) . Then the LMR test to test H against K maximizes the expression

$$\min(A_\psi^1, A_\psi^2)$$

The proofs of theorems are not included in this paper,however in general they follow the lines of proofs in Rudnicki (1984).

Determination of the critical function of the LMR test on the basis of this theorem is in general rather complicated. It could be performed numerically, but the form of this test is too complicated for applications.Therefore it is desirable to derive the LMR test in a class of rank tests of reasonable form. We restrict our considerations to the set of rank tests based on a linear combination of the locally most powerful test statistics for one parameter location and scale alternatives, i.e.

$$(1) \quad \psi_\nu(\underline{R}) = \begin{cases} 1 & \text{if } T_\nu = \nu\, S_1(\underline{R}) + (1-\nu)\, S_2(\underline{R}) > k_\alpha^\nu \\ 0 & \text{otherwise} \end{cases}$$

where $0 \le \nu \le 1$. The problem now consists in the determination of the optimal number ν^*.

Notice, that $A^1_{\psi_\nu}$ is a nondecreasing function of ν and $A^2_{\psi_\nu}$ is a nonincreasing function of ν. As a consequence of this assertion and of Theorem 1 we have

Theorem 2.

There exists $\varepsilon_o > 0$ such that for every $0 < \varepsilon < \varepsilon_o$ it holds:

a. There exists ν^* (0,1) such that
$\lim_{\nu\to\nu^*_-} M_{\psi_\nu}(\varepsilon,0) < \lim_{\nu\to\nu^*_-} M_{\psi_\nu}(0,\varepsilon)$ and
$\lim_{\nu\to\nu^*_+} M_{\psi_\nu}(\varepsilon,0) > \lim_{\nu\to\nu^*_+} M_{\psi_\nu}(0,\varepsilon)$
Then one of the tests: $\lim_{\nu\to\nu^*_-}\psi_\nu$, ψ_{ν^*}, $\lim_{\nu\to\nu^*_+}\psi_\nu$ is the LMR test of the form (1).

b. If $M_{\psi_\nu}(\varepsilon,0) > M_{\psi_\nu}(0,\varepsilon)$ for all $\nu\in(0,1)$ then the test ψ_1 is the LMR test of the form (1) .

c. If $M_{\psi_\nu}(\varepsilon,0) < M_{\psi_\nu}(\varepsilon,0)$ for all $\nu\in(0,1)$ then the test ψ_o is the LMR test of the form (1) .

These statements are true if both $A^1_{\psi_\nu}$ and $A^2_{\psi_\nu}$ are nonnegative for the determined test.

It follows from Theorem 2, that ν^* corresponds to the point of sign change of the expression $M_{\psi_\nu}(\varepsilon,0) - M_{\psi_\nu}(0,\varepsilon)$ for ε's sufficiently small. If $M_{\psi_\nu}(\varepsilon,0)$ and $M_{\psi_\nu}(0,\varepsilon)$ were continuous functions of ν, the LMR test would correspond to the point of intersection of curves $M_{\psi_\nu}(\varepsilon,0)$ and $M_{\psi_\nu}(0,\varepsilon)$.

THE ASYMPTOTICALLY LOCALLY MAXIMIN RANK TEST

For any finite sample functions $M_{\psi_\nu}(\varepsilon,0)$ and $M_{\psi_\nu}(0,\varepsilon)$, as functions of ν, are not continuous. However, by the theorem of Noether (1955), under suitable regularity conditions, these functions tend to continuous limits as $N \to \infty$ and $\varepsilon \sim N^{-1/2}$. This assertion leads to the following definition of the test optimality.

Definition 2.

The level α test ψ^* is the asymptotically locally maximin rank (ALMR) test if for any level α rank test ψ it holds

$$\lim_{N\to\infty} \inf_{\beta^2+\theta^2=\varepsilon^2/N} M_{\psi^*}(\beta,\theta) \geq \lim_{N\to\infty} \inf_{\beta^2+\theta^2=\varepsilon^2/N} M_{\psi}(\beta,\theta)$$

for any $\varepsilon>0$.

If the assumptions of the theorem of Noether (1955) are satisfied, the limit power functions $\lim_{N\to\infty} M_{\psi_\nu}(\varepsilon N^{-1/2},0)$ and $\lim M_{\psi_\nu}(0, \varepsilon N^{-1/2})$ are continuous functions of ν. Thus, by Theorem 2, ν^* is uniquely determined and the ALMR test of the form (1) is given by

$$\nu_{opt} = \begin{cases} 1 & \text{if } \nu^* > 1 \\ \nu^* & \text{if } 0 \leq \nu^* \leq 1 \\ 0 & \text{if } \nu^* < 0 \end{cases}$$

where ν^* is the solution of the equation

(2) $$\lim_{N\to\infty} M_{\psi_{\nu^*}}(\varepsilon N^{-1/2},0) = \lim_{N\to\infty} M_{\psi_{\nu^*}}(0,\varepsilon N^{-1/2})$$

In order to apply the theorem of Noether (1955) one must establish the asymptotic normality of statistics T_ν of the form (1) . Assume that

$$J_k(u) = \phi_k\,(F^{-1}(u)) \qquad k=1,2.$$

are nondecreasing, nonconstant , twice differentiable functions satisfying the condition

$$|J_k^{(i)}(u)| \le K[\,u\,(1-u)\,]^{-i-1/2+\delta} \qquad k=1,2.$$

for $i=0,1,2$ and some constants $K,\delta > 0$.Here $J^{(i)}$ denotes the i-th derivative of J.

The above assumptions are obviously satisfied also by the function $\nu J_1 + (1-\nu)J_2$, which is the score generating function to the statistic T_ν . Thus, by Theorem VI.5.1. and Lemma VI.5.1.a. of Hájek and Šidák (1967) , statistics T_ν are asymptotically normal with asymptotic mean

$$\text{(3)} \qquad \begin{aligned} \mu^\nu_{m,n}(\beta,\Theta) &= \nu n \int J_1\left[\tfrac{m}{N} F(\mu\beta + (\Theta+1)x) + \tfrac{n}{N} F(x)\right] dF(x) + \\ &+ (1-\nu) n \int J_2\left[\tfrac{m}{N} F(\mu\beta + (\Theta+1)x) + \tfrac{n}{N} F(x)\right] dF(x) \end{aligned}$$

By the theorem of Noether (1955) equation (2) is equivalent to

$$\Phi\left(\lambda_\alpha - \varepsilon \cdot \lim_{N\to\infty} \frac{\frac{\partial}{\partial\beta}\,\mu^\nu_{m,n}(0,0)}{\sqrt{N}\,\sigma_N(0,0)}\right) =$$

$$= \Phi\left(\lambda_\alpha - \varepsilon . \lim_{N\to\infty} \frac{\frac{\partial}{\partial\Theta}\,\mu^\nu_{m,n}(0,0)}{\sqrt{N}\,\sigma_N(0,0)}\right)$$

where λ_α is the α-quantile of the standard normal distribution. Since Φ is a bijective function we obtain finally

$$\text{(4)} \qquad \lim_{N\to\infty} \frac{\partial}{\partial\beta}\,\mu^\nu_{m,n}(0,0) = \lim_{N\to\infty} \frac{\partial}{\partial\Theta}\,\mu^\nu_{m,n}(0,0)\ .$$

From (3) we obtain

$$\frac{\partial}{\partial\beta}\mu^{\nu}_{m,n}(0,0) = -\nu\frac{m\cdot n}{N}\mu^2\int\left[\frac{f''(x)}{f(x)} - \left(\frac{f'(x)}{f(x)}\right)^2\right]f(x)dx +$$

$$-(1-\nu)\frac{m\cdot n}{N}\mu\int\left[\frac{f'(x)}{f(x)} + x\frac{f''(x)}{f(x)} - x\left(\frac{f'(x)}{f(x)}\right)^2\right]f(x)\,dx,$$

$$\frac{\partial}{\partial\Theta}\mu^{\nu}_{m,n}(0,0) = -\nu\frac{m\cdot n}{N}\mu\int x\left[\frac{f''(x)}{f(x)} - \left(\frac{f'(x)}{f(x)}\right)^2\right]f(x)\,dx + \tag{5}$$

$$-(1-\nu)\frac{m\cdot n}{N}\int\left[x\frac{f'(x)}{f(x)} + x^2\frac{f''(x)}{f(x)} - x^2\left(\frac{f'(x)}{f(x)}\right)^2\right]f(x)dx.$$

Putting (5) to the equation (4) we can obtain the formula for ν^*.

Theorem 3.

The ALMR test of the form (1) corresponds to

$$\nu_{opt} = \begin{cases} 1 & \text{for } \nu^* > 1 \\ \nu^* & \text{for } 0 \le \nu^* \le 1 \\ 0 & \text{for } 0 > \nu^* \end{cases}$$

where

$$\nu^* = \frac{\int\left[x(x-\mu)\frac{f''(x)}{f(x)} + (x-\mu)\frac{f'(x)}{f(x)} - x(x-\mu)\left(\frac{f'(x)}{f(x)}\right)^2\right]dF(x)}{\int\left[(x-\mu)^2\frac{f''(x)}{f(x)} + (x-\mu)\frac{f'(x)}{f(x)} - (x-\mu)^2\left(\frac{f'(x)}{f(x)}\right)^2\right]dF(x)}.$$

EXAMPLES

Consider the Weibull model with the cumulative distribution function

$$F(x) = 1-\exp(-x^p) \qquad \text{for } x \geq 0$$

The necessary assumptions are satisfied if $p > 2$.
The ALMR test of the form (1) is based on the statistic

$$\nu_{opt}\ S_1\ (\underline{R}) + (1-\nu_{opt})\ S_2\ (\underline{R})$$

where $S_2(\underline{R})$ is the well known Savage statistic and $S_1(\underline{R})$ is the linear rank statistic with the scores

$$a_1(i) = N\binom{N-1}{i-1} \sum_{j=0}^{i-1} \binom{i-1}{j}(-1)^j\Gamma(1-1/p)(N-i+1-j)^{1/p-2}(j+i-N-1/p)$$

The optimal weight ν_{opt} depends on the shape parameter p in the way presented in Fig.1.

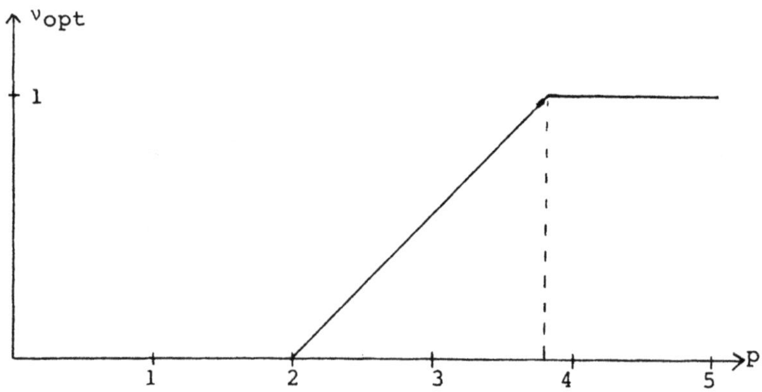

Fig.1.

If F is the gamma distribution with the density function

$$f(x) = x^{a-1}[\Gamma(a)]^{-1} \exp(-x) \quad \text{for } x \geq 0, \quad a>1.$$

then $\nu^* = 0$. Consequently the ALMR test of the form (1) in the gamma model is also the locally most powerful rank test of the scale parameter.

REFERENCES

Hájek J. and Šidák Z. (1967) : Theory of Rank Tests, Academia , Praque.

Noether G.E. (1955) : On a Theorem of Pitman, Ann.Math. Statist. 26, No 1; 64 -68.

Rudnicki J. (1984) : Locally Optimal Rank Tests for Independence, Statistics & Probability Letters 2 , No 2, 83-89.

Polish Academy of Sciences
Systems Research Institute
Newelska 6
01-447 WARSZAWA
P O L A N D

AN ALGORITHM FOR AN ELECTRONIC NETWORK

STATISTICAL ANALYSIS

C.Sambotin, L.Ciocoiu

B u c h a r e s t

A B S T R A C T

A general purpose network analysis algorithm for performing nonlinear transient dc and ac analysis provides statistical simulation to determine the distribution of circuit output due to parameter variations. The algorithm combines a user-oriented input language capable to describe completely general nonlinear devices with the latest advances in numerical and programming techniques: implicit integration of variable order, tableau formulation and sparsematrix solution methods.

Circuit-analysis program is used in computer-aided design of electronic circuit. Dc, ac and transient analysis uses a single program. Furthermore, it`s necessary to pass automatically **the results of one type of analysis as input to another type of analysis, for example, using a prior dc analysis to establish the operating point of the nonlinear devices for a subsequent small-signal ac analysis. Also, there is a demand for statistical simulation to determine the distribution of circuits outputs due to input parameter variation such as component tolerances, aging of temperature effects. This paper describes the techniques used in the algorithm.**

MODIFIED TABLEAU FORMULATION WITH IMPLICIT INTEGRATION

Consider an electrical network constructed by interconnecting n two-terminal elements. Clearly, 2n equations are required to determine currents and voltages in each elements. The equivalent formulations of Kirchoff's curent and voltage laws used here, expresses these one in term's of f-circuit equation. To obtain these equations, one first selects a tree (or a forest of trees if the network graph consists of several maximally connected subgraphs). Network elements that are members of a tree are called branches; network elements that are not in a tree are called chords. There are $n_B = n_n - n$ branches and $n_C = n - n_n + n_s$ chords, where n_n is the number of nodes and n_s is the number of maximally subgraphs in the network graph. Each chord, together with the tree path that connects the two nodes of the chord, forms an f-circuit with respect to the selected tree.

If the $n_c \times n_b$ matrix B is defined by

$$B_{iy} = \begin{cases} +1 & \text{if branch } y \text{ is in the f-circuit defined by chord } i \text{ and the orientation of branch } y \text{ is the same as that of chord } i \\ -1 & \text{if branch } y \text{ is in the f-circuit defined by chord } i \text{ and the orientation of branch } y \text{ is opposite to that of chord } i \\ 0 & \text{if branch } y \text{ is not in the f-circuit defined by chord } i \end{cases}$$

and the $n_B \times n_C$ matrix Q is defined by

$$(1)\quad Q = -B^T$$

where the superscript T denotes the matrix transpose, then the f-cut set equations are

$$(2)\quad \begin{bmatrix} U_B, & Q \end{bmatrix} \begin{bmatrix} I_B \\ I_C \end{bmatrix} = 0$$

and the f-circuit equations are

$$(3)\quad \begin{bmatrix} -Q^T, & U_C \end{bmatrix} \begin{bmatrix} V_B \\ V_C \end{bmatrix} = 0$$

In (2) U_B is a unit matrix of dimension n_B.

I_B, I_C are vectors of branch currents and, respectively, chords currents.

In (3) U_C is a unit matrix of dimension n_c.

V_B, V_C are vectors of branch voltages and, respectively, chords voltages.

The remaining n equations describe the current-voltage characteristics of network elements.

The reactive elements i

(4) $I_i = C_i \frac{dV_i}{d_t}$

(5) $V_i = L_i \frac{dI_i}{dt}$

The nonreactive elements

(6) $I_i = J_i(t)$

(7) $V_i = E_i(t)$

The dependent sources

(8) $I_i = f_i(V_1, V_2 \dots V_n, I_{i-1}, I_{i+1} \dots I_n)$

(9) $V_i = f_i(V_1, V_2 \dots V_{i-1}, V_{i+1}, \dots V_n, I_1, I_2 \dots I_n)$

For a Newton-like iteration procedure

(10) $I_i = J_i + \sum_{y=1}^{n} G_{iy}V_y + \sum_{y=1}^{n} K_{iy}I_y$

(11) $V_i = E_i + \sum_{y=1}^{n} K_{iy}V_y + \sum_{y=1}^{n} R_{iy}I_y$

The 2n equations are to be solved approximately at a time sequence $t_1, t_2 \dots t_n$ given the solution at the initial time t_o. The implicit integration formulas [4] are used for the numerical integration of the differential equations.

$y`(t) = f(y(t), t)$ becomes

(12) $y`(t_n) \approx a_r(y(t_n) - \sum_{k=1}^{r} b_{rk}y(t_n - k))$

With the application of (12) to (4) the differential equation for a capacitor produces

$$G_{iy} = \begin{cases} a_rC_i & i = y \qquad K_{iy} = 0 \\ 0 & i \neq y \qquad y_i = -a_rC_i \sum_{k=1}^{r} b_{rk}V_i(t_n - k) \end{cases}$$

Equations (2), (3), (10), (11) can be collected into a single matrix as shown in (13).

$$\text{(13)} \quad \begin{bmatrix} M \\ G(S(t_n)) \end{bmatrix} S(M_n) = \begin{bmatrix} 0 \\ E(S(t_n)) \end{bmatrix}$$

where $S(t_n)$ is the solution vector of currents and voltages at time t_n M is the matrix for branches and chords of equations 10, 11. $G(S(t_n))$ is a conductances and resistances matrix of branches and chords and $E(S(t_n))$ is a matrix for dependent and independent sources of currents and voltages

To solve (13), an initial estimate S_o of the solution is made by passing a polynomial of degree r, where r is the order of the integration formula, through the known solution points at time $t_{n-1}, t_{n-2}, \ldots, t_{n-r-1}$. Given the initial estimate S_o, a sequence $S_1, S_2, S_3 \ldots$ of approximate solutions is constructed iteratively using the formula

$$\text{(14)} \quad \begin{bmatrix} M \\ G(S_k) \end{bmatrix} S_{k+1} = \begin{bmatrix} 0 \\ E(S_k) \end{bmatrix}$$

The solution of the nonlinear system (13) is replaced by a sequence of solutions of the linear system (14). The convergence rate of the iteration can be controlled by changing the integration time step.

Solution of DC Problem

The procedure can be described as follows. Equation (13) can be decomposed into 2 partitions:

$$\text{(15)} \quad \begin{aligned} MS &= 0 \\ GS &= E \end{aligned}$$

Suppose that (15) describes a dc problem (no reactive elements, all independent sources constant). Now consider the system of equations

$$\text{(16)} \quad \begin{aligned} MS &= 0 \\ GS + C\frac{ds}{dt} &= E \end{aligned}$$

where C is a matrix having the same dimension as G. The choice of matrix C is not unique each choice representing a "strategy" for solving the dc problem. A strategy which has proven is to add a constant pseudocapacitor in parallel with each nonlinear current source and a constant pseudoinductor in series with each nonlinear voltage source.

If reactive elements are present in a network convergence of the iteration described above usually can be achieved by making the time step sufficiently small. However, the iteration procedure is quite likely to fail for pure dc problems. Fortunately, dc problems can be converted to transient problems by the addition of pseudoreactive elements. The resulting transient network can be integrated until dc levels are reached.

Solution of AC Problems

To analyze a nonlinear circuit in the frequency domain (ac), the network is first linearized about some desired operating point. At that point, small sinusoidal signals are applied to the linearized network for certain specified frequencies. All derivatives are to be evaluated at the given operating point.

The dependence will be on frequency instead of time. The time derivative operator will be replaced by yw where y is the complex unit element and w is the angular frequency. The equations (13) are solved for the real and imaginary parts of each small-signal current and voltage.

Statistical Simulation

Statistical simulation may be applied to a network using ac, dc or transient analysis. The basic Monte Carlo method is used to compute the probability distributions of the output parameters of a system from known distribution of the inputs. Nonlinearities are entered as statistical tables which give a tolerance and distribution to the nonlinear curve. The Monte Carlo method makes no assumptions concerning the distribution shapes themselves on the linearity of input/output relations.

The basic output of statistical simulation consists of histograms of specified output parameters. A scatter diagram correlating one output to another can be requested.

Sparse Matrix and Control of Fill-In

The fundamental problem in analyzing networks using the formulation described above is rapid solution of the linearized network equations.

The solution of the linear system (13) is found by LU decomposition followed by forward and backward substitution. Fill-in elements arise because of the presence of nonzero elements in the submatrices Q and Q^{-T} of the upper-right-hand quarter of the network matrix. The mechanism for controlling the growth of fill-in elements is to pick a tree from the network graph so the associate Q-matrix has a minimum number of nonzero elements.

To control growth of fill-in elements during LU decomposition requires a fill-in-reducing pivot-selection strategy [3]. At each stage of LU decomposition a weight is assigned to each nonzero element in the partition that does not lie in a row or column that has already been used as a pivot row or column. The weight is assigned according to the rule

$$w_{iy} = (R_i - 1)(C_y - 1)$$

where w_{iy} is the weight of the nonzero element at the intersection of row i and column y, R_i is the number of nonzero elements in row i which lie in columns that have not yet been used as pivot columns, and C_y is the number of nonzero elements in column y that lie in rows that have not yet been used as pivot rows. The element with the lowest weight is found and its row and columns are used for the next pivot columns, respectively W_{iy} is an upper bound on the number of fill-ins that will be created if row i and column y are used as a pivot row and pivot column, respectively, for the next step of the LU decomposition. Thus, the algorithm picks a least upper bound on the number of fill-ins at each step of LU decomposition. To protect against loss the numerical precision, matrix elements that are the partial derivatives of dependent sources or are multiples of such derivatives are forbidden as pivot elements.

Weight-Tree Selection and Determination of Q

The tree-selection algorithm assigns element to branches of a tree on a first-encounter basis. Prior to selecting a tree, each element is assigned a weight equal to the seam of the number of network elements incident at the initial and final nodes of the given element. Then, the elements are sorted by decreasing weight. Elements with the highest weights are at the beginning of the list; they have priority for selection as tree branches.

There are the valid networks, however, for which the pivot-selection algorithm fails to find a pivot order. These networks are characterized by a large number of nonlinear devices that generate element in the network matrix which are forbidden as pivot elements in order to preserve numerical precision. To handle these cases, the weighted-tree selection scheme is modified so independent voltage sources are forced to be branches and independent current sources are forced to be chords. Then, all the elements on the diagonal of the network matrix are nonzero.

The determination of the Q-submatrix given the tree is quite straight forward. The technique used is to construct the $n_n \times n$ incidence matrix A, where n_n is the number of nodes and n number of elements in the network. In terms of the incidence matrix, Kirchhoff's law can be stated as:

$$A \cdot I = 0$$

So, equation (3) canbe solved for n_B branch current in term of the n_C chord currents by applying the Gauss-Jordan elimination:

$$\begin{bmatrix} V_B & Q \\ 0 & 0 \end{bmatrix} \begin{bmatrix} I_B \\ I_C \end{bmatrix} = 0$$

where Q is the desired submatrix.

C O N C L U S I O N

The combined used of variable-order implicit integration formulas, tableau formulation of network equation and sparse-matrix techniques lead to the rapid analysis.

The use of the fixed second-order formula with halving and doubling of the time step for truncation error control requires from 2.0 to 2.5 times as many passes (iterations) as the use of the first- and second-order formula in a variable order combination.

The sparse-tableau formulation of network equations does not require the selection of a tree from the network graph. In the formulation we describe the selection of an appropiate tree which leads to a system of linear equations for which LU decomposition is simple.

REFERENCES

1. S.W.Director, R.A.Rohrer - Automated network design - The frequency-domain case, IEEE Trans Circuit Theory CT - 16 pg.330-337

2. W.J.McCalla, D.O.Pederson - Elements of computer-aided circuit analysis, IEEE Trans Circuit Theory

CT-18 pg.14-26

3. H.M.Markowitz - The elimination form of the inverse and its application to linear programming, IEEE Trans Circuit Theory CT-16 pg.205-221

4. C.W.Gear - The control of parameters in the automatic integration of ordinary differential equations, IEEE Trans Circuit Theory CT-17 pg.14

5. G.D.Hachtel, R.K.Brayton - The sparse tableau approach to network analysis and design, IEEE Trans Circuit Theory CT-18 pg.101-113

OPTIMAL CONTROL AND P-TH MEAN STABILIZABILITY OF STOCHASTIC LINEAR SYSTEMS

Tetsushi Sasagawa

Tokyo

Key words: stochastic linear system, optimal control, stabilizability

ABSTRACT

For the deterministic case, the controlled system is always stable as long as we use the control law which is obtained as the solution of so-called LQ-problem. For the stochastic case, however, the controlled system is not always p-th mean stable for large p even if we use the LQ-optimal control law. Hence, it is meaningful to solve the LP-optimal control problem (i.e., linear system - p-th order cost function) for each p. For the scalar case, the LP-optimal control problem is completely solved. For the multidimensional case, we consider the p-th mean stabilizability problem only and give a sufficient condition for the existence of p-th mean stabilizing control by using the contraction mapping method in a Hilbert space. Some examples are also given.

MOTIVATION

We shall explain the motivation of this paper by using the scalar system. Let us consider the following scalar control system:

$$dx_t = [ax_t + bu_t] + \sigma x_t \, dw_t \, . \tag{1}$$

$$(x_0 (\neq 0)\text{: given a.s., } b \neq 0,\ t \geq 0)$$

Here, $u_t = u(t,x_t)$, $u(t,0) \equiv 0$ and the set of admissible controls $\mathcal{U}$ is defined as the set of functions $u(\cdot,\cdot)$ such that the equation (1) with $u_t = u(t,x_t)$ has a unique solution on $[0,\infty)$.
We present the following problem.

LP-Optimal Control Problem.

⌈For the control system (1), find the optimal control $\hat{u}_t(p) \in \mathcal{U}$ such that

$$\text{(2)} \qquad \underset{u.\in\mathcal{U}}{\text{Minimize}} \int_0^\infty \mathbb{E}_{x_0}[q|x_t|^p + r|u_t|^p]dt. \qquad (q,r>0)$$ ⌋

The case of $p = 2$ is well-known as the LQ-problem. For simplicity, we assume through this paper that p is a positive even number. We call such a problem simply the "LP-problem" as there is no confusion from the context with Linear Programming problems.

Is it so useful to solve the LP-problem?

For the deterministic case (i.e., $\sigma = 0$), even if we apply the solution $\hat{u}_t$ (2) of the LQ-problem which is linear to the system (1) (with $\sigma = 0$), the system is exponentially stable and the integral of (2) does not diverge for any p, i.e., the cost function (2) (without the expectation operator $\mathbb{E}_{x_0}$) is well-defined in spite of the higher order of the integrand because we have $|x_t|^{2n} \leq \alpha^n \exp(-n\beta t)$ provided $|x_t|^2 \leq \alpha\exp(-\beta t)$. Of course, for this case the cost function is not minimized but we may say that the cost function has a kind of robustness in the deterministic case. Hence, for the deterministic case it may be enough in some sense for the LQ-problem to be solved if one does not deeply pursue the optimality.

For the stochastic case, the situation is not so simple.

The solution of the LQ-problem is given by

$$\text{(3)} \qquad \hat{u}_t(2) = -r^{-1}bS_+x_t ,$$

where

$$S_+ = rb^{-2}[\hat{a} + \sqrt{\hat{a}^2 + r^{-1}b^2q}] , \quad \hat{a} = a + 2^{-1}\sigma^2 .$$

We call $\hat{u}_t(2)$ the LQ-optimal control. The LQ-optimal control system is

$$dx_t = [a - r^{-1}b^2 S_+]x_t dt + \sigma x_t dw_t$$

and this system is exponentially p-stable iff $\lambda(p) < 0$, where

$$\lambda(p) = [a - r^{-1}b^2 S_+] + 2^{-1}(p-1)\sigma^2 .$$

Naturally, $\lambda(2) < 0$ but we have

Proposition 1.1.

For the system (1) controlled by the LQ-optimal control $u_t = \hat{u}_t(2)$, the cost function (2) *diverges* iff

$$p \geq 2 [\![2 + \sigma^{-2}\sqrt{a^2 + r^{-1}b^2 q}]\!]$$

where $[\![x]\!]$ implies the maximum integer less than or equal to x.

□

From this proposition, it follows that the integral (2) diverges for some large p and hence we must not be contented to use the solution of the LQ-problem for the case where the cost function (2) would be more suitable. Namely, we have to solve the LP-problem for *each* p.

For the scalar case, the LP-problem is easily solved by using the dynamic programming approach.

Proposition 1.2.

The LP-optimal control for the system (1) is given by

(4) $$\hat{u}_t(p) = -\sqrt[p-1]{r^{-1}b}\; S_+(p)x_t \in \mathcal{U} ,$$

where $S_+(p)$ is the "unique" positive solution of

(5) $$-(p-1)\, r^{-\frac{1}{p-1}}\, b^{\frac{p}{p-1}}\, S^p + [a + \tfrac{p-1}{2}\sigma^2]p\, S^{p-1} + q = 0$$

□

One can easily verify that the combined system of (1) and (4) is exponentially p-stable for any positive even number p. It is worth-while to note that the LP-optimal control is "linear". This fact does not give rise to the question as to the existence and uniqueness of the solution of the combined system equation.

P-STABILIZABILITY

For the multidimensional case, the LP-optimal control problem has not been solved as long as the author knows. We shall consider the exponential p-stabilization problem in this section. Consider the system :

$$(6) \quad \begin{aligned} dx(t) &= [\, Ax(t) + Bu(t)\,]dt + \sum_{r=1}^{k} D_r x(t)dw_r(t) , \\ x(o) &= x_0 \quad \text{a. s.} , \qquad\qquad (t \in R^+) \end{aligned}$$

where $x(\cdot): R^+ \times \Omega \to R^n$.

Let us say the control system (6) to be *exponentially p-stabilizable* if there exist some control $u(\cdot) \in U$ and positive constants α, β such that

$$E \,\|x^u(t;x_0)\|^p \le \alpha \,\|x_0\|^p \exp(-\beta t) \qquad \text{a.s.}$$

for any initial condition $x_0 \in R^n \backslash \{ o \}$ and $t \in R^+$.

The solution of the LP-problem for the scalar case was given by a *linear* feedback control. Here, for simplicity, we consider only the class of linear feedback controls. Therefore, letting $u(t) = -Kx(t)$, we obtain

$$(7) \quad \begin{aligned} dx(t) &= \hat{A}x(t)dt + \sum_{r=1}^{k} D_r x(t)dw_r(t) , \\ x(o) &= x_0 \quad \text{a. s.} , \end{aligned}$$

where $\hat{A} = A - BK$. Here, for the solution process $x(t;x_0)$ of (7), we define the (p-th mean) Lyapunov exponent:

$$\lambda(x_0(\omega), p) := \limsup_{t \to \infty} \frac{1}{t} \log E \,\|x(t;x_0)\|^p .$$

From the definition, the fact that the top Lyapunov exponent is almost surely negative for p implies that the trivial solution $x(t;o) \equiv o$ is exponentially p-stable. Define the scalar positive process

$$\xi(t;\xi_0) := \langle Sx(t;x_0), x(t;x_0) \rangle ,$$

where $\xi_0 = \langle Sx_0, x_0 \rangle$, $S = S^T > o$. Note that the Lyapunov exponents of the process $[\xi(t;\xi_0)]^{p/2}$ and $x(t;x_0)$ coincide.

By using the transformation

$$y : x \to \sqrt{S}\, x \,/ < Sx, x >^{1/2},$$

we have finally

$$\lambda(x_0(\omega),p) = \lim_{t\to\infty} \sup \frac{1}{t} \log E_{x_0} \exp\{ \frac{p}{2} \int_0^t [2<\sqrt{S}^{-1}\hat{A}^T\sqrt{S}\, y, y>$$

$$(8) \qquad + \sum_{r=1}^{k} <\sqrt{S}^{-1}D_r^T S D_r \sqrt{S}^{-1} y, y> + \sum_{r=1}^{k} <\sqrt{S}^{-1}D_r^T\sqrt{S}\, y, y>^2]dt$$

$$- \frac{p^2}{2} \sum_{r=1}^{k} \int_0^t <\sqrt{S}^{-1}D_r^T\sqrt{S}\, y, y>^2 dt + p \sum_{r=1}^{k} \int_0^t <\sqrt{S}^{-1}D_r^T\sqrt{S}\, y, y> dw_r(t)\}.$$

Here, we assume that, for some given symmetric positive definite matrix Q, the algebraic equation

$$(9) \qquad \hat{A}^T S + S\hat{A} + (p-1)\sum_{r=1}^{k} D_r^T S D_r = -Q$$

has a positive definite solution S. Then

$$\lambda(x_0(\omega),p) \le -\frac{p}{2}\hat{\lambda} \quad (p \ge 2) \quad \text{for almost all } x_0 \in R^n \backslash \{ o \}$$

where $\hat{\lambda} = \lambda_{min}(Q)/\lambda_{max}(S)$ (> o).

Define the norm of a matrix F by $|F| := \max \{ \|Fx\| \,;\, \|x\| = 1 \}$.

Lemma 1. Assume that $\hat{A}$ is a stable matrix and that the inequality

$$(10) \qquad \left| \sum_{r=1}^{k} \int_0^\infty [D_r \exp(\hat{A}t)]^T [D_r \exp(\hat{A}t)] dt \right| < \frac{1}{p-1} \qquad (p \ge 2)$$

is valid. Then, linear matrix equation (9) has a unique positive definite solution S for any given symmetric positive definite matrix Q. [See Sasagawa (1982)]. □

Proposition 2.1. Let $\hat{A} = A - BK$. Assume that there exists a matrix K, say $\hat{K}$, such that $\hat{A}$ is stable and, moreover, the inequality (10) is satisfied.

Then, the system (6) is exponentially p-stabilizable (p ≥ 2) by using the linear feedback control $u = -\hat{K}x$. □

The next example shows that the inequality (10) is not always satisfied even if (A, B) is controllable.

Example 1. Consider the system with k = 2 and

$$A = \begin{bmatrix} 0 & 1 \\ 0 & 0 \end{bmatrix}, \quad B = \begin{bmatrix} 0 \\ 1 \end{bmatrix}, \quad D_1 = \gamma \begin{bmatrix} 1 & 0 \\ 0 & -1 \end{bmatrix}, \quad D_2 = \gamma \begin{bmatrix} 0 & 1 \\ 1 & 0 \end{bmatrix}.$$

The pair (A, B) is controllable (i.e., rank [B, AB] = 2) and we can make the matrix $\hat{A} = A - BK$ stable. In fact, letting $K = [k_1 \;\; 2k_2]$, we obtain $\lambda_{1,2} = -k_2 \pm \sqrt{k_2^2 - k_1}$ as the eigenvalues of $\hat{A}$. $\hat{A}$ is stable iff $k_1 > 0$, $k_2 > 0$.

The computation yields us

$$\left| \sum_{r=1}^{2} \int_0^\infty [D_r \exp(\hat{A}t)]^T [D_r \exp(\hat{A}t)] dt \right| > 2\gamma^2.$$

Hence, the inequality (10) is not satisfied for sufficiently large $|\gamma|$.

SPECIAL CASE

Let $\bar{k} := \{1, 2, \ldots, k\}$. We note that

$$\langle \sqrt{S}^{-1} D_r^T \sqrt{S}\, y, y\rangle = 2^{-1} \langle \sqrt{S}^{-1} (D_r^T S + SD_r) \sqrt{S}^{-1} y, y\rangle .$$

Then, the terms from the third up to the last of the exponent of (8) are equal to zero if $D_r^T S + SD_r = 0$ for all $r \in \bar{k}$. This condition is satisfied if D_r, $r \in \bar{k}$, are skew-symmetric and we take $S = I$ (identity matrix).

Proposition 3.1. Let D_r, $r \in \bar{k}$, be skew-symmetric matrices.
If there exists some matrix K such that all the eigenvalues of

$$\hat{A}^T + \hat{A} + \sum_{r=1}^{k} D_r^T D_r , \qquad \text{where } \hat{A} = A - BK ,$$

are negative (i.e., the above matrix is negative definite), then the system (6) is exponentially p-stabilizable for *all* $p \geq 2$ by using $u = -Kx$. □

Remark. The condition of the above proposition does not depend on p.
Hence, if a feedback matrix K which exponentially 2-stabilizes the system can be found, then the system is also exponentially p-stabilizable with the same feedback gain K.

Example 2. Consider the exponential p-stabilization problem of the 2-dimensional system, namely, we want to determine a feedback gain K under which the system is exponentially p-stabilized.

Let

$$A = \begin{bmatrix} a_1 & a_2 \\ a_3 & a_4 \end{bmatrix}, \quad B = \begin{bmatrix} b_1 \\ b_2 \end{bmatrix}, \quad K = [k_1 \;\; k_2], \quad D_r = \begin{bmatrix} 0 & \sigma_r \\ -\sigma_r & 0 \end{bmatrix},$$

where $|b_1| + |b_2| \neq 0$ and $|k_1| + |k_2| \neq 0$, and let

$$\alpha = a_1 + 2^{-1}\sigma^2, \quad \beta = 2^{-1}(a_2 + a_3), \quad \gamma = a_4 + 2^{-1}\sigma^2, \quad \sigma^2 = \sum_{r=1}^{k} \sigma_r^2 .$$

From Prop. 3.1, what we have to do is to determine k_1, k_2 such that

$$\hat{A}^T + \hat{A} + \sum_{r=1}^{k} D_r^T D_r = 2 \begin{bmatrix} \alpha - b_1k_1 & \beta - \dfrac{b_1k_2 + b_2k_1}{2} \\ \beta - \dfrac{b_1k_2 + b_2k_1}{2} & \gamma - b_2k_2 \end{bmatrix} < 0 .$$

By Sylvester's theorem, the principal minors, say Δ_1 and Δ_2, should satisfy the conditions :

$$\Delta_1 = \alpha - b_1k_1 < 0 ,$$

$$\Delta_2 = (\gamma - b_2k_2) \Delta_1 - (\beta - \frac{b_1k_2 + b_2k_1}{2})^2 > 0 .$$

(i) For the case $b_1 = 0$, under the condition $\alpha < 0$, the system is exponentially p-stabilizable if we take

$$k_1 = 2 \beta b_2^{-1} \;; \qquad k_2 \gtrless \gamma b_2^{-1} \quad (b_2 \gtrless 0) .$$

(ii) For the case $b_2 = 0$, under the condition $\gamma < 0$, the system is exponentially p-stabilizable if we take

$$k_1 \gtrless \alpha b_1^{-1} \quad (b_1 \gtrless 0) \;; \qquad k_2 = 2 \beta b_1^{-1} .$$

(iii) For the case $b_1b_2 \neq 0$, the system is exponentially p-stabilizable if

$$[b_2^2\alpha - 2b_1b_2\beta + 2b_1^2\gamma] < 0 ,$$

$$(b_1b_2)^2 \det[A^T + A + \sum_{r=1}^{k} D_r^T D_r] > - [b_2^2\alpha - 2b_1b_2\beta + 2b_1^2\gamma]^2$$

and the feedback gains are given by

$$k_1 = -2b_2^{-2}[\ b_1\gamma - b_2\beta\ ] \ ; \quad k_2 = -2b_1^{-2}[\ b_2\alpha - b_1\beta\ ] \ .$$

ACKNOWLEDGEMENT

This work was written while the author was a visitor at the " Forschungsschwerpunkt Dynamische Systeme ", Universität Bremen. The author is grateful to Professor L. Arnold for providing interesting seminars and exellent working conditions during his stay. The financial support given by the Alexander-von-Humboldt Foundation during author's stay is also gratefully acknowledged.

REFERENCES

Has'minskii R. Z. (1980): Stochastic stability of differential equations, Sijthoff & Noordhoff, Alphen aan den Rijn, The Netherlands.

Sasagawa T. (1982): Sufficient conditions for the exponential p-stability and p-stabilizability of linear stochastic systems, INT. J. SYSTEM SCI., Vol.13, No.4, 399-408.

Department of Mechanical Engineering
Sophia University
Kioi-cho 7-1, Chiyoda-ku
102 Tokyo, Japan

A SEQUENTIAL LEBESGUE-RADON-NIKODYM THEOREM
AND THE LEBESGUE DECOMPOSITION OF MARTINGALES

Klaus D. Schmidt

Mannheim

Key words: Lebesgue decomposition, generalized Radon-Nikodym derivatives, martingales, a.s. convergence

ABSTRACT

The present paper contains generalizations of Neveu's sequential Lebesgue-Radon-Nikodym theorem and of Baez-Duarte's Lebesgue decomposition of martingales. The results are based on Darst's extension of the classical Lebesgue-Radon-Nikodym theorem and on the properties of the generalized Radon-Nikodym operator.

1. INTRODUCTION

Let $(\Omega,\mathcal{F},\lambda)$ be a probability space, let $\{ \mathcal{F}_n \mid n \in \mathbb{N} \}$ be an increasing sequence of sub-σ-algebras of $\mathcal{F}$, and let $\mathcal{F}_\infty$ denote the smallest σ-algebra satisfying $\mathcal{F}_n \subseteq \mathcal{F}_\infty \subseteq \mathcal{F}$ for all $n \in \mathbb{N}$. If $\mathcal{G}$ is any sub-σ-algebra of $\mathcal{F}$, $\mu : \mathcal{G} \longrightarrow \mathbb{R}$ is an additive set function, and $X : \Omega \longrightarrow \mathbb{R}$ is a $\mathcal{G}$-measurable λ-integrable random variable, then, for all $m \in \mathbb{N} \cup \{\infty\}$ satisfying $\mathcal{F}_m \subseteq \mathcal{G}$, let $R_m\mu$ denote the restriction of μ to $\mathcal{F}_m$ and let E_mX denote the conditional expectation of X with respect to $\mathcal{F}_m$.

The following result is well-known: If $\mu : \mathcal{G} \longrightarrow \mathbb{R}$ is a countably additive set function, then there exists a random variable $X \in L^1(G,\lambda)$ and a λ-singular countably additive set function $\psi : G \longrightarrow \mathbb{R}$ satisfying $\mu = \int X \, d\lambda + \psi$. This classical Lebesgue-Radon-Nikodym theorem was extended by Darst [4] to the case where μ is only additive and bounded. While Darst's proof is elementary but lengthy, a shorter and also more informative proof can be given by using vector lattice theory and will be outlined in Section 2.

A sequential version of the classical Lebesgue-Radon-Nikodym theorem was proven by Neveu [5]: If $\mu : F \longrightarrow \mathbb{R}$ is a positive countably additive set function and if, for $m \in \mathbb{N} \cup \{\infty\}$, $R_m\mu = \int X_m \, d\lambda + \psi_m$ is the Lebesgue-Radon-Nikodym decomposition of $R_m\mu$, then $\{ X_n \mid n \in \mathbb{N} \}$ is a positive supermartingale satisfying $\lim X_n = X_\infty$ a.s. In Section 3, this sequential Lebesgue-Radon-Nikodym theorem will be extended to the case where the positive set function μ is only additive, and a similar result will be obtained in the case where μ is only additive and bounded.

A martingale version of the classical Lebesgue-Radon-Nikodym theorem was proven by Baez-Duarte [2]: If $\{ X_n \in L^1(F_n,\lambda) \mid n \in \mathbb{N} \}$ is an L^1-bounded martingale such that there exists a countably additive set function $\mu : F_\infty \longrightarrow \mathbb{R}$ satisfying $R_n\mu = \int X_n \, d\lambda$ for all $n \in \mathbb{N}$, then $\{ X_n \mid n \in \mathbb{N} \}$ is the sum of two martingales $\{ Y_n \mid n \in \mathbb{N} \}$ and $\{ Z_n \mid n \in \mathbb{N} \}$ such that there exists a random variable $Y \in L^1(F_\infty,\lambda)$ and an $R_\infty\lambda$-singular countably additive set function $\psi : F_\infty \longrightarrow \mathbb{R}$ satisfying $E_nY = Y_n$ and $R_n\psi = \int Z_n \, d\lambda$ for all $n \in \mathbb{N}$. In Section 4, this Lebesgue decomposition of martingales will be extended to general L^1-bounded martingales.

2. ON THE GENERAL LEBESGUE-RADON-NIKODYM THEOREM

In this section we outline a short proof of Darst's extension of the classical Lebesgue-Radon-Nikodym theorem to bounded additive set functions. Further details may be found in [6; Section 2.1].

Let G be an algebra of subsets of some set Ω , and let $ba(G)$ denote the class of all bounded additive set functions $G \longrightarrow \mathbb{R}$.

Under the pointwise defined addition, scalar multiplication, and order relation, $ba(G)$ is a vector lattice. In particular, each $\mu \in ba(G)$ has a Jordan decomposition $\mu = \mu^+ - \mu^-$, with $\mu^+ := \mu \vee 0$ and $\mu^- := (-\mu) \vee 0$, and the modulus $|\mu| := \mu \vee (-\mu)$ agrees with the variation of μ. For $\lambda \in ba(G)$, let $ba^\lambda(G)$ denote the class of all λ-continuous set functions in $ba(G)$ and let $\{\lambda\}^\perp$ denote the class of all λ-singular set functions in $ba(G)$. Then $ba^\lambda(G)$ and $\{\lambda\}^\perp$ are projection bands of $ba(G)$, and $ba(G)$ is the direct sum of $ba^\lambda(G)$ and $\{\lambda\}^\perp$. In particular, each $\mu \in ba(G)$ has a unique Lebesgue decomposition $\mu = \varphi + \psi$, with $\varphi \in ba^\lambda(G)$ and $\psi \in \{\lambda\}^\perp$.

Let now $\lambda \in ba(G)$ be a countably additive set function, and let $\overline{\lambda}$ denote the extension of λ to the σ-algebra $\overline{G}$ generated by G. Then each $\varphi \in ba^\lambda(G)$ is countably additive and thus has a unique extension $\overline{\varphi}$ to $\overline{G}$, and by the Radon-Nikodym theorem there exists an a.s. unique random variable $X \in L^1(\overline{G},\overline{\lambda})$ satisfying $\overline{\varphi} = \int X \, d\overline{\lambda}$. This random variable is said to be the Radon-Nikodym derivative of φ with respect to λ, and the map $ba^\lambda(G) \longrightarrow L^1(\overline{G},\overline{\lambda})$ associating with each set function in $ba^\lambda(G)$ its Radon-Nikodym derivative is a vector lattice isomorphism which is called the Radon-Nikodym operator with respect to λ. It then follows that the map

$$D^\lambda : ba(G) \longrightarrow L^1(\overline{G},\overline{\lambda})$$

associating with each set function in $ba(G)$ the Radon-Nikodym derivative of its λ-continuous part is a vector lattice homomorphism, and that it is the unique vector lattice homomorphism which extends the Radon-Nikodym operator to the whole of $ba(G)$. Therefore, D^λ is called the generalized Radon-Nikodym operator with respect to λ, and for each set function $\mu \in ba(G)$ the random variable $D^\lambda\mu$ is said to be the generalized Radon-Nikodym derivative of μ with respect to λ.

By the preceding construction, there exists for each set function $\mu \in ba(G)$ an a.s. unique random variable $X \in L^1(\overline{G},\overline{\lambda})$ and a unique set function $\psi \in \{\lambda\}^\perp$ satisfying $\mu(A) = \int_A X \, d\overline{\lambda} + \psi(A)$ for all $A \in G$. This is Darst's result [4]. We remark, however, that the existence of a Lebesgue-Radon-Nikodym decomposition only allows for the formulation of the results of this paper whereas their proofs will depend on the fact that the generalized Radon-Nikodym operator is linear and preserves suprema.

Throughout this paper, let F be an algebra of subsets of some set Ω, let $\{ F_n \mid n \in \mathbb{N} \}$ be an increasing sequence of subalgebras of F, let F_∞ be the smallest algebra satisfying $F_n \subseteq F_\infty \subseteq F$ for all $n \in \mathbb{N}$, and let $\lambda \in ba(F)$ be a countably additive set function. In the sequel, we shall not distinguish between λ, its extension $\overline{\lambda}$ to $\overline{F}$, and the restrictions of $\overline{\lambda}$ to sub-σ-algebras of $\overline{F}$. For all $m \in \mathbb{N} \cup \{\infty\}$, let

$$D_m : ba(F_m) \longrightarrow L^1(\overline{F}_m, \lambda)$$

denote the generalized Radon-Nikodym operator with respect to $R_m\lambda$, and, if G is any algebra satisfying $F_m \subseteq G \subseteq F$, $\mu : G \longrightarrow \mathbb{R}$ is an additive set function, and $X : \Omega \longrightarrow \mathbb{R}$ is a $\overline{G}$-measurable λ-integrable random variable, let $R_m\mu$ denote the restriction of μ to F_m and let E_mX denote the conditional expectation of X with respect to $\overline{F}_m$. For all $n \in \mathbb{N}$, define $\mathbb{N}(n) := \{ m \in \mathbb{N} \mid n \leq m \}$.

3. A SEQUENTIAL LEBESGUE-RADON-NIKODYM THEOREM

The following result extends the sequential Lebesgue-Radon-Nikodym theorem of Neveu [5; Proposition III-2-7], who considered positive additive set functions which are countably additive:

<u>3.1. Theorem.</u>
Let $\mu : F \longrightarrow \mathbb{R}$ be a positive additive set function.
Then $\{ D_nR_n\mu \mid n \in \mathbb{N} \}$ is a positive supermartingale satisfying $\lim D_nR_n\mu = D_\infty R_\infty\mu$ a.s. In particular, $\lim D_nR_n\mu = 0$ a.s. holds if and only if $R_\infty\mu$ is $R_\infty\lambda$-singular.
Moreover, $\{ D_nR_n\mu \mid n \in \mathbb{N} \}$ is a martingale if and only if, for all $n \in \mathbb{N}$ and $m \in \mathbb{N}(n)$, the restriction of the $R_m\lambda$-singular part of $R_m\mu$ to F_n is $R_n\lambda$-singular, and it is a uniformly integrable martingale if and only if, for all $n \in \mathbb{N}$, the restriction of the $R_\infty\lambda$-singular part of $R_\infty\mu$ to F_n is $R_n\lambda$-singular.

Proof. For all $m \in \mathbb{N} \cup \{\infty\}$, let $R_m\mu = \varphi_m + \psi_m$ be the Lebesgue decomposition of $R_m\mu$, with $R_m\lambda$-continuous part φ_m and $R_m\lambda$-singular part ψ_m. Then we have, for all $n \in \mathbb{N}$ and $m \in \mathbb{N} \cup \{\infty\}$,

$$0 \leq E_nD_mR_m\mu = E_nD_m\varphi_m = D_nR_n\varphi_m \leq D_nR_n\varphi_m + D_nR_n\psi_m = D_nR_nR_m\mu = D_nR_n\mu \quad ,$$

which means that $\{ D_nR_n\mu \mid n \in \mathbb{N} \}$ is a positive supermartingale. The preceding inequality also reveals that each $R_n\psi_m$ is $R_n\lambda$-singular if and only if $E_nD_mR_m\mu = D_nR_n\mu$ holds for all $n \in \mathbb{N}$ and $m \in \mathbb{N}(n)$, which means that $\{ D_nR_n\mu \mid n \in \mathbb{N} \}$ is a martingale, and that each $R_n\psi_\infty$ is $R_n\lambda$-singular if and only if $E_nD_\infty R_\infty\mu = D_nR_n\mu$ holds for all $n \in \mathbb{N}$. In the latter case, $\{ D_nR_n\mu \mid n \in \mathbb{N} \}$ is a uniformly integrable martingale, by [5; Proposition V-1-2] or Corollary 4.7 below. Conversely, if $\{ D_nR_n\mu \mid n \in \mathbb{N} \}$ is a uniformly integrable martingale, then there exists some $X \in L^1(F_\infty,\lambda)$ satisfying $D_nR_n\mu = E_nX$ for all $n \in \mathbb{N}$, and we have

$$\int_A X\, d\lambda = \int_A D_nR_n\mu\, d\lambda \leq \mu(A)$$

for all $n \in \mathbb{N}$ and $A \in F_n$. Define $\eta \in ba(F_\infty)$ by letting

$$\eta(A) := \int_A X\, d\lambda$$

for all $A \in F_\infty$. Then we have $\eta \leq R_\infty\mu$, hence

$$X = D_\infty\eta \leq D_\infty R_\infty\mu = D_\infty\varphi_\infty \quad ,$$

and thus

$$E_nX \leq E_nD_\infty\varphi_\infty = D_nR_n\varphi_\infty \leq D_nR_n\varphi_\infty + D_nR_n\psi_\infty = D_nR_nR_\infty\mu = D_nR_n\mu = E_nX \quad ,$$

which yields $D_nR_n\varphi_\infty = 0$, for all $n \in \mathbb{N}$.

Furthermore, by the convergence theorem for positive supermartingales [5; Théorème II-2-9], $\lim D_nR_n\mu$ exists and the inequality

$$\int_A \lim D_nR_n\mu\, d\lambda \leq \int_A D_mR_m\mu\, d\lambda \leq \mu(A)$$

holds for all $m \in \mathbb{N}$ and $A \in F_m$. Define $\gamma \in ba(F_\infty)$ by letting

$$\gamma(A) := \int_A \lim D_nR_n\mu\, d\lambda$$

for all $A \in F_\infty$. Then we have $\gamma \leq R_\infty\mu$, and thus

$$\lim D_nR_n\mu = D_\infty\gamma \leq D_\infty R_\infty\mu \quad .$$

Conversely, using the identification of the limit of successive conditional expectations [5; Proposition II-2-11], we obtain

$$D_\infty R_\infty\mu = \lim E_nD_\infty R_\infty\mu \leq \lim D_nR_n\mu \quad .$$

Therefore, we have $\lim D_nR_n\mu = D_\infty R_\infty\mu$, and it is then obvious that $\lim D_nR_n\mu = 0$ holds if and only if $R_\infty\mu$ is $R_\infty\lambda$-singular. □

<u>3.2. Corollary.</u>
Let $\mu : F \longrightarrow \mathbb{R}$ be a bounded additive set function.
Then $\{ D_nR_n\mu \mid n \in \mathbb{N} \}$ is an L^1-bounded quasimartingale satisfying $\lim D_nR_n\mu = D_\infty R_\infty\mu$ a.s.

This is immediate from the Jordan decomposition.

4. THE LEBESGUE DECOMPOSITION OF MARTINGALES

In the sequel we assume that the algebras F_∞ and F agree.

For a sequence $\{ X_n \in L^1(\overline{F}_n,\lambda) \mid n \in \mathbb{N} \}$ and all $n \in \mathbb{N}$, define set functions $\mu_n \in ba(F_n)$ by letting

$$\mu_n(A) \quad := \quad \int_A X_n \, d\lambda$$

for all $A \in F_n$. Then $X_n = D_n\mu_n$ holds for all $n \in \mathbb{N}$. Moreover, if $\{ X_n \mid n \in \mathbb{N} \}$ is a martingale, then

$$\mu(A) \quad := \quad \lim \mu_n(A)$$

exists for all $A \in F_\infty$, and this defines an additive set function $\mu : F_\infty \longrightarrow \mathbb{R}$ which is called the limit measure of the martingale. Obviously, a martingale is positive if and only if its limit measure is positive, and it is easy to see that a martingale is L^1-bounded if and only if its limit measure is bounded; see [6; Theorem 2.3.2]. As a consequence of Corollary 3.2 we obtain the following extension of a result of Baez-Duarte [2], who considered L^1-bounded martingales which have a countably additive limit measure:

4.1. Corollary.
Each L^1-bounded martingale converges a.s. to the Radon-Nikodym derivative of the λ-continuous part of its limit measure.

The previous result in turn yields the following extension of a result which was recently proven by Austin, Bellow, and Bouzar [1], who considered L^1-bounded martingales which are positive:

4.2. Corollary.
Let $\{ X_n \mid n \in \mathbb{N} \}$ be an L^1-bounded martingale with limit measure μ. Then $\lim X_n = 0$ a.s. if and only if μ is λ-singular.

A sequence $\{ X_n \mid n \in \mathbb{N} \}$ is singular if there exists a λ-singular set function $\eta \in ba(F_\infty)$ satisfying $|\mu_n|(A) = \int_A |X_n| \, d\lambda \leq \eta(A)$ for all $n \in \mathbb{N}$ and $A \in F_n$. In this case, $|\mu_n| \leq R_n\eta$ and hence

$$|X_n| = |D_n\mu_n| = D_n|\mu_n| \leq D_nR_n\eta$$

holds for all $n \in \mathbb{N}$. Now the following result which was recently proven by Baez-Duarte [3] is immediate from Theorem 3.1 when applied to η:

4.3. Corollary.
Each singular positive supermartingale converges a.s. to zero.

We remark that the previous results are also evident from the convergence theorems proven in [6; Chapter 2].

A martingale $\{ X_n \mid n \in \mathbb{N} \}$ is hereditary if there exists a random variable $X \in L^1(\mathcal{F}_\infty, \lambda)$ satisfying $X_n = E_n X$ for all $n \in \mathbb{N}$.

4.4. Lemma.
An L^1-bounded martingale is singular if and only if its limit measure is λ-singular, and it is hereditary if and only if its limit measure is λ-continuous.

Proof. Let $\{ X_n \mid n \in \mathbb{N} \}$ be an L^1-bounded martingale. Then

$$|\mu|(A) = \sup_{\mathbb{N}} |\mu_n|(A) = \sup_{\mathbb{N}} \int_A |X_n| \, d\lambda$$

holds for all $A \in \mathcal{F}_\infty$, by [6; Theorem 2.3.2]. Since $\{\lambda\}^\perp$ is a projection band in $ba(\mathcal{F}_\infty)$, this yields the assertion concerning singularity. If μ is λ-continuous, then $X_n = D_n\mu_n = D_n R_n \mu = E_n D_\infty \mu$ holds for all $n \in \mathbb{N}$, hence $\{ X_n \mid n \in \mathbb{N} \}$ is hereditary. The converse is obvious. □

The following result extends the Lebesgue decomposition of martingales of Baez-Duarte [2], who considered L^1-bounded martingales which have a countably additive limit measure:

4.5. Theorem.
Each L^1-bounded martingale is the sum of a hereditary and a singular martingale. The decomposition is unique.

Proof. Let $\{ X_n \mid n \in \mathbb{N} \}$ be an L^1-bounded martingale and let $\mu = \varphi + \psi$ be the Lebesgue decomposition of its limit measure μ, with λ-continuous part φ and λ-singular part ψ. For all $n \in \mathbb{N}$, define $Y_n := D_n R_n \varphi$ and $Z_n := D_n R_n \psi$. Then

$$X_n = D_n\mu_n = D_n R_n \mu = D_n R_n \varphi + D_n R_n \psi = Y_n + Z_n$$

holds for all $n \in \mathbb{N}$, and $Y_n = D_n R_n \varphi = E_n D_m R_m \varphi = E_n Y_m$ holds for all $n \in \mathbb{N}$ and $m \in \mathbb{N}(n)$. Then $\{ Y_n \mid n \in \mathbb{N} \}$ is a martingale with limit measure φ, hence $\{ Z_n \mid n \in \mathbb{N} \}$ is a martingale with limit measure ψ . The assertion now follows from Lemma 4.4 and the uniqueness of the Lebesgue decomposition of μ . □

To complete the discussion of martingales and their limit measures, we include two further results:

4.6. Lemma.
An L^1-bounded martingale is uniformly integrable if and only if its limit measure is λ-continuous.

Proof. This follows from [6; Theorem 2.3.5]. □

4.7. Corollary.
A martingale is hereditary if and only if it is uniformly integrable.

The previous result is of course well-known, but usually a proof is given which relies on the convergence theorem for martingales.

REFERENCES

[1] Austin, D., Bellow, A., and Bouzar, N. (1985): The Fatou inequality revisited - variations on a theme by A. Dvoretzky. In: Probability in Banach Spaces V. Lecture Notes in Mathematics, vol. 1153, pp. 17-39. Springer, Berlin - Heidelberg - New York.

[2] Baez-Duarte, L. (1968): Another look at the martingale theorem. J. Math. Anal. Appl. 23, 551-557.

[3] Baez-Duarte, L. (1982): Divergence of L^1-bounded martingales. Repub. Venezuela Bol. Acad. Cienc. Fis. Mat. Natur. 42, 39-72.

[4] Darst, R.B. (1962): A decomposition of finitely additive set functions. J. Reine Angew. Math. 210, 31-37.

[5] Neveu, J. (1972): Martingales à Temps Discret. Masson, Paris.

[6] Schmidt, K.D. (1983): Amarts - a measure theoretic approach. In: Amarts and Set Function Processes. Lecture Notes in Mathematics, vol. 1042, pp. 51-236. Springer, Berlin - Heidelberg - New York.

Seminar für Statistik
Universität Mannheim
A 5
6800 Mannheim
West Germany

A NOTE ON POTENTIAL ESTIMATION FOR THE STATIONARY GIBBS MODEL ON THE LATTICE Z^d

Arthur Sieders

DELFT

Keywords: point estimation, random fields, Gibbs model

ABSTRACT

An estimation procedure for parameters of the potential function in a weakly coupled Gibbs model on Z^d is proposed which can be performed even on small computers. Using results of Sylvester (1979) its statistical properties are investigated. It is near consistent and conjectured to be asymptotically normally distributed.

1. Introduction.

The Gibbs model originated in statistical physics and is commonly assumed to give a correct description of thermodynamical equilibrium states of physical particles under potential forces. It has also found application in other branches of science where experimental data may be considered as stochastic subsets of the Z^2 - lattice, e.g. biology, anatomy and geography; seeBesag (1974).

It is well known that the Gibbs model, in contrast to its intuitive appeal, is very hard to treat mathematically. Therefore it is not surprising that the associated statistical problem of estimating the potential function can be considered intractable as well. Recently, Mase (1984) showed that, under plausible conditions, the maximum likelihood estimator has asymptotically optimal properties but, again, this estimator is impossible to compute in any realistic situation and therefore is of theoretical interest only.

For this reason, Mase (1984) also proposed a moment-type estimator but its statistical properties cannot be evaluated at present. In this paper, we propose an estimator which is not optimal but at least amenable to analysis by the methods of mathematical statistics. Using results of Sylvester (1979) and some heuristic arguments we show its near consistency and conjecture its asymptotic normality under the assumption that the Gibbs measure is weakly coupled. The crux is that this estimator is not based on the likelihood but on a conditional likelihood, which is tractible because the 'monstrous' grand partition function is divided out.

This note was inspired by the papers of Mase (1984) and Sylvester (1979). To tease the reader, we use a mixture of their notations.

2. Preliminiaries.

Sylvester (1979) gives a very general definition of lattice models. Although here we consider much simpler models we briefly set out his description because we shall refer to his results. A lattice model consists of the following quintuple:

L	is a denumerable set consisting of the vertices i of some lattice;
$\{(X_i,B_i,\nu_i); i\in L\}$	is a family of probability spaces indexed by L;
G	is a hypergraph on L, i.e. a family of finite non-empty subsets E of L, which is assumed to be of finite degree;
$\{p_E; E \in G\}$	is a family of weights, indexed by G; each weight p_E is an element of $[1,\infty]^E$ and is assumed to be conformable for integration;
$\{g_E; E \in G\}$	is a family of complex functions in $L^{p_E}(\prod_E X_i; \prod_E \nu_i)$, indexed by G; the functions g_E are called the Gibbs factors.

The intuitive meaning of these concepts is understood most clearly when one uses the traditional physical terminology for e.g. a spin system or a lattice gas, see e.g. Ruelle (1969). The set L is called the lattice, the elements i of L represent possible locations for particles or up-spins. The spaces (X_i, B_i,ν_i) account for a random behaviour of these particles or spins. The Gibbs factors g_E account for the multi-particle interaction between the sites $i \in E$, i.e. the interaction remaining if all lower order (self-, binary-, etc.) interactions are removed. The hypergraph G embodies all nontrivial interactions.

Like Mase (1984), we shall restrict ourselves to translation invariant models describing random subsets of Z^d. In this context, we call these models *simple*. This restriction allows us to make some simplifications: (1) L will always denote the vertex set of Z^d; (2) all probability spaces (X_i,B_i,ν_i) are copies of a probability space $(\{0,1\},B,\nu)$ where ν is a probability measure on $\{0,1\}$ and B is the σ-algebra

3

$\{\emptyset,\{0\},\{1\},\{0,1\}\}$, so X_i will henceforth denote the set $\{0,1\}$; (3) as we shall only consider finite functions on $\prod_E X_i$, we may set all weights p_E equal to ∞ so that conformability for integration is assured; (4) there exists a set of fundamental edges which generates G by translation; (5) our functions g_E are of a very simple form: if $\xi_E \in \prod_E X_i$ then

$$(2.1) \qquad g_E(\xi_E) = \begin{cases} \exp - \Phi(E) & \text{if } \xi_i = 1 \text{ for all } i \in E, \\ 1 & \text{otherwise,} \end{cases}$$

where the 'potential' function Φ maps $G \to \mathbb{R}$ (and may be set $= 0$ outside G).

We define the *spread* of a family of finite subsets $A_1, A_2, \ldots$ of L by

$$(2.2) \qquad \rho(A_1,A_2,\ldots) = \min\{\, |\Lambda| : \Lambda \subset G \text{ and } \Lambda \cup \{A_j\} \text{ is connected}\,\}.$$

The spread is a metric on the subsets of L. The ball $B_r(A)$ around an $A \subset L$ is defined as the subhypergraph $\{\, E \subset G : \rho(A,E) \leq r \,\}$.
Throughout we shall identify subsets ξ of L and their characteristic functions $I_\xi : L \to \{0,1\}^L$. If S is a subset of L then ξ_S denotes the restriction $\xi \cap S$ of ξ to S. Under the technical assumption that each $i \in L$ is covered by at least one $E \subset G$, we may also identify a subset A of L and the subhypergraph $\Lambda(A)$ of G describing the interactions inside A. Thus

$$(2.3) \qquad G \supset \{\, E \subset G : E \subset A \,\} =: \Lambda(A) \;\leftrightarrow\; A := \cup \{\, E \subset G : E \subset \Lambda(A) \,\} \subset L.$$

We wish to consider random subsets of L. Let $(\Omega, F, \mathbb{P})$ be a probability space; a random subset Ξ of L is a measurable map $(\Omega,F) \to (\prod_L X_i, \sigma(\prod_L B_i))$. Let μ be the measure corresponding to Ξ, i.e. let $\mu(B) := \mathbb{P}(\Xi \in B)$ for all $B \in \sigma(\prod_L B_i)$. In the case that the Ξ_i are expected to behave independently we may obviously write

$$(2.4) \qquad d\mu = \prod_L d\nu_i =: d\nu.$$

Dependent Ξ_i may be modelled by introducing the so-called Gibbs factors g_E: let $\{\Lambda\}$ be a sequence of finite subhypergraphs of G which increases to G; then one defines

$$(2.5) \qquad d\mu_\Lambda := Z(\Lambda)^{-1} \prod_{E \in \Lambda} g_E \, d\nu$$

where

$$(2.6) \qquad Z(\Lambda) := \int_{\prod_L X_i} \prod_{E \in \Lambda} g_E \, d\nu$$

and defines

(2.7) $$d\mu := \lim_{\Lambda \to G} d\mu_\Lambda$$

provided the limit exists. Sylvester (1979) has shown (his theorem 4.2) that, under weak coupling conditions (see (3.1)), the limit exists in the following sense: let A be a finite subset of L and f_A be a finite function on $\prod_A X_i$. Denote by $\mathrm{IE}_\Lambda f_A$ the expectation of f_A with respect to $d\mu_\Lambda$ (provided it exists) and, in particular, let $\mathrm{IE}_0 f_A$ denote the expectation with respect to $d\mu_0 := d\nu$ (i.e. the case where Λ is empty). Then the limit of $\mathrm{IE}_\Lambda f_A$ exists for all possible f_A as $\Lambda \to G$: one writes $\lim \mathrm{IE}_\Lambda f_A =: \mathrm{IE}_\infty f_A$. From the expectations IE_∞ the measure μ may be recovered. The measures μ_Λ and μ are called local and global Gibbs measures, respectively.

For a 'simple' lattice model, it is convenient to rewrite definitions (2.5) and (2.6) in terms of the potential function Φ. Moreover, we wish to consider a potential function which depends also on some parameter θ. Let Λ be a finite subhypergraph of G. One has

(2.8) $$d\mu_{\theta,\Lambda}(\xi) = Z_\theta(\Lambda)^{-1} \left[\exp - \sum_{E \in \xi_\Lambda} \Phi_\theta(E) \right] d\nu(\xi),$$

(2.9) $$Z_\theta(\Lambda) = \sum_{\xi_\Lambda} \left[\exp - \sum_{E \in \xi_\Lambda} \Phi_\theta(E) \right] \prod_{i \in E \in \Lambda} \nu_i(\xi_i).$$

The limit μ_θ , if it exists, is defined pointwise in θ.

3. Parametric potential estimation.

For a given lattice model of the 'simple' type defined in section 2, we wish to consider the following statistical problem. Let $\{ \Phi_\theta;\ \theta \in \Theta \}$ be a known family of potential functions indexed by some open subset Θ of IR. To each Φ_θ corresponds a global Gibbs measure μ_θ. A random configuration Ξ is modelled by the probability measure μ_t ,where t, the true value of the parameter, is unknown and has to be estimated from the observation of Ξ on some finite subset S of L. I.e. we want to construct a functional $\hat{\theta}(\Xi_S)$ which is close to t in some sense.

Maximum likelihood estimation (MLE) with respect to the global measure μ_θ is, in general, out of the question. Mase (1984) performs MLE with respect to the local measure $\mu_{\theta,S}$, i.e. all interactions in the complement S^c and between S and S^c are neglected. Still, for any S of realistic size, the evaluation of $Z_\theta(S)$ can be considered a hopeless task because of the extremely large number of summands. See Baxter (1982) for details.

In the sequel, we shall use the notion 'computable'. An expression is called computable if a sufficiently accurate numerical value can be obtained within a reasonable amount of computer time. The obvious subjectivity of this notion is of less importance in this context. Our aim is to propose a computable estimator. The idea is the following. (The bounds are mere indications.)
We assume that

(1) μ is weakly coupled (see (3.1)) and that

(2) the number m and the range r of the fundamental edges is small ($m<100$, $r<10$).

We next choose a subset $A \subset S$, and

(3) translate A into the sets A_1 (= A), A_2, ..., $A_n \subset S$ which are

(4) so far apart that, by (1), the Ξ_{A_i} are almost uncorrelated.

Now we construct events P and $P\cap Q$ on $\{0,1\}^A$ (with translates P_i and $(P\cap Q)_i$)

(5) which have probability at least $10/n$, so that

(6) based on the observed numbers of occurences $n_P := |\{ i\colon \Xi_{A_i} \in P_i \}|$ of P and $n_{P\cap Q}$ of $P\cap Q$, a reasonable estimate $\hat{\mu}(\theta)$ of the probability $\mu_\theta\{ \Xi_A \in Q | \Xi_A \in P\}$ can be made (with known distribution), using (4).

(7) Put $B:= B_s(A)$, where $s \simeq 2r$ (so that, by (2), $s < 20$).

Now assume that we have chosen the events P and $P\cap Q$ such that they are also

(8) of limited cardinality (i.e. $|\xi_A\colon \xi_A \in P| < 10^5$ and the same for $P\cap Q$)

so that, using (2.8), (2), (7) and (8),

(9) the probability $\mu_{\theta,B}\{ \Xi_A \in Q | \Xi_A \in P\}$ (note that Z divides out!) is a calculable† function of θ.

We remark that often the amount of calculation can be reduced by at least a factor $|A|$ by choosing clever A, P and Q and using symmetry and stationarity.

†Provided one approximates the sum over all N configurations on the 'edge' $B_s(A)\backslash A$ by N/n_P times the sum over the realisations of Ξ on the edges of those A_i where P occurs (and for $P\cap Q$ idem).

6

Finally, we have by (1) and (3) that $\mu_{\theta,B} \simeq \mu_\theta$. This, together with (6), suggests the estimate $\hat{\theta} := \hat{\mu}^{-1}(n_{P\cap Q} / n_P)$, which is calculable by (9). We make this partly rigorous in section 4.

4. Asymptotic behaviour of the estimator.

We make the following assumptions:

A1. The lattice model is simple (see section 2).

A2. G is generated by a finite set of fundamental edges $E_1, E_2, \ldots, E_m$ and every vertex is covered by at least two edges of G.

A3. The Gibbs measure is weakly coupled, uniformly over Θ, i.e. for some chosen δ

$$(3.1) \qquad |\Phi_\theta(E_i)| < \delta \qquad \theta \in \Theta,\ i=1,2,\ldots,m.$$

Assumptions A1,A2 and A3 imply all the assumptions of Sylvester(1979). In particular, A2 implies his technical assumption on p.102, (1.2) and (2.21). It also implies finite range r of the potential, where $r := \max_E \max_{i,j \in E} |i-j|$. Hence, by theorem 4.2 of Sylvester (1979), $\mu_{\theta,\Lambda} \to \mu$ uniformly over Θ in the sense discussed.

The following theorems support the heuristics set out in section 3. We shall write $\mu(P)$ for $\mu\{\Xi_A \in P\}$, $\mu(Q|P)$ for $\mu\{\Xi_A \in Q \mid \Xi_A \in P\}$ and $I(P)$ for $I(\{\Xi_A \in P\})$.

Theorem 1. Let A be a finite subset of L. Then, for all events P and Q in $\prod_A X_i$ with $\inf_\Theta \mu_\theta(P) > 0$, there exists a constant C such that, for all c>0, there exists a $\delta(c,m)$ such that, under the assumptions A1, A2, A3 ($\delta=\delta(c,m)$ one has, for all $\theta \in \Theta$,

$$(3.2) \qquad |\mu_\theta(Q|P) - \mu_{\theta,B_s(A)}(Q|P)| \leq C \exp(-cs).$$

Proof. Apply theorem 4.2 of Sylvester (1979), with $r_A = \infty$, to $f_A = I(P)$ and take the limit $\Lambda \to \infty$. Do the same for $f_A = I(P\cap Q)$. Combine the resulting inequalities, using that $\mu_{B_s(A)}(P)$ is bounded away from 0 by the first inequality and $\mu(P) > 0$.□

Of course, C may be large. However, intuition suggests that the left-hand side of (3.2) is small for $s \simeq 2r$.

Theorem 2. Let (I_n) be an increasing sequence of finite subsets $I_n := \{i_1, i_2, \ldots, i_n\}$ of L and let A be a finite subset of L, with translates A_i, $i \in I_n$. Let P and Q be events on $\prod_A X_i$ with translates P_i and Q_i, $i \in I_n$. Put $n_P := |\{i \in I_n: \Xi_{A_i} \in P_i\}|$ and define $n_{P\cap Q}$ likewise. Finally, let V_n be the increasing sequence of random sets defined by

7

(3.3) $$V_n := \{ i \in I_n : \Xi_{A_i} \in P_i \}.$$

Then there exists a $\delta(m)$ such that, if the assumptions A1,A2, A3 hold, with $\delta=\delta(m)$ one has, uniformly over Θ,

(3.4) $$n_{P\cap Q} / n_P \to \mu_\theta \text{ (Q|P) in } \mu_\theta\text{-probability as } n \to \infty.$$

Conjecture. Under the above conditions, also the following hold:

(3.5) $$n_P^{\frac{1}{2}} [(n_{P\cap Q} / n_P) - \mu_\theta \text{ (Q|P) }] \to N(0, \sigma^2_{\theta,P,Q}),$$

where $\sigma^2_{\theta,P,Q} := \sum_{i\in V_\infty} u_{\theta,\infty}(I_i(Q), I_i(Q)$, u denotes the cumulant and $i\in V_\infty$;

moreover, one has

(3.6) $$\sigma^2_{\theta,P,Q} \to \mu_\theta(Q|P) - \mu_\theta(Q|P)^2 \text{ exponentially as } \rho(V_i;\ i\in I_n) \to 0.$$

Proof of theorem 2.
We merely sketch the proof. On applies theorem 4.6 of Sylvester (1979): take all weights = ∞ and put f_A = I(P) and I(P$\cap$Q), respectively. It is seen that both $n_{P\cap Q}/n$ and n_P/n converge in μ_θ-probability.□

Application of the same theorem to the conjecture is not immediate, as the sets V_n are random. Still, one might use the same technique as in the proof of this theorem 4.6 provided one shows that a conditional version of theorem 4.3 holds, i.e. that

(3.7) $$|u_\infty(f_{A_i};\ i \in I_n \mid P_{I_n}) - u_0(f_{A_i};\ i \in I_n \mid P_{I_n})| \to 0 \text{ exponentially in } \rho(A_i;\ i \in I_n)$$

where P_{I_n} is the event that all $\Xi_{A_i} \in P_i$. We illustrate the difficulty by the following example. Let Y_i, $i \in Z$, be a stationary centered real stochastic process with exponentially decreasing cumulants, for instance $u(Y_i,Y_j) < C \exp - c|i-j|$. Let Y_{i_k} be the subsequence defined by $Y_{i_k} \in P$, where $\mu\ (Y_i \in P) > 0$.

Can one show that the cumulants of Y_{i_k} also decrease exponentially?

We remark that condition (4.15) of Sylvester (1979) seems to be insufficient for his theorem 4.6 to hold: it does not easily imply finiteness of the finite-volume generating function, as the author suggests. This can probably be repaired by the use

s

of coroll. 3.5, equation (3.40). The well-known Cramer's condition

$$\mathbb{E} \exp \lambda f < \infty \qquad \text{for all } \lambda \in \mathbb{R}$$

real λ, is certainly sufficient and is satisfied for our choice of f_A. Another thing which strikes us as surprising is that Sylvester (1979) does not impose regularity conditions on the growth of the sets V_n; compare Ruelle (1969), chapter 2.4, and Bolthausen (1982).

Acknowledgment of support. This research was supported by the Netherlands organisation for the advancement of pure research (Z.W.O.).

references:

Baxter, Rodney J. (1982). Exactly solved models in statistical mechanics. Academic Press, New York.

Bolthausen, E. (1982). On the central limit theorem for stationary mixing random fields. Annals of Probability, Vol. 10, No. 4, 1047-1050.

Mase, Shigeru (1984). Locally asymptotic normality of Gibbs models on a lattice. Advances in Applied Probability 16, 585-602.

Ruelle, D. (1969). Statistical mechanics. Benjamin, New York.

Sylvester, Garrett S. (1979). Weakly coupled Gibbs measures. Zeitschr. Warsch. Th. Verw. Gebiete, 50, 97-118.

Delft University of Technology
Dept. of Mathematics and Informatics
Julianalaan 132
2628 BL Delft
The Netherlands

RANDOM GENERATOR OF FORMULAS OF PROPOSITIONAL CALCULUS

Jan Šindelář

Prague

Key words: probabilistic algorithm, propositional calculus, binary tree

ABSTRACT

Probabilistic algorithm for the generating formulas of propositional calculus is presented. Some of its properties are analyzed. Generated formulas contain propositional indeterminates $x_1, x_2, x_3, \ldots$, logical constants T (true) and F (false), and any of binary logical connectives. Instead of $\neg\phi$ subformulas $\phi \to F$ are used. As an auxiliary notion is used the notion of binary tree.

§1. Propositional indeterminates are denoted by $x_1, x_2, x_3, \ldots$, the set $\{x_1, x_2, x_3, \ldots\}$ of propositional indeterminates by X. Logical constants are denoted by T (true) and F (false). The set of all binary propositional connectives is denoted by BIN. Conjunction is denoted by $\vee$, disjunction by $\wedge$, implication by $\to$, equivalence by $\sim$, negation by $\neg$. Hence, $\{\vee, \wedge, \to, \sim\} \subseteq \text{BIN}$, $\neg \notin \text{BIN}$.

§2. Now we define formulas of propositional calculus (PC) in the form used below. It is a well-known fact that formulas $\neg\phi$ and $\phi \to F$ are equivalent. In the following $\phi \to F$ is used instead of $\neg\phi$. In our definition of PC only binary logical connectives are used.

T, F, $x_1, x_2, x_3, \ldots$ are formulas of PC. If ν_L and ν_R are formulas of PC and $b \in \text{BIN}$, then $\nu_L\ b\ \nu_R$ is formula of PC. The set of formulas of PC will be denoted by $\mathcal{F}$.

A bijection (i. e. one-to-one mapping) from X to X will be called a permutation of X. Let Π be a permutation of X, ν a formula of PC. Let $x_{i_1}, x_{i_2}, \ldots, x_{i_n}$ be all propositional indeterminates occurring in ν. Then by $\nu[\Pi]$ we denote the formula arising from ν by substituting (simultaneously) all occurrences of x_{i_j} by $\Pi(x_{i_j})$ ($i = 1,\ldots,n$).

§3. As an auxiliary notion we use the notion of binary tree, i. e. a tree where each node is either a leaf, or has two immediate successors - the left successor and the right successor. Left successor of a node w will be denoted by $\Gamma_L(w)$, right successor by $\Gamma_R(w)$. Moreover, we suppose that the empty set $\emptyset$ is not a node of any binary tree used below. If w is a leaf of a binary tree S, we put $\Gamma_L(w) = \Gamma_R(w) = \emptyset$. The root of a binary tree will be denoted by s. Hence, binary tree S is a triple $\langle V, \Gamma_L, \Gamma_R \rangle$, where V is set of nodes of S, Γ_L and Γ_R are as above. An example of binary tree is on Fig. 1.

§4. In §4 and §5 we describe two special cases of procedure generating formulas of PC. In the sequel the general case is described.

Let H be nonempty set of formulas of PC, i. e. $\emptyset \neq H \subseteq F$. Let $B = H \cup \mathrm{BIN}$. First, we ascribe to each node w of a finite binary tree S, a member b(w) of the set B. Hence, b(w) is either a formula of PC, or a binary logical connective. An example is given on Fig. 2. Members of B are written at the left hand side of corresponding nodes. For the sake of simplicity let us suppose that to each leaf a formula of PC is ascribed.

Now, for each node w of S, we generate a formula ν_w of PC. We will start from the leaves and proceed to the root S. If w is a leaf, then $\nu_w = b(w)$. Let us suppose that w is not a leaf. If $b(w) \in H$ then $\nu_w = b(w)$. Let $b(w) \in \mathrm{BIN}$. If $\nu_{\Gamma_L(w)}$ and $\nu_{\Gamma_R(w)}$ were already generated, then $\nu_w = \nu_{\Gamma_L(w)}\ b(w)\ \nu_{\Gamma_R(w)}$. E. g., if $\nu_{\Gamma_L(w)} = (x_1 \vee x_3)$, $\nu_{\Gamma_R(w)} = x_2 \rightarrow F$, $b(w) = \wedge$, then $\nu_w = (x_1 \vee x_3) \wedge (x_2 \rightarrow F)$.

When the formula ν_s is obtained the procedure terminates and the result is ν_s. An example is given on Fig. 2. Formulas are written at the right-hand side of the corresponding nodes.

If, for example, $H = \{T, F, x_1, x_2, \ldots\}$, then each formula of PC may be generated.

§5. Let t_1, t_2 be two mappings from F to F such that for each $\nu \in F$ $t_1(\nu) \sim \nu$ and let there be a permutation Π of X such that for each $\nu \in F$ $t_2(\nu) \sim \nu[\Pi]$.

The first step is the same as in §4. To each node w of S a member b(w) from $BIN \cup H$ is ascribed. If w is a leaf, then $b(w) \in H$. To each node w we ascribe a formula ν_w. As in §4, our procedure starts from the leaves and proceeds to the root s. If $b(w) \in H$, then $\nu_w = t_2(b(w))$. If $b(w) \in BIN$ and $\nu_{\Gamma_L(w)}$ and $\nu_{\Gamma_R(w)}$ were already generated, then $\nu_w = t_1(\nu_{\Gamma_L(w)}\ b(w)\ \nu_{\Gamma_R(w)})$. When the formula ν_s is obtained the procedure terminates and the result is ν_s.

§6. Now, a general version of our procedure is described. A binary tree $S = \langle V, \Gamma_L, \Gamma_R \rangle$ is called complete if each node w has left successor, $\Gamma_L(w)$ and right successor $\Gamma_R(w)$.

Let $K \in \{0,1,2,\ldots\}$. For each $i = 0,1,2,\ldots,K$ let $\langle t_1^i, t_2^i, H^i \rangle$ be a triple satisfying

1. t_1^i, t_2^i are mappings from F to F,
2. for each $\nu \in F$, $t_1^i(\nu) \sim \nu$,
3. $H^i \subseteq F$,
4. there is a permutation Π^i of X such that, for each $\nu \in H^i$, $t_2^i(\nu) \sim \nu[\Pi^i]$.

Moreover, for $i = 0,1,\ldots,K-1$, let

$$t_2^0 \circ t_2^1 \circ \ldots \circ t_2^i(H^0) \subseteq H^{i+1}$$

where $f \circ g$ denotes the composition of mappings f, g (i. e. $f \circ g(t) = g(f(t))$).

Let b^0 be a mapping from V to F. For each node $w \in V$ we ascribe either the formula ν_w^0 of PC by a procedure analogical to that one described in §5, or an empty set $\emptyset$. Here b^0, t_1^0, t_2^0 and H^0 play the role of b, t_1, t_2 and H from §5. We have not started from leaves, because complete binary tree has no leaves. Procedure starts from nodes satisfying $b^0(w) \in H^0$. For such nodes we put

$\nu_w^o = t_2^o(b^o(w))$. Now we proceed from nodes satisfying $b^o(w) \in H^o$ to the root. If $w \in V$ and, for nodes $\Gamma_L(w)$, $\Gamma_R(w)$, formulas $\nu^o_{\Gamma_L(w)}$, $\nu^o_{\Gamma_R(w)}$ were generated, then we put $\nu_w^o = t_1^o(\nu^o_{\Gamma_L(w)} b^o(w) \nu^o_{\Gamma_R(w)})$.

If to a node w a formula from F is not ascribed, we ascribe to w the empty set $\emptyset$, i. e. we put $\nu_w^o = \emptyset$.

Let $i \in \{0,1,\dots,K-1\}$, and to each $w \in V$ a member ν_w^i from $F \cup \{\emptyset\}$ has been ascribed. We define a mapping b^{i+1} from V to BIN $\cup$ $\cup\, H^{i+1}$ as follows. Let $w \in V$. If $\nu_w^i \in H^{i+1}$, we put $b^{i+1}(w) = \nu_w^i$. If $\nu_w^i \notin H^{i+1}$ (in such a case $b^o(w) \in$ BIN), we put $b^{i+1}(w) = b^o(w)$.

Now, b^{i+1}, t_1^{i+1}, t_2^{i+1}, H^{i+1} play the role of b, t_1, t_2, H from §5. If $b^{i+1}(w) \in H^{i+1}$, we put $\nu_w^{i+1} = t_2^{i+1}(b^{i+1}(w))$. Now we proceed from nodes satisfying $b^{i+1}(w) \in H^{i+1}$ to the root S. If formulas $\nu^{i+1}_{\Gamma_L(w)}$, $\nu^{i+1}_{\Gamma_R(w)}$ were generated, we put $\nu_w^{i+1} = t^{i+1}(\nu^{i+1}_{\Gamma_L(w)} b^{i+1}(w) \nu^{i+1}_{\Gamma_R(w)})$.

If, by the procedure just described, no formula is ascribed to a node w, we put $\nu_w^{i+1} = \emptyset$.

If a formula ν_s^k is obtained, it is a result of our procedure. If $\nu_s^k = \emptyset$, our procedure fails.

The following can be proved.

(a) The fact that ν_s^k is a formula of PC depends only on the mapping b^o and the set H^o. Δ

This leads to the definition

Definition. A pair $\langle b^o, H^o\rangle$ is generable if $\nu_s^k \in F$. Δ

Hence, ν_s^o is a formula of PC iff ν_s^k is a formula of PC. In Šindelář (1984) is proved

Theorem 1. If $\langle b^o, H^o\rangle$ is generable, then there is a permutation Π_k of X such that

$$\nu_s^o \sim \nu_s^k[\Pi_k]. \qquad \Delta$$

§7. Till now, the mapping b^o has been deterministic. In this §7 we proceed to a probabilistic algorithm generating formulas of PC by randomizing b^o.

First, for each $n \in \{1,2,3,\dots\}$ we define levels H_L^n, H_R^n and a level H^o of nodes of complete binary tree $S = \langle V, \Gamma_L, \Gamma_R\rangle$. We put

$$H^o = \{s\}.$$

Let $n \in \{1,2,3,\ldots\}$ and H^{n-1} be defined. Then

$$H^n_L = \{\Gamma_L(w) \mid w \in H^{n-1}\}$$

$$H^n_R = \{\Gamma_R(w) \mid w \in H^{n-1}\}$$

$$H^n = H^n_L \cup H^n_R.$$

Let $\langle \Omega, A, P\rangle$ be a probability space. To each node $w \in V$ and each $\omega \in \Omega$ we ascribe $b^o(w,\omega) \in H^o \cup \text{BIN}$. We suppose, that

1. Random variables $b^o(w,.)$ $(w \in V)$ are statistically independent.
2. For each $n \in \{1,2,3,\ldots\}$, random variables $b^o(w,.)$ $(w \in H_L)$ are identically distributed. For each $n \in \{1,2,3,\ldots\}$, random variables $b^o(w,.)$ $(w \in H^n_R)$ are identically distributed.

Now, the main problem is to find or estimate the probability that by procedure described above a formula of PC is generated, i. e. the probability that v^k_s is a formula of PC. This probability is, by (a) and Definition 1, equal to the probability that a pair $\langle H^o, b^o(.,\omega)\rangle$ is generable.

For each $n \in \{1,2,3,\ldots\}$ and $w \in H^n_L$, let

$$p^n_L = P(b^o(w,\omega) \in H^o),$$

for each $n \in \{1,2,3,\ldots\}$ and $w \in H^n_R$, let

$$p^n_R = P(b^o(w,\omega) \in H^o).$$

Moreover, let

$$p^o = P(b^o(s,\omega) \in H^o).$$

Let, for $n = 1,2,3,\ldots$, $V_n = \bigcup^n_{i=0} H^i$. Moreover, for $w \in \bigcup^{n-1}_{i=0} H^i_L$, let $\Gamma_{nL}(w) = \Gamma_L(w)$, for $w \in \bigcup^{n-1}_{i=0} H^i_R$ let $\Gamma_{nR}(w) = \Gamma_R(w)$. For $w \in H^n_L$ let $\Gamma_{nL}(w) = \emptyset$, for $w \in H^n_R$ let $\Gamma_{nR}(w) = \emptyset$. Then $S_n = \langle V_n, \Gamma_{nL}, \Gamma_{nR}\rangle$ is a finite binary subtree of S, where $S = \langle V, \Gamma_L, \Gamma_R\rangle$ is a complete binary tree.

Let $S' = \langle V', \Gamma'_L, \Gamma'_R\rangle$ be a binary tree. A sequence $w_o, w_1, \ldots, w_n$ of nodes is called a path from the root to leaves if $w = S$, for each $i = 0,1,\ldots,n-1$, $w_{i+1} \in \{\Gamma'_L(w_i), \Gamma'_R(w_i)\}$ and w_n is a leaf of S'.

Let $S = \langle V, \Gamma_L, \Gamma_R\rangle$ be a complete binary tree. A sequence of nodes $w_o, w_1, w_2, \ldots$ is called a path from the root if $w_o = s$ and for each $i \in \{0,1,2,\ldots\}$, $w_{i+1} \in \{\Gamma_L(w_i), \Gamma_R(w_i)\}$.

In Šindelář (1984), the following theorem is proved.

Theorem 2. A pair $\langle b^o(.,\omega), H^o \rangle$ is generable iff
(a) for every path $w_0, w_1, w_2, \ldots$ from the root s (in complete binary tree S) there is $i \in \{0,1,2,\ldots\}$ such that $b^o(w_i,\omega) \in H^o$. Δ

Let us denote the event (a) as A. Moreover, for each $n \in \{1,2,3,\ldots\}$, let A_n be the following event: for every path $w_0, w_1, \ldots, w_n$ from the root to leaves (in binary tree S_n) there is $i \in \{0,1,\ldots,n\}$ such that $b^o(w_i,\omega) \in H^o$. Then

$$P(A_n) \nearrow P(A).$$

In Šindelář (1984) the following theorems are proved.

Theorem 3. Let $n \in \{1,2,3,\ldots\}$. Let us consider a system of n+1 equations

$$x_n = p_L^n \cdot p_R^n,$$

for $i \in \{1,2,\ldots,n-1\}$

$$x_j = [p_L^i + (1-p_L^i) \cdot x_{i+1}] \cdot [p_R^i + (1-p_R^i) \cdot x_{i+1}],$$

$$x_0 = p^o + (1-p^o) \cdot x_1.$$

Then

$$P(A_n) = x_0. \qquad \Delta$$

Theorem 4. For each $n \in \{1,2,3,\ldots\}$,

$$P(A) \geq P(A_n) \geq 1 - (1-p^o) \cdot \left[\prod_{i=1}^{n-1} (2-p_L^i \cdot p_R^i) \right] \cdot (1-p_L^n \cdot p_R^n),$$

where for n = 1 we put $\prod_{i=1}^{n-1} (2-p_L^i \cdot p_R^i) = 1$. Δ

A detailed description of the probabilistic algorithm generating formulas of PC and based on the procedure described in this paper is given in Šindelář (1984).

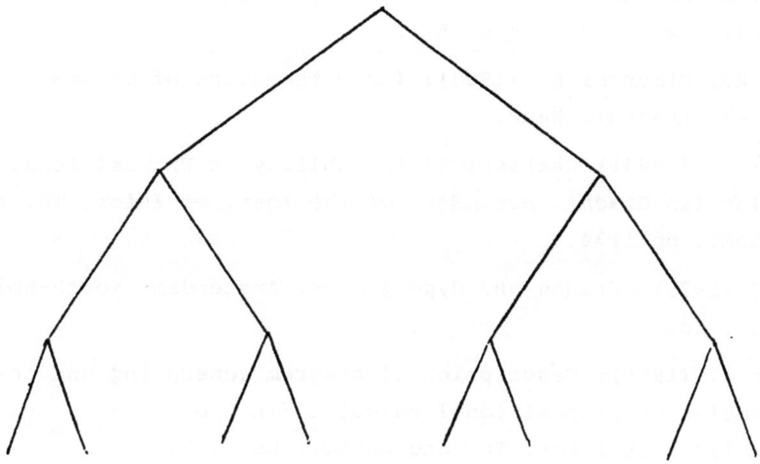

Fig. 1

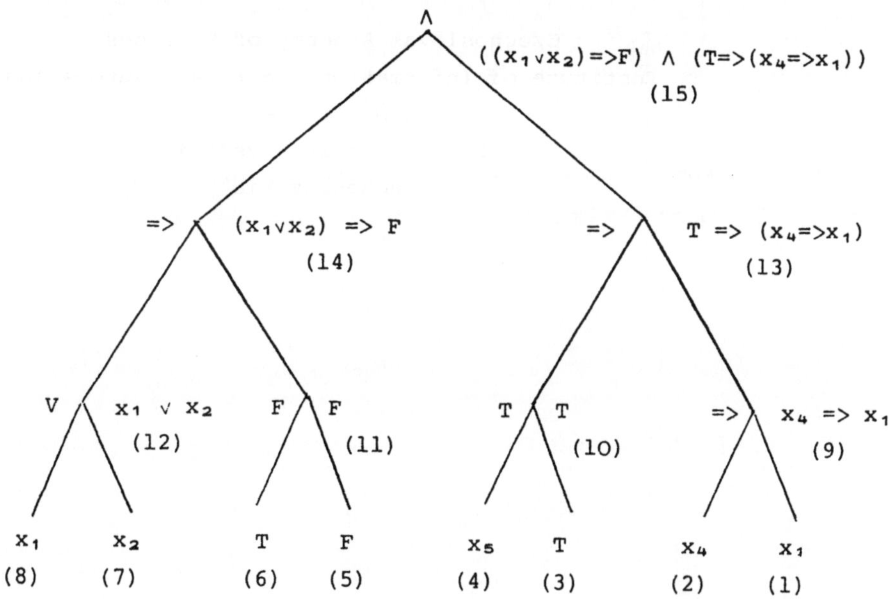
∧
((x₁∨x₂)=>F) ∧ (T=>(x₄=>x₁))
(15)
=>
(x₁∨x₂) => F
(14)
=>
T => (x₄=>x₁)
(13)
V
x₁ ∨ x₂
(12)
F
F
(11)
T
T
(10)
=>
x₄ => x₁
(9)
x₁
(8)
x₂
(7)
T
(6)
F
(5)
x₅
(4)
T
(3)
x₄
(2)
x₁
(1)

Fig. 2

REFERENCES

Bledsoe, W. W.; Loveland, D. W. (Editors)(1984): Automated theorem proving after 25 years. AMS.

Raisowa H., Sikorski R. (1963): The mathematics of metamathematics. Polska Akademia Nauk.

Šindelář J. (1983): Statistical Provability in Propositional Calculus (in Czech). Res. Rep. of the Inst. of Infor. Th. and Autom., no 1174.

Berge C. (1973): Graphs and Hypergraphs. Amsterdam, North-Holland Publ. Co.

Šindelář J. (1984): Description of program generating and testing formulas of propositional calculus (in Czech). Res. Rep. of the Inst. of Infor. Th. and Autom., no. 1311.

Czechoslovak Academy of Sciences
Institute of Information Theory and Automation
182 08 Prague 8
Pod vodárenskou věží 4
Czechoslovakia

SENSITIVE GROWTH ANALYSIS OF CONTROLLED MULTIPLICATIVE SYSTEMS

Karel Sladký

Prague

Key words: Dynamic programming, nonnegative matrices, Markov programming.

ABSTRACT

We consider at discrete time points a dynamic system whose utility vector at time n is a product of n nonnegative matrices (taken from a given finite set) postmultiplied by the initial utility vector. Under certain "product property" of the considered matrix set the maximum utilities (i. e. components of the utility vector) at each time point are computable by a dynamic programming recursion. It can be shown that the utilities at time n are bounded by the n-th power of (nonnegative) characteristic values (of a suitable admissible matrix) possibly multiplied by some polynomials in n. For fixed initial conditions we can also calculate coefficients of these polynomials by policy iterations or linear programming as a solution to some nested functional equations.

1. INTRODUCTION AND NOTATIONS

We shall consider at discrete time points $n=0,1,\dots$ a controlled multiplicative systems with finite state space $E=\{1,\dots,N\}$ whose utility vector at time n is a product of n nonnegative matrices postmultiplied by the nonnegative initial utility vector $x(0)$. The maximum utility vector at time n, denoted $x(n)$, can be calculated by the following dynamic programming recursion

(1) $$x(n+1) = \max_{f\in F} Q(f)\, x(n) = Q(\hat{f}^{(n)})\, x(n), \quad x(0) > 0.$$

In (1) $Q(f)$ is an $(N\times N)$-nonnegative matrix depending on a decision vector $f\in F$, i. e. an N-vector whose i-th component $f(i)$ specifies the i-th row of $Q(f)$ taken from a given finite set of nonnegative N-vectors. Observe that the set of all $Q(f)$'s possesses an important "product property", i. e. if $f_1, f_2 \in F$ then there exists also $f\in F$ such that for each pair $i_1, i_2 \in E$ $[Q(f_1)]_{i_1} = [Q(f)]_{i_1}$, $[Q(f_2)]_{i_2} = [Q(f)]_{i_2}$ (here $[A]_i$ denotes the i-th row of the matrix A), and hence vectorial maximum in (1) always exists. Symbol $\hat{f}^{(n)}$ denotes the maximizer over all $f\in F$ at time n.

Controlled multiplicative systems generalize controlled Markov chains (intensively studied in the literature) and possess many other interesting and useful interpretations (e. g. supervised linear economic models, controlled branching processes and models of social processes). In particular, recursive relation (1) generalizes the functional equations for maximum expected rewards of a controlled Markov chain; more precisely, considering a Markov decision chain with transition probability matrices $P(f)$ and one-stage reward vectors $r(f)$ (both depending on $f\in F$), the vector of maximum expected rewards to be obtained in the n next transitions satisfies (1) for $Q(f) = \begin{bmatrix} P(f) & r(f) \\ 0 & 1 \end{bmatrix}$, $x(n) = \begin{bmatrix} v(n) \\ 1 \end{bmatrix}$. Observe that the N-th element of $x(n)$ is a dummy variable and for $i = 1,\dots,N-1$ the i-th element of $x(n)$ is the value of expected total reward to be obtained in the n next transitions if the chain starts in state i.

In contrast with the total expected rewards of Markov decision chains (that grow linearly in n) the growth of $x(n)$ is given by an exponential as well as a polynomial parts. More precisely, for the i-th component of $x(n)$ we have $[x(n)]_i \approx \sigma^n n^{-\nu+1}$, where the number $\sigma > 0$ is equal to the spectral radius of an appropriate submatrix of some $Q(f)$ and integer $\nu \geq 1$ denotes the number of such (successively accessible) submatrices. These facts as well as the proofs and a more detailed discussion of the results presented in Theorems 1, 2 can be found in Sladký (1980), Zijm (1982), Rothblum and Whittle (1982) and in Chapter 35 of Whittle (1983). A more detailed analysis shows (cf. Sladký (1980), Zijm (1982)) that if $\sigma > 0$ and some aperiodic assumptions of $Q(f)$'s are satisfied, letting $n \to \infty$ σ^{-n} $[x(n)]_i$ converges to some polynomial of degree $\nu - 1$ whose coefficients also depend on initial condition $x(0)$.

In what follows we shall denote by $\sigma(f)$, resp. $\sigma_i(f)$, the spectral radius of $Q(f)$, resp. of its diagonal submatrix $Q_{ii}(f)$, and (cf. the Perron-Frobenius theorem) by $u(f)$, $u_i(f)$ the nonnegative right eigenvector corresponding to $\sigma(f)$, $\sigma_i(f)$ respectively. Supposing that $Q_{ii}(f)$'s are the biggest submatrices of $Q(f)$ having spectral radius $\sigma_i(f)$ and strictly positive right eigenvectors $u_i(f)$, let $\nu_i(f) \geq 1$ denote the number of diagonal submatrices of $Q(f)$ having spectral radius $\sigma_i(f)$ that are accesssible from the class $Q_{ii}(f)$. We write $u(f) > 0$, resp. $u(f) \gg 0$, if all elements of $u(f) \neq 0$ are nonnegative, resp. positive. The symbol I, resp. e, will denote unit matrix, resp. unit vector, of appropriate dimensions.

Now we are in a position to formulate the following important fact on "uniform block-triangular decomposition" of any $Q(f)$ with $f \in F$.

Theorem 1. There exists a (fixed) permutation of rows and corresponding columns and an (independent of $f \in F$ and obtainable by a policy iteration method) block-triangular decomposition of any $Q(f)$ with the following properties:

(i) Any $Q(f)$ with $f \in F$ is upper block-triangular, i. e.

$$(2) \qquad Q(f) = \begin{bmatrix} Q_{11}(f) & Q_{12}(f) & \cdots & Q_{1s}(f) \\ 0 & Q_{22}(f) & \cdots & Q_{2s}(f) \\ & & \ddots & \vdots \\ 0 & 0 & \cdots & Q_{ss}(f) \end{bmatrix}$$

(ii) There exists $\hat{f} \in F$ such that for all $i = 1,\dots,s$ and any $f \in F$

$$(3) \qquad \sigma_i(\hat{f}) \geq \sigma_i(f), \quad \sigma_i(\hat{f}) \geq \sigma_{i+1}(\hat{f}) \quad \text{with}$$

$$(3') \qquad \sigma_i(\hat{f}) = \sigma_{i+1}(\hat{f}) \Rightarrow \nu_i(\hat{f}) = \nu_{i+1}(\hat{f}) + 1, \quad \text{and}$$

$$(4) \qquad \sigma_i(\hat{f})u_i(\hat{f}) = Q_{ii}(\hat{f})\, u_i(\hat{f}) \geq Q_{ii}(f)\, u_i(\hat{f}), \text{ where } u_i(\hat{f}) \gg 0.$$

From now on the same decomposition as in (2) will be considered for any $Q(f)$. Similarly, for any N-dimensional vector, say k, k_i denotes the subvector of k whose elements correspond to the elements of $Q_{ii}(f)$. Applying this "uniform block-triangular dècomposition" in dynamic programming recursion (1) we get for $i = 1,\dots,s$

(5) $$x_i(n+1) = Q_{ii}(\hat{f}^{(n)})\, x_i(n) + \sum_{j=i+1}^{s} Q_{ij}(\hat{f}^{(n)})\, x_j(n).$$

In virtue of (3), (4), (5) we can evaluate the first-order approximation on the growth of $x(n)$. The results are summarized as Theorem 2 (we abbreviate $\sigma_i(f)$, $\nu_i(f)$, $u_i(f)$ by σ_i, ν_i, u_i respectively).

Theorem 2. There exists vectors $k_i \gg 0$ such that if $\sigma_i > 0$

(6) $$x_i(n) \le k_i\, \sigma_i^n\, n^{\nu_i - 1} \qquad \text{for } n = 0,1,\dots .$$

Moreover, if $\sigma_i = 0$ then $x_i(n) \le k_i$ for $n = 0,\dots,\nu_i - 1$, but $x_i(n) = 0$ for all $n \ge \nu_i$.

Throughout the paper we make

Assumption GA. $\sigma_s > 0$ and each $Q_{ii}(f)$ with $\sigma_i(f) = \sigma_i$ $(i = 1,\dots,s)$ is aperiodic (i. e., $\sigma_i > 0$ and $Q^*_{ii}(f) = \lim_{n\to\infty} \sigma_i^{-n}(Q_{ii}(f))^n$ always exists; note that $\sigma_i > \sigma_i(f) \Rightarrow Q^*_{ii}(f) = 0$).

2. POLYNOMIAL BOUNDS ON UTILITIES

In what follows we shall investigate polynomial bounds on $\{x_i(n)\}$. Introducing for any $j = i,\dots,s$ $\bar{x}_j(n) = \sigma_i^{-n} x_j(n)$ and $\bar{Q}_{ij}(.) = \sigma_i^{-n} Q_{ij}(.)$ by our definition of the index (cf. (4), (4')) we get

$$\sigma_j = \sigma_i, \quad \nu_j = \nu_i + i - j \qquad \text{for } j = i,\dots,i+\nu_i-1$$

$$\sigma_j < \sigma_i \qquad \text{for any } j \ge i+\nu_i.$$

Hence by (5), (6) we immediately get

(7) $$\bar{x}_i(n+1) = \sum_{j=i}^{i+\nu_i-1} \bar{Q}_{ij}(\hat{f}^{(n)})\, \bar{x}_j(n) + o_i(\hat{f}^{(n)};n)$$

where $o_i(.;n) \to 0$ as $n \to \infty$ and the convergence is geometrical, i. e.

(7') $$c_i'\, \lambda^n \le o_i(f;n) \le c_i''\, \lambda^n$$

for some vectors $c_i' \ll 0$, $c_i'' \gg 0$, number $\lambda \in (0,1)$, and any $f \in F$. In case that $\nu_i = 1$ $\{\bar{x}_i(n)\}$ is bounded by Theorem 2, and under Assumption GA (aperiodicity of $Q_{ii}(f)$'s) the following stronger property holds (cf. Sladký (1979), Zijm (1982)).

Theorem 3. Let $\nu_i = 1$. Then $\bar{x}_i(n) \to x_i$ as $n \to \infty$ and the convergence is geometrical.

Similarly, if $\nu_i > 1$ by Theorem 2 $\{n^{-\nu+1}\bar{x}_i(n)\}$ is bounded for $\nu = \nu_i$, and also the following stronger result holds (cf. Sladký (1981) for details).

Theorem 4. Let $\nu_i > 1$ and for $j = i+1,\dots,i+\nu_i-1$ the (vector) coefficients a_{jk} (with $k = 0,\dots,\nu_j-1$) be selected such that

$$y_j(n) = \bar{x}_j(n) - \sum_{k=0}^{\nu_j-1} \binom{n}{k} a_{jk} \to 0 \quad \text{as } n \to \infty \tag{8}$$

and the convergence is geometrical. Then there exist (computable) a_{ik}'s ($k=\nu_i-1,\dots,1$) such that $\{y_i(n)\}$ (defined in (8)) is bounded.

Remark 1. Under the aperiodicity condition on $Q_{ii}(f)$'s (cf. Assumption GA) it can be shown that for suitably selected a_{io} also $y_i(n) \to 0$ and the convergence is geometrical (i. e. (8) holds for $i = j$). Hence Assumption GA guarantees convergence of any $\bar{x}_i(n)$ to some polynomial of degree ν_i-1.

Remark 2. In case that $Q(f) = \begin{bmatrix} P(f) & r(f) \\ 1 & 0 \end{bmatrix}$ where $P(f)$ is a transition probability matrix (i. e. $P(f)\, e = e$) and $r(f) > 0$ is a vector of one-stage rewards of a Markov decision chain, $\sigma_1 = \sigma_2 = 1$, $\nu_1 = 1$ and, obviously, $Q_{11}(f) = P(f)$; $Q_{12}(f) = r(f)$ and the vector coefficient a_{11} of the bounding polynomial is the vector of maximum average rewards of the considered Markov decision chain.

To show how to construct the vector coefficients a_{ik}, define (for yet unspecified a_{ik}'s) the vectors $b_{ik}(f)$ by

$$\begin{aligned} b_{ik}(f) &= (\bar{Q}_{ii}(f) - I)\, a_{ik} && \text{for } k=\nu_i-1, \quad \text{and} \\ b_{ik}(f) &= (\bar{Q}_{ii}(f) - I)\, a_{ik} - a_{i,k+1} + r_{ik}(f) && \text{for } k=\nu_i-2,\dots,0 \end{aligned} \tag{9}$$

where $r_{ik}(f) = \sum_{j=i+1}^{i+\nu_i-k-1} \bar{Q}_{ij}(f)\, a_{jk}$ with a_{jk}'s selected such that condition (8) holds for $j = i+1,\dots,i+\nu_i-1$. In Sladký (1981) it was shown (constructively by the policy iteration method) that a_{ik}'s can be selected such that the matrix

$$\begin{aligned} B_i(f) &= \left[b_{i,\nu_i-1}(f),\dots,b_{io}(f)\right] \preceq 0 && \text{for any } f \in F \\ B_i(f) &= 0 && \text{at least for one } f = f^* \end{aligned} \tag{9'}$$

(the symbol $\preceq$ means lexicographically nonpositive).
Inserting from (8) into (7) after some manipulations we have

$$(10)\quad y_i(n+1) = \sum_{j=i}^{i+\nu_i-1} \bar{Q}_{ij}(\hat{f}^{(n)})\, y_j(n) + \sum_{k=0}^{\nu_i-1} \binom{n}{k} b_{ik}(\hat{f}^{(n)}) + o_i(n).$$

Hence if (9) holds the second term on the right-hand side of (10) is non-positive for sufficiently large n, and as for $j > i$ $y_j(n) \to 0$ geometrically, we immediately get $y_i(n+1) \le \bar{Q}_{ii}(\hat{f}^{(n)})\, y_i(n) + o_i(n)$ (for $n > n_o$); similarly, $y_i(n+1) \ge \bar{Q}_{ii}(f^*)\, y_i(n) + o_i(n)$ where $o_i(n) \to 0$ geometrically. Iterating the above inequalities we get

Theorem 5. Let the assumptions of Theorem 4 be fulfilled and a_{ik}'s (with $k = \nu_i-1,\dots,1$) satisfy (9), (9'). Then $\{y_i(n)\}$ (defined by the first part of (8)) is bounded.

3. COMPUTING OF POLYNOMIAL BOUNDS

First we show that equation (9) fulfilling condition (9') will also arise if sensitive discount optimal policies are considered for a Markov decision chain whose one-stage rewards are polynomials in interest rate ρ. To this order let us consider a Markov decision chain with transition rate matrices $P(f) = \bar{Q}_{ii}(f)$ and one-stage reward vectors $r^\rho(f) = \sum_{k=0}^{\nu-2} \rho^k r_k(f)$, where $r_k(f) = r_{i,\nu-2-k}(f)$ (for $k = 0,\dots,\nu-2$), $r_k(f) = 0$ (for $k > \nu-2$), and $\rho = (1-\alpha)/\alpha$ is the interest rate associated with discount factor α. Observe that the matrix $P(f)$ need not be stochastic, e. g. its row sum can exceed 1, but (cf. (4)) using the similarity transformation given by the non-singular matrix $U = \operatorname{diag} u_i$, $U\,P(f)\,U^{-1}$ is (sub)-stochastic.

The vector $v^\alpha(f)$ of α-discounted rewards if stationary policy $\pi = (f)$ is followed, i. e. $v^\alpha(f) = \sum_{n=0}^{\infty} (\alpha P(f))^n r^\rho(f)$, has the Laurent expansion (cf. Miller, Veinott (1969) for the case $r^\rho(f) = r_o(f)$)

$$(11)\qquad v^\alpha(f) = (I - \alpha P(f))^{-1}\, r^\rho(f) = (\rho+1) \sum_{m=-1}^{\infty} \rho^m u_m(f)$$

where

$$(11')\qquad u_m(f) = -\sum_{k=0}^{m} (-H(f))^{k+1}\, r_{m-k}(f) + P^*(f)\, r_{m+1}(f).$$

Recall that (we omit the argument f in $P(f)$, $H(f)$ etc.) $(I-\alpha P)^{-1} = (\rho+1)\left[\rho^{-1}P^* - \sum_{m=0}^{\infty} \rho^m (-H)^{m+1}\right]$, the deviation matrix $H = (I-P+P^*)^{-1} - P^*$ where for P^* (the Césaro limit of P^n) under

Assumption GA $P^n \to P^*$ as $n \to \infty$. Moreover, $u_m(f)$'s can be calculated as a solution to the set of equations

(12)
$$(P(f) - I)\, u_{-1}(f) = 0, \quad \text{and for } m = 0,1,\ldots$$
$$(P(f) - I)\, u_m(f) - u_{m-1}(f) + r_m(f) = 0$$

and $[u_{-1},\ldots,u_m]$ maximizing lexicographically $[u_{-1}(f),\ldots,u_m(f)]$ can be found (together with the corresponding decision vector $f_m^* \in F$) by policy iterations as $u_i(f)$'s such that (cf. (9), (9'))

(13)
$$0 = B_{m+1}(f_m^*) \succeq B_{m+1}(f) = [b_{-1}(f), b_o(f),\ldots,b_{m+1}(f)]$$

where for any $f \in F$

(13')
$$b_{-1}(f) = (P(f) - I)\, u_{-1}, \quad \text{and for } m = 0,1,\ldots$$
$$b_m(f) = (P(f) - I)\, u_m - u_{m-1} + r_m(f).$$

To formulate finding of the decision vector f_m^* as a linear programming problem, first observe that for any N-dimensional row vector $\beta = [\beta_1,\ldots,\beta_N]$ such that $\beta_j > 0$, $\sum_j \beta_j = 1$, and any $f \in F$

(14)
$$v^\alpha(f_\alpha) \ge v^\alpha(f) \qquad \text{iff} \qquad v(\alpha,\beta,f_\alpha) \ge v(\alpha,\beta,f)$$

where $v(\alpha,\beta,f) = z^\rho(f)\, r^\rho(f)$ with $z^\rho(f) = \beta(I-\alpha P(f))^{-1}$ (i. e., $v(\alpha,\beta,f)$ is the value of total expected α-discounted reward if the stationary policy $\pi = (f)$ is followed and the initial distribution is given by β).

Moreover, we shall denote by $\tilde{F}$ (with generic $\tilde{f}$) the set of all randomized decision vectors, i. e. the rules selecting in state $i \in E$ action $k = \{1,\ldots,K\}$ with given probability $d(i,k;\tilde{f}) \ge 0$ (obviously, $\sum_k d(i,k;f) = 1$). The (stationary) policy maximizing $v(\alpha,\beta,f)$ can be found as $f \in \tilde{F}$ such that

(P)
$$\max_{f \in \tilde{F}} z^\rho(f)\, r^\rho(f) \quad \text{s.t.} \quad z^\rho(f)\,[I-\alpha P(f)] = \beta$$

Denoting by $z^\rho(i;k) = [z^\rho(f)]_i\, d(i,k;\tilde{f})$ (similarly $p(i,j;k)$ is reserved for the transition probability from state i to state j if action k is selected, and $r^\rho(i;k)$ for the respective one-stage reward) problem (P) can be explicitly written as the following linear program (δ_{ij} is the Kronecker symbol)

$$\text{(P') s.t.} \quad \max \sum_{i=1}^{N} \sum_{k=1}^{K} z^{\rho}(i;k)\; r^{\rho}(i;k)$$
$$\sum_{i=1}^{N} \sum_{k=1}^{K} z^{\rho}(i;k)[\delta_{ij} - p(i,j;k)] = \beta_j \qquad (j = 1,\dots,N)$$

Letting $\alpha \to 1$ (or equivalently $\rho \to 0$) by the Laurent expansion of $(I-\alpha P(f))^{-1}$ we get

$$z^{\rho}(f) = (\rho+1) \left[\rho^{-1}\, x(f) + \sum_{m=0}^{\infty} \rho^{m}\, y_m(f)\right] \tag{15}$$

where

$$x(f) = \beta P^{*}(f), \quad \text{and for } m = 0,1,\dots$$
$$y_m(f) = -\beta(-H(f))^{m+1}. \tag{15'}$$

Inserting (15) into the constraint of problem (P) we get

$$\left[\rho^{-1}\, x(f) + \sum_{m=0}^{\infty} \rho^{m}\, y_m(f)\right] \left[I - P(f) + \rho I\right] = \beta$$

and hence, by comparing terms of the same power of ρ, we conclude that

$$x(f)\,[I - P(f)] = 0$$
$$x(f) + y_0(f)\,[I - P(f)] = \beta \tag{16}$$

and for $m = 1,2,\dots$

$$y_{m-1}(f) + y_0(f)\,[I - P(f)] = 0. \tag{16'}$$

Similarly, for the Laurent expansion of $v(\alpha,\beta,f)$ we have

$$v(\alpha,\beta,f) = (\rho+1) \sum_{m=-1}^{\infty} \rho^{m} \beta u_m(f) \tag{17}$$

where

$$\beta\, u_{-1}(f) = x(f)\; r_0(f), \quad \text{and for } m = 0,1,\dots$$
$$\beta\, u_m(f) = \sum_{k=0}^{m} y_k(f)\; r_{m-k}(f) + x(f)\; r_{m+1}(f). \tag{17'}$$

Hence finding stationary policy $\pi_{-1} = (f^{*}_{-1})$ maximizing $u_{-1}(f)$ over all $f \in \tilde{F}$ can be formulated as

$$\max_{f \in \tilde{F}} x(f)\; r_0(f)$$
$$(P_{-1}) \qquad \text{s.t.} \qquad x(f)\,[I - P(f)] = 0$$
$$x(f) + y_0(f)\,[I - P(f)] = \beta$$
$$y_0(f) + y_1(f)\,[I - P(f)] = 0$$

Similarly, policy $\pi_o = (f_o^*)$ maximizing lexicographically the pair $[u_{-1}(f), u_o(f)]$ can be found as a solution to

$$(P_o) \quad \text{s.t.} \quad \begin{array}{l} \max\limits_{f \in \tilde{F}} [y_o(f)\ r_o(f) + x(f)\ r_1(f)] \\ x(f)\ r_o(f) = x(f_{-1}^*)\ r_o(f_{-1}^*) \\ \text{and the constraints of } (P_{-1}) \end{array}$$

Repeating this reasoning we can also formulate problem (P_m) for finding decision vector $f_m^* \in F$ maximizing lexicographically $[u_{-1}(f),\dots,u_m(f)]$.

Remark. Observe that for any $f \in \tilde{F}$ $x(f) \geq 0$, $[x(f),y(f)] \succ 0$ (note that in $(P_{-1}) \in y_o(f)$ is unique for the selected f F). Ignoring the last equation in the constraints of (P_{-1}) we must add condition $y(f) \geq 0$ to guarantee that $[x(f),y(f)] \succ 0$. Introducing (cf. problems (P), (P')) $x(i;k) = [x(f)]_i\ d(i,k;f)$, $y(i;k) = [y(f)]_i \times d(i,k;f)$, we can write (P_{-1}) explicitly as a linear programming problem. Having found the solution of (P_{-1}), we can choose $f_{-1}(i) = k$ if $[x(i;k), y(i;k)] \succ 0$. For a more detailed analysis of linear programming methods in classical Markov decision chains see Denardo, Fox (1968), Derman (1970) and Hordijk, Kallenberg (1979), as well as Denardo (1970) for discussion on maximizing lexicographically $[u_{-1}(f), u_o(f)]$ in a classical Markov decision chain.

REFERENCES

Denardo E. V. (1970): Computing a bias-optimal policy in a discrete-time Markov decision problem. Oper. Res. 18, No. 2, 279-289.

Denardo E. V. and Fox B. L. (1968): Multichain Markov renewal programs. SIAM J. Appl. Math. 16, No. 3, 468-487.

Derman C. (1970): Finite State Markovian Decision Processes. Academic Press, New York.

Hordijk A. and Kallenberg L. C. M. (1979): Linear programming and Markov decision chains. Manag. Sci. 25, No. 4, 352-362.

Miller B. L. and Veinott A. F. (1969): Discrete dynamic programming with a small interest rate, Ann. Math. Statist. 40, No. 2, 366-370.

Rothblum U. G. and Whittle P. (1982): Growth optimality for branching Markov decision chains. Mathem. Oper. Res. 7, No. 4, 582-601.

Sladky K. (1979): Successive approximation methods for dynamic programming models. In: Proc. of the Third Formator Symposium, Academia, Prague.

——— (1980): Bounds on discrete dynamic programming recursions I, models with nonnegative matrices. Kybernetika 16, No. 2, 526-547.

——— (1981): Bounds on discrete dynamic programming recursions II, polynomial bounds on problems with block-triangular structure. Kybernetika 17, No. 4, 310-328.

Whittle P. (1983): Optimization over Time - Dynamic Programming and Stochastic Control, Vol. II. Wiley, Interscience, London - New York.

Zijm W. H. M. (1982): Nonnegative Matrices in Dynamic Programming. Mathematical Centre Tract, Amsterdam.

Czechoslovak Academy of Sciences
Institute of Information Theory and Automation
Pod vodárenskou věží 4
182 08 Praha 8 - Libeň
Czechoslovakia

RIGHTS AND SIMILARITY

OF INDIVIDUALS

Honorata Sosnowska

Warsaw

Key words: group decision making, theory of rights.

ABSTRACT

We investigate the problem whether there exists respecting rights choice function if all individuals are similar. We show, that, generally, similarity of individuals does not imply the existence of respecting rights choice function. We also show, that for a set of alternatives consisting of two elements, similarity of individuals implies the existence of such a function.

TERMS AND NOTATION

For a set X, $\#X$ denotes the cardinality of X. We consider binary relations P,Q,R defined on a nonempty set X. A relation $Q \subset X^2$ is identified with a directed graph having the set of vertices X and the set of edges Q. We say that a relation Q has a cycle if there exists a sequence $C = (c_o, c_1, \ldots, c_k) \in X^{k+1}$ such that $c_i Q c_{i+1}$ for $i = 0,\ldots,k-1$ and $c_o = c_k$. A relation having no cycles is called acyclical. A transposition of the set $\{1,\ldots,n\}$ defines a bijection from X^n to X^n:

$$t_{ij}(x_1,\ldots,x_i,\ldots,x_j,\ldots,x_n) = (x_1,\ldots,x_j,\ldots,x_i,\ldots,x_n).$$

COMMON CHOICE RESPECTING RIGHTS

We descibe the problem in the following way. There are n individuals (members of society for example). They are labelled by numbers 1,...,n. They have a common set of alternatives $X \neq \emptyset$. Each individual i has his preference relation $P_i \subset X^2$. We assume that preference relations are acyclical. For $a,b \in X$ aP_ib means that individual i prefers b than a. A mapping $W : 2^X \to 2^X$ such that for every nonempty and finite $A \subset X$, $\emptyset \neq W(A) \subset A$ is called common choice function. The set $W(A)$ can be treated as a set of the best alternatives chosen from A by a common choice. Usually, connections between the individual preferences and common choice function are wanted. We follow Gibbard (1974) and use a concept of a common choice function respecting rights. Each individual has his rights relation $U_i \subset X^2$, which describes those personal matters which ought to be reflected by a common choice. We say that a common choice function W respects rights if aP_ib and aU_ib together imply that $a \notin W(A)$ for all $A \subset X$ such that $a,b \in A$.
The above condition can be interpreted as follows: If i prefers b than a and has a right to choose b, than a common choice does not prefer a than b. In what follows the relation $\bigcup_{i=1}^{n} (P_i \cap U_i)$ is denoted for short by V. It is easy to check.

Lemma 1. There exists a common choice function respecting rights iff the relation V is acyclical.

Gibbard (1974) and Hammond (1982) consider special form of the rights relation in the case if the set of alternatives is a Cartesian product, $X = X_1 \times \ldots \times X_n$ (the number of factors is equal to the number of individuals). X_i is interpreted as the set of personal features of an individual i. Alternatives are n-tuples, $a = (a_1,\ldots,a_n) \in X$. The rights relations U_i are called strongly independent of

$$aU_ib \Leftrightarrow (a_1 = b_1) \wedge \ldots \wedge (a_{i-1} = b_{i-1}) \wedge (a_{i+1} = b_{i+1}) \wedge \ldots \wedge (a_n = b_n).$$

Under strong independence, individual i is decisive over all issues affecting just his component of X. The following example shows that a respecting rights common choice function may not exist, even if the rights are strongly independent.

Example 1. We consider two individuals. The set of alternatives can be treated as subset of the plane. The x_1-axis is the component of the individual 1, the x_2-axis of the individual 2. Horizontal lines represent the relation U_1, vertical U_2. If the horizontal (vertical) arrow is directed from a to b, it means that $bP_1 \cap U_1a$ $(bP_2 \cap U_2a)$. Alternatives are marked by points. The preference relations and rights are given in the Fig.1. The relation V consists of a cycle. The example is essentialy due to Gibbard.

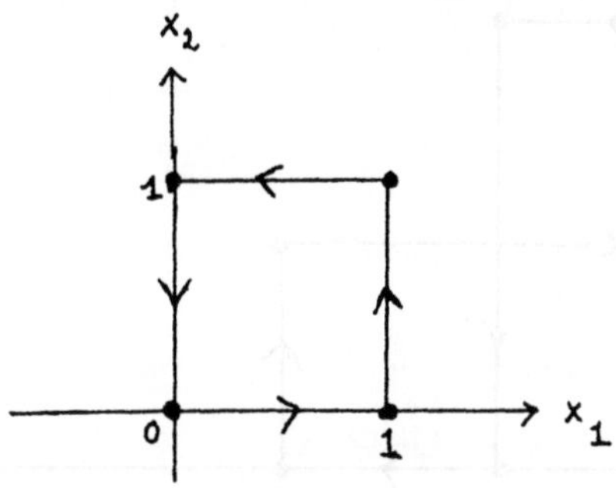

Fig.1.

SIMILARITY OF INDIVIDUALS

We shall consider strongly independent rights and a set of alternatives $X = X_o^n$. Let us note, that if xU_iy then $t_{ij}(x)U_jt_{ij}(y)$. So, transposition t_{ij} is an isomorphism between relations U_i and U_j. We shall say that individuals i and j are similar if transposition t_{ij} is an isomorphism between preference relations P_i and P_j (i.e. if xP_iy then $t_{ij}(x)P_jt_{ij}(y)$. Such similarity was considered by Hildrecht (1953) who studied another problem. Intuitionally, i and j are similar if an alternative x means for i the same as $t_{ij}(x)$ for j. Isomorphism transforms individual i into individual j. Let us see that if individuals are identical $(P_i=P_j=P,\ U_i=U_j=U)$, then they are similar. In this case $V = P \cap U \subset P$ is acyclical since P is acyclical. Now, we shal try to answer the question whether there exists a respecting rights common choice function in the case of all individuals similar. By Lemma 1 this question is equivalent to the following one: Is the relation V acyclical if all individuals are similar ? The following example shows that the answer is negative.

Example 2. We consider two individuals and use the same notation as in Example 1. The preference relations and rights are given in the Fig.2.

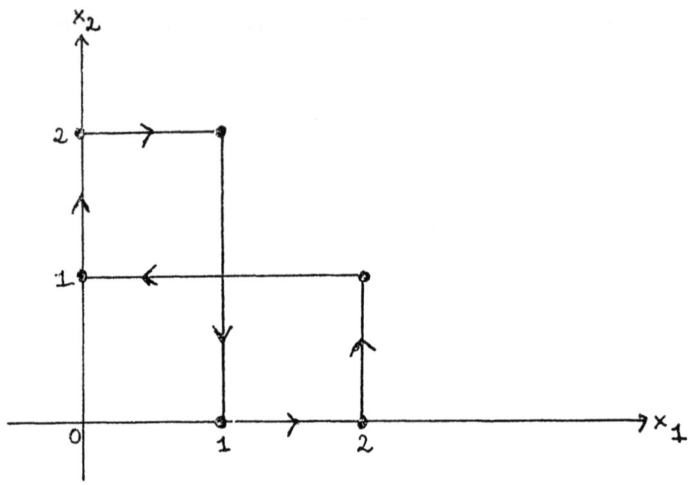

Fig.2

Relation V consists of a cycle. Individuals are similar because the line $x_1 = x_2$ is the symmetry axis of the diagram of V.

There is a case in which similarity implies the acyclicity of V. The following theorem holds.

Theorem 1. Let $X = X_o^n$. If $\#X_o = 2$ and all individuals are similar, then the relation V is acyclical.

Proof is complicated and is not included for editorial reasons. The full theorem is available from the author on request. The assumptions $\#X_o = 2$ is essential (see Example 2).

REFRENCES

Gibbart A.(1974): A Pareto consistent liberation claim. J. of

Economic Th. 7, 388-410.
Hammond P.J.(1982): Liberalism, independent rights and Pareto principle. In: Logic Mathodology and Philosophy of Science VI, North Holland, 607-620.
Hildrecht C.(1953): Alternative conditions for social orderings. Econometrica 2, 81-94.

Central School of
Planning and Statistics
Institute of Econometrics
Al.Niepodległości 162
Poland

STADIA

A STATISTICAL DIALOGUE SYSTEM

Jan Špitálský, Miloslav Vošvrda, Algimantas Senkus

Prague, Vilnius

Key words: Interactive systems, statistical analysis and modelling, robust estimators, testing of hypothesis.

ABSTRACT

In this paper is described the STADIA, the interactive system whose programme equipment is intent on solving problems from the areas of mathematical statistics and statistical analysis of stochastic processes. The dialogue is based on choice of a row in the hierarchical structure of commands that are coupled with relevant required system activity and on diagnostic and informative messages, which are given to user by the system in good time.

The STADIA contents available forms of input data structures and output results of statistical evaluation and inference, linear space arithmetic, different estimators of a distribution function and its parameters, different estimators of a spectral density, estimators of coefficients in multiple linear regression analysis, usually used tests of hypothesis and means for construction of models.

The STADIA is a dialogue system whose programme equipment is focused on solving of tasks of mathematical statistics or tasks of statistical analysis of univariate random processes. A solving belongs both to a statistical inference and to a modelling of random events. User having been allowed both to formulate task and to interprete obtained results may be able to use the system. That system does not suppose any programme knowledges for its using.

User's activity provides with the system by the dialogue. The aim is compiling of the command /or commands/ for solving user's task. A command constitutes one branch in a hierarchical structure. The hierarchical structure is consequently offered to user on computer display. One command corresponds to one alternative activity of the system. A base of the dialogue is the hierarchical structure with diagnostic and different informative messages, which are yield to user on time.

Any command is constituted by a sequence of key parts and key words. These key parts and key words are building-stones of the hierarchical command structure.

We shall not go in to very detailed forming of commands. At this level, we are sufficiently telling to be naturally.

The STADIA is constituted both the superviser system, the user's data space for deposite data and the system of files, which is formed by system, user, functional and auxiliary files. The system files contain commands both working which they are facilitating user's work with system and functional which they assure input/output operations and computing operations at all. Also, there is the list of diagnosis and the informative messages here. The user's files contain both the list of users with their security code and the list of data labels. The auxiliary files are used to easy depositing or to erasing files or to compiling of STADIA.

The functional files are constituted by segments, modules and programme units for a computation. The system is partially segmented to make unified adressing and to economize using of a operating memory. If a user launches any command it will be activated available segment and in this one available module. The moduls safeguard STADIA from termination by the following reasons:

a/ fail in a logical rule,

b/ uncorrect text and/or digital variable in STADIA standard form.

In this case, each module issues message. Further, any module assures formal connection of a data organization among STADIA and programme units. In this way, we are able quickly to carry out connection of different prepared subroutines to STADIA. Each module executes an activation of programme units for solving user's task.

Programme units assure system execution activity i.e. an input of data structures, an estimating, a testing, a modelling and an output of data structures. These activities are started by following key words:

a/ READ (for an input)

b/ COMPUTE (for an estimating and a modelling)

c/ TEST (for a testing)

d/ WRITE (for an output).

The key word <u>READ</u> assures an input of data structure both of a keybord and of a disc and of different transformations of a realization from a random numbers generator. STADIA enables to a user to read of data structures of a disc both integer digits and real digits, both formatted and unformatted, both direct access and sequential access. An input to STADIA of a generator of sequences of random numbers is assured with following possibilities of realizations of random variable: uniform, normal, lognormal, exponential, Weibull, χ^2, gamma, Student, Rayleigh, Rice, logistic, Cauchy, Tukey's λ-family, Johnson's system, binomical, Poisson, geometrical.

The key word <u>COMPUTE</u> assures an estimating of location and scale parameters, an estimating of density both a distribution function and a spectral function, an estimating of regression coefficients, an estimating of parameters of a model, a forming of arithmetical operations over files and a computing of critical values.

An estimating of a location parameter may be executed by

- a sample mean
- Takeuchi's adaptive location estimator

- Jackel's estimator
- an adaptive minimum divergence estimator
- a sample median.

An estimating of a scale parameter may be executed by
- a sample mean square
- an adaptive minimum divergence estimator
- Downton's estimator.

An estimating of a covariance structure may be executed by a sample covariance structure.

An estimating of a density function has the following framework:

- density function estimator
  - of a distribution function by
    - a histogram
    - a Epanechnikov's kernel
    - a robust estimator
  - of a spectral function by
    - an estimator of Blackman-Tukey type
      - triangle window
      - Hamming's window
      - Kaiser's window
      - Black.-Har.'s window
      - Parzen's window
      - Riesz's window
      - Poisson's window
    - a maximum entropy estimator
    - an autoregressive estimator
    - Prony's estimator for
      - a line spectrum
      - a continue spectrum

An estimating of regression coefficients may be executed both by methods of a linear regression and by a method of regression quantile (Bassett-Koenker's estimator).

An estimating of model parameters may be executed by hidden periodicities method (Fisher's test).

STADIA contains following arithmetical operations over files:
- a sum of two files
- a product by a scalar
- an inversion of files (provided that it exists)
- an inner product
- a shift of vector
- a collect of vectors
- a discollect of vectors.

A computing of critical values may be executed in an empirical distribution by a sample quantile and in a theoretical distribution for normal, χ^2, Student's one.

The key-word TEST assures a testing of hypothesis:
- about an empirical distribution by Kolmogorov-Smirnov's test
- outlier test
- Bartlett's test for a significance of a difference between variances
- about a significance of correlation coefficient by Spearmen's, Kendall's test and t-test for normal sample.

The key-word WRITE assures an output of data structure both on a digital form and on a graphic form. A digital output may be carried out either on a printer or terminal display and graphic output may be carried out either on a graphic display or by Graphic Function System'82 on a graphic printer, on a x-y plotter, on a TV display.

STADIA is now implemented on a minicomputer ADT 4500/operating system DOS IV/. STADIA has been programmed in FORTRAN 4X/i.e. HP version of FORTRAN 77/. STADIA has been supported by extent statistical programme package made in the Institute of Information Theory and Automation mainly. The framework of STADIA makes it possible to extend it (but and to enlarge it) in a simple and quick way. User's activity with the system is efficient and the dialoque is telling and easy. That system may be transfered on-to different computer systems with FORTRAN'77 translator only and with an operating memory outnumbering 57KB including.

REFERENCES

Havlová H., Špitálský J., Vošvrda M. (1986): User's Manual STADIA. Publication of the Institution of Information Theory and Automation.

Senkus A.(1986): Collection of Programms of Statistical Analysis and Random series modelling STADIA - system part. Publication of the Institution of Mathematics and Cybernetics.

Institution of Information Theory
and Automation,
Czechoslovak Academy of Science
Pod vodárenskou věží 4
182 08 Prague 8
Czechoslovakia

Institution of Mathematics and
Cybernetics,
Lith. Academy of Science
K. Poželos 54
232600 Vilnius
USSR

ON THE COMPUTATION OF HAUSDORFF DIMENSION AND ITS RELATION TO CHANNEL CAPACITY

Ludwig Staiger

Berlin

Key words: Hausdorff dimension, channel capacity, Minkowski dimension

ABSTRACT

The paper derives combinatorial conditions on subsets of the unit interval which guarantee that their Hausdorff dimension equals the channel capacity (or Minkowski dimension) of their corresponding sets of r-adic expansions.

INTRODUCTION

Many important spatial patterns of noture are irregular in such a degree that "smooth" models are not suitable for describing their form. As appropriate means for solving this problem it turned out to consider so-called fractals (Mandelbrot, 1977), i. e. sets whose topological dimension is smaller than their Hausdorff dimension. In this respect it is of interest to derive methods for calculating the Hausdorff dimension of sets.
A great deal of this work was done by looking for a connection between dimension theory and information theory by making infinite sequences of symbols correspond to r-adic expansions of points of the unit interval (cf. Billingsley, 1965).
For practical purposes it is sometimes easier to estimate an approximation of the Hausdorff dimension of a set: its so-called Minkowski dimension. The latter being nothing else than Shannon's channel capacity for the corresponding set of r-adic expansions. Here we develop several sufficient conditions under which Hausdorff dimension and lower or upper Minkowski dimension of sets of r-adic expansions coincide. We consider the space X^{ω} of infinite sequences on a finite alphabet X of cardinality $\#X = r$. These sequences may be considered as r-adic expansions of real numbers in the unit interval, and thus our results develop-

ed in X^ω may be also applied to point sets in $[0,1]$ (neglecting the ambiguity in the countably many points of the form ir^{-j}). By X^* we denote the set of finite strings (words) on X including the empty word e. For $w \in X^*$ and $b \in X^* \cup X^\omega$, let wb be their concatenation, and $|w|$ denotes the length of the word w ($|e| = 0$). Moreover, $w \sqsubseteq b$ means that w is an initial segment (prefix) of b.

For a set $B \subseteq X^* \cup X^\omega$ we call $B/w := \{b : wb \in B\}$ the <u>state</u> of B derived by the word $w \in X^*$, and we call $B \subseteq X^* \cup X^\omega$ <u>finite-state</u> provided $\{B/w : w \in X^*\}$ is a finite set.

We consider X^ω as a metric space with the metric ρ defined by

$$\rho(\beta, \xi) := \inf\{r^{-|w|} : w \sqsubseteq \beta \ \& \ w \sqsubseteq \xi\}$$

(where $r = \#X$). This metric ρ satisfies the ultrametric inequality

$$\rho(\beta, \xi) \leq \max\{\rho(\beta, \eta), \rho(\xi, \eta)\} ,$$

and, since X is finite, (X^ω, ρ) is a compact space homeomorphic to the Cantor discontinuum. The open (and simultaneously closed) balls in (X^ω, ρ) are the sets $w \cdot X^\omega := \{\beta : w \sqsubseteq \beta\}$ $(w \in X^*)$.

COVERINGS AND HAUSDORFF DIMENSION

An <u>r^{-n}-covering</u> of a set $F \subseteq X^\omega$ is a family $(v \cdot X^\omega)_{v \in V}$ of balls satisfying $\bigcup_{v \in V} v \cdot X^\omega \supseteq F$ and whose diameters diam $(v \cdot X^\omega)$ do not exceed r^{-n}, i. e. $|v| \geq n$. In our investigations we can confine ourselves to minimal coverings $(v \cdot X^\omega)_{v \in V}$ of F, i. e. to those coverings all of whose proper subfamilies are not coverings of F.

Since $v \sqsubseteq w$ implies $v \cdot X^\omega \supseteq w \cdot X^\omega$ (and vice versa), the index set V of a minimal covering is <u>prefix-free</u>, i. e. V does not contain words $v \neq w$ such that $v \sqsubseteq w$. Moreover, from the equivalence $F/w = \emptyset$ iff $w \cdot X^\omega \cap F = \emptyset$ we obtain that the index set V of a minimal covering does not contain words w with $F/w = \emptyset$.

Consequently, a set $V \subseteq X^*$ is an index set of a minimal covering of a set F iff V is prefix-free, $V \subseteq \{w : F/w \neq \emptyset\}$, and for every $\beta \in F$ there is a $v \in V$ such that $v \sqsubseteq \beta$. In what follows we shall abbreviate this fact by $V \triangleleft F$.

Let for $W \subseteq X^*$: $\underline{l}(W) := \inf\{|w| : w \in W\}$, and

$\overline{l}(W) := \sup\{|w| : w \in W\}$.

From diam $(v \cdot X^\omega) = r^{-|v|}$ we obtain that V is the index set of an r^{-n}-covering if and only if $\underline{l}(V) \geq n$.

According to Billingsley (1965), Section 14, the Hausdorff dimension (Hausdorff, 1919) of a set $F \subseteq X^{\omega}$ is defined in the following way:

Let $L_{\alpha}(F, V) := \sum_{v \in V} (\operatorname{diam} v \cdot X^{\omega})^{\alpha} = \sum_{v \in V} r^{-\alpha \cdot |v|}$

when $V \lhd F$ and $0 \leq \alpha \leq 1$.

Then $L_{\alpha}(F) := \liminf_{n \to \infty} \{ L_{\alpha}(F, V) : V \lhd F \ \& \ \underline{l}(V) \geq n \}$ is the α-dimensional outer measure of F.

Consider $L_{\alpha}(F)$ as a function of α. Then there is an $\alpha_0 \in [0,1]$ such that $L_{\alpha}(F) = \infty$ if $\alpha < \alpha_0$, and $L_{\alpha}(F) = 0$ if $\alpha > \alpha_0$.

This "change-over point" α_0 is called the <u>Hausdorff-dimension</u> dim F of the set $F \subseteq X^{\omega}$, i.e.

$\dim F = \sup\{ \alpha : L_{\alpha}(F) = \infty \} = \inf\{ \alpha : L_{\alpha}(F) = 0 \}$.

CHANNEL CAPACITY AND MINKOWSKI DIMENSION

Now, we investigate the connection of Hausdorff dimension to a combinatorial parameter of subsets $F \subseteq X^{\omega}$ called the <u>structure function</u> s_F of F

$$s_F(n) := \# \{ w : |w| = n \ \& \ F/w \neq \emptyset \}.$$

Observe that $V_n = \{ w : |w| = n \ \& \ F/w \neq \emptyset \}$ is the index set of the coarsest r^{-n}-covering of F, and this covering has minimum cardinality among all r^{-n}-coverings of F.

Following Shannon (1948) we call

$$H_F = \limsup_{n \to \infty} \frac{\log s_F(n)}{n}$$

the <u>entropy</u> (or <u>channel capacity</u>) of F and

$$\underline{H}_F := \liminf_{n \to \infty} \frac{\log s_F(n)}{n}$$

the <u>lower entropy</u> of F. (All logarithms are taken to the base $r = \# X$).

Observe that $\underline{H}_F$ and H_F are the respective "change-over points" of the functions

$$L'(F) := \liminf_{j \to} \{ L_{\alpha}(F, V_n) : n \geq j \}, \text{ and}$$

$$L''(F) := \limsup_{j \to} \{ L_{\alpha}(F, V_n) : n \geq j \}.$$

This means that $\underline{H}_F$ and H_F are the <u>lower</u> and <u>upper Minkowski dimension</u> (Minkowski, 1901) of the set F in the space (X^{ω}, ρ).

Thus, we have the following.

Property 1. dim $F \leq \underline{H}_F \leq H_F$.

We are interested in conditions under which dim $F = \underline{H}_F$ (or even $= H_F$), i. e. when the estimation of the Hausdorff dimension can be reduced to the simpler case of the estimation of the Minkowski dimension. To this end we mention the obvious fact that the Minkowski dimension $\underline{H}_F (H_F)$ of a set $F \subseteq X^\omega$ equals the Minkowski dimension $\underline{H}_{C(F)} (H_{C(F)})$ of its closure $C(F)$.

Therefore, our first results concern closed subsets of X^ω.

We call a subset $F \subseteq X^\omega$ balanced provided there is a function $f: N \rightarrow N$ of subexponential growth such that for all $w \in X^*$ and all $n \in N$

$$s_{F/w}(n) \leq f(|w|) \frac{s_F(n+|w|)}{s_F(|w|)}$$

holds true.

(A function $f: N \rightarrow N$ has subexponential growth iff for every $\varepsilon > 0$ we have $\lim_{j \rightarrow \infty} f(j) \cdot r^{-\varepsilon j} = 0$)

Remark. Since $s_F(n+m) = \sum_{|v|=m} s_{F/v}(n)$, the above definition requires that the structure function $s_{F/w}$ of an arbitrary state does not exceed "too much" the average over all structure functions

$$s_{F/v} \quad \text{where} \quad F/v \neq \emptyset \quad \text{and} \quad |v| = |w| .$$

The following result is shown in (Staiger, 1985a).

Theorem 2. If F is balanced and closed then dim $F = \underline{H}_F$.

In the same paper another condition guaranteeing dim $F = H_F$ is shown. To this end we say that a subset F has uniformly bounded growth iff there is a function $f: N \rightarrow N$ having subexponential growth such that

$$s_{F/w}(n) \leq f(|w|) \cdot s_F(n)$$

holds for every state F/w ($w \in X^*$) and every $n \in N$.

Theorem 3. If F is closed and has uniformly bounded growth then dim $F = H_F$.

Remark. In both theorems the hypothesis that F is closed is essential. Consider e. g. a set F countable and dense in X^ω.

APPLICATIONS TO FINITE-STATE SUBSETS

As a generalization of the concept of self-similarity (cf. Hutchinson, 1981) one may see finite-state subsets of X^{ω} where for all $n \in N$ in the r^n-partition $\{F \cap w \cdot X^{\omega} : |w| = n\}$ of F each member $F \cap w \cdot X^{\omega} = w \cdot (F/w)$ is congruent to one sample from the finite set $\{F/v : v \in X^*\}$.

As one verifies easily from the definition and Theorem 3 the following holds true.

Property 4. Every finite-state subset $F \subseteq X^{\omega}$ has uniformly bounded growth.

Corollary 5. Every finite-state closed subset $F \subseteq X^{\omega}$ satisfies $\dim F = H_F$.

A particular case of Corollary 5, namely, when F/w depends only on the last K letters of w, was proved by Conner (1970).

The entropy (or channel capacity) of finite-state subsets of X^{ω}, especially its relation to topological density, was investigated in (Staiger, 1983 and 1985b). Utilizing these results we can prove the equality $\dim F = H_F$ also for some non-closed finite-state subsets of X^{ω}.

To this end let V be a prefix-free subset of X^*. Consider

$V^{\omega} := \{ v_1 v_2 \ldots v_i \ldots : v_i \in V \ \& \ v_i \neq e \} \subseteq X^{\omega}$.

If V is a finite-state subset of X^*, it is easily verified that V^{ω} as well as its closure $C(V^{\omega})$ are finite-state subsets of X^{ω}.

We obtain the following result.

Theorem 6. Let $V \subseteq X^*$ be prefix-free and finite-state. Then

$$\dim V^{\omega} = H_{V^{\omega}} = \dim C(V^{\omega}).$$

This theorem enables us to reduce the computation of the Hausdorff dimension to the computation of the channel capacity in some cases when the set $F \subseteq X^{\omega}$ is constructively given (cf. Eilenberg, 1974), e. g. in the simplest case as a union of sets of the form $W \cdot V^{\omega}$ where $W, V \subseteq X^*$, and V is prefix-free and finite-state.

REFERENCES

Billingsley P. (1965): Ergodic Theory and Information. Wiley, New York.

Conner W.M. (1970): The capacity and ambiguity of a transducer. Ann. Math. Stat. 41, No. 6, 2093-2104.

Eilenberg S. (1974): Automata, Languages and Machines, Vol. A. Academic Press, New York, Ch. XIII and XIV.

Hausdorff F. (1919): Dimension und äußeres Maß. Math. Ann. 79, 157-179.

Hutchinson J.E. (1981): Fractals and self similarity. Indiana Univ. Math. J. 30, 713-747.

Mandelbrot B.B. (1977): Fractals. Form, Chance, and Dimension. Freeman, San Francisco.

Minkowski H. (1901): Über die Begriffe Länge, Oberfläche und Volumen. Jahresbericht d. Dt.Math. Vereinigung 9, 115-121.

Shannon C.E. (1948): A mathematical theory of communication. Bell System Tech. J. 27, No. 3, 379-423, and No. 4, 623-656.

Staiger L. (1983): On the relative density of sources. In: Trans. of the ninth Prague Conference, 1982, Academia Prague, 185-188.

Staiger L. (1985a): Combinatorial properties of the Hausdorff dimension. In: Proc. Geobild' 85, Wiss. Beitr. Friedrich-Schiller-Universität, Jena, 43-48.

Staiger L. (1985b): The entropy of finite-state ω-languages. Probl. Control and Inform. Theory 14, No. 5, 383-392.

Akademie der Wissenschaften der DDR
Zentralinstitut für Kybernetik
und Informationsprozesse

Kurstraße 33
PSF 1298

Berlin

DDR-1086

LIMIT THEOREMS FOR LAG PROCESSES

Josef Steinebach

Marburg

Key words: Strong theorems, lag processes, strong approximations, law of iterated logarithm, increments of Wiener processes

ABSTRACT

Consider a stochastic process $\{Z(t)\}_{t\geq 0}$ (e.g. a partial sum process or a renewal process), satisfying certain almost sure limit relations. Such limit relations will change if the corresponding "lag" process (or "delayed" process) is observed, i.e. the process $\{Z(t) - Z(t-a_t)\}_{t\geq 0}$, where $a_t \to \infty$ (as $t \to \infty$) at a given rate. Based upon strong invariance principles, certain asymptotics of the lag processes can be derived from corresponding properties of Wiener processes. Some law of iterated logarithm type results are discussed as well as necessity and sufficiency of appropriate generalized moment conditions in special situations.

1. INTRODUCTION

Hanson and Russo (1981) studied extended versions of the strong law of large numbers (SLLN), motivated by the following statistical problem: Suppose $\{X_i\}_{i=1,2,...}$ is an independent sequence of random variables whose distributions approach (in some appropriate sense) a fixed distribution (say F) with mean μ . Then, under mild regularity conditions, $S_n/n \to \infty$ (as $n \to \infty$) almost surely (a.s.), where $S_n = X_1 + \dots + X_n$. But there will usually be bias associated with the earlier observations, and one might hope to reduce this bias by considering "lag" sums (or "delayed" sums) $(S_n - S_{n-k_n})/k_n$. Under appropriate moment conditions, Hanson and Russo (1981) answered the question of how many X_i's must be kept (how fast must k_n grow) in order to retain the SLLN, i.e. $(S_n - S_{n-k_n})/k_n \to \mu$ (a.s.), or even an extended version of it, i.e. $\max\{(S_n - S_{n-k} - k\mu)/k : k_n \leq k \leq n\} \to 0$ a.s. (as $n \to \infty$).

More recently, based upon the Komlós-Major-Tusnády (1975/76) and Major (1976) strong approximation techniques, Hanson and Russo (1983 a,b) improved on their earlier results by stating certain law of iterated logarithm (LIL)-type results and thus also provided convergence rate statements for their lag sums. In addition,via a useful duality argument, the results turned out to be closely connected to the Erdös-Rényi (197o) law of large numbers for moving averages and related work in that area.

In a recent note (see Steinebach (1986 a)), we further investigated on the relationships between the limiting behaviour of lag sums, moving averages and strong invariance principles. A unified approach to LIL-type results for lag sums has been presented, where in particular,necessity of the involved moment conditions could also be proved. For earlier results in this direction we refer to Lai (1974).

The purpose of this paper is to discuss analogues of the forementioned statements in a rather general framework. Consider, for instance, a real-valued stochastic process $\{Z(t)\}_{t \geq 0}$ satisfying a strong invariance principle, e.g.

$$(1) \qquad \sup_{0 \leq s \leq t} \left| \frac{Z(s) - \mu s}{\sigma} - W(s) \right| = O(r(t)) \qquad \text{a.s.},$$

where $\{W(t)\}_{t \geq 0}$ is a standard Wiener process, μ and $\sigma > 0$ are real constants, and $\{r(t)\}_{t \geq 0}$ is an increasing positive-valued function. If $r(t) = o(t)$ (as $t \to \infty$), then by the SLLN for $\{W(t)\}_{t \geq 0}$,

(2) $Z(t)/t \to \mu$ a.s. (as $t \to \infty$).

Moreover, if $r(t) = o(\{t \text{ loglog } t\}^{1/2})$, by the LIL also

(3) $\limsup_{t \to \infty} |Z(t) - \mu t| / (t \text{ loglog } t)^{1/2} = \sigma$ a.s. (as $t \to \infty$).

The question is what kind of related statements could be retained for the "lag process" $\{Z(t) - Z(t-a_t)\}_{t \geq 0}$. The answer will depend upon the relationship between the "lag function" $\{a_t\}$ and the strong approximation rates $\{r(t)\}$ given in (1). Some general results are presented in Section 2 and illustrated by specific examples. Section 3 concentrates on discussing necessity of certain (generalized) moment conditions being involved. The main ideas of proofs are based upon those developed by Hanson and Russo (1983 a,b) or Steinebach (1986 a,b).

2. ALMOST SURE LIMITING BEHAVIOUR OF LAG PROCESSES

Let $\{Z(t)\}_{t \geq 0}$ be a real-valued stochastic process as introduced before, satisfying a strong invariance principle of type (1). For sake of notational convenience set

$$
\begin{aligned}
d(t,s) &= (2s\{\log(t/s) + \text{loglog } s\})^{1/2} \\
L(t,s) &= |Z(t) - Z(t-s) - s\mu| \\
L^*(t,s) &= \sup\{|Z(t) - Z(t-r) - r\mu| : 0 \leq r \leq s\}, \\
M(t,a) &= \sup\{L(t,s)/d(t,s) : a \leq s \leq t\}, \\
M^*(t,a) &= \sup\{L^*(t,s)/d(t,s) : a \leq s \leq t\}, \\
I(t,a) &= \sup\{L(s+a,a) : 0 \leq s \leq t-a\} / d(t,a), \\
I^*(t,a) &= \sup\{L^*(s+a,a) : 0 \leq s \leq t-a\} / d(t,a),
\end{aligned}
$$

where $0 \leq s \leq t$, $0 < a \leq t$ are real numbers and μ is the constant appearing in (1).

Based upon (1) and the Hanson and Russo (1983 a,b) results on increments of the Wiener process $\{W(t)\}_{t \geq 0}$ (cf. also Csörgő and Révész (1981), Ortega and Wschebor (1984)), one can immediately state a number of LIL-type results for the lag process

$\{Z(t)-Z(t-a_t)\}_{t\geq o}$. For sake of brevity we confine ourselves to some specific examples.
As in Hanson and Russo (1983 a,b), it will be convenient to introduce the notations

$$\log x \;=\; \log_e(\max\{x,1\}) \quad \text{and}$$
$$\log\log x = \log_e\log_e(\max\{x,e\}),$$

so that $\log x \geq o$ and $\log\log x \geq o$ for all x. Moreover, we assume that the strong approximation rate appearing in (1) satisfies the general condition

(R) $\quad r(t) = O(t^p) \quad$ for some $p < 1/2$.

The case of Strassen-type strong invariance rates (cf. Strassen (1964)), i.e. $r(t) = o(\{t \log\log t\}^{1/2})$ could also be dealt with, but the methods would have to be modified slightly. Under (R), however, we have a kind of uniform approach leading to the following extensions of earlier results (see e.g. Hanson and Russo (1983 a, Section 5), Steinebach (1986 a,b)):

THEOREM 1. Suppose a_t is measurable, $0<a_t\leq t$ $(t>0)$ and

(A1) $\quad a_t\, t^{\alpha} \to \infty$ (as $t \to \infty$) for every $\alpha>0$.

a) Under (1), (R) and

(A2) $\quad a_t \,/\, \dfrac{r^2(t)}{\log t} \to \infty \quad (\text{as } t \to \infty)$,

we have

(4) $\quad \limsup_{t\to\infty} L(t,a_t)/d(t,a_t) = \limsup_{t\to\infty} L^*(t,a_t)/d(t,a_t) = \sigma$ a.s.

(5) $\quad \limsup_{t\to\infty} M(t,a_t) = \limsup_{t\to\infty} M^*(t,a_t) = \sigma \quad$ a.s.

b) If $O(\cdot)$ in (1) can be replaced by $o(\cdot)$ and (R) holds, then assumption (A2) can be weakened to

(A3) $\quad \liminf_{t\to\infty} a_t \,/\, \dfrac{r^2(t)}{\log t} > 0$

in order to retain (4) and (5).

REMARK. In all typical situations one has $r^2(t)/\log t \to \infty$ (as $t \to \infty$), so that assumption (A1) becomes superfluous in view of (A2) or (A3).

THEOREM 2. Suppose a_t is measurable, $0 < a_t \leq t$ $(t > 0)$ and

(A4) $a_t \to \infty$ (as $t \to \infty$).

a) Under (1), (R) and (A2), we have

(6) $$\limsup_{t \to \infty} I(t,a_t) = \limsup_{t \to \infty} I^*(t,a_t) = \sigma \quad \text{a.s.}$$

b) If $O(\cdot)$ in (1) can be replaced by $o(\cdot)$ and (R) holds, then assumption (A2) can be weakened to (A3) in order to retain (6).

Next we consider some examples:

EXAMPLE 1. Let $\{X_i\}_{i=1,2,\ldots}$ be an i.i.d. sequence with $EX_1 = \mu$, $\mathrm{Var}(X_1) = \sigma^2 > 0$, and $EH(|X_1|) < \infty$, where $H(x) \geq 0$ $(x \geq 0)$ is a nondecreasing continuous function such that $x^{-2-\gamma}H(x)$ is nondecreasing for some $\gamma > 0$ and $x^{-1} \log H(x)$ is nonincreasing. Define

$$Z(t) = \sum_{i \leq t} X_i, \quad 0 \leq t < \infty.$$

By the Komlós, Major and Tusnády (1975, 1976) and Major (1976) construction the process $\{Z(t)\}_{t \geq 0}$ can be redefined on a suitable probability space, which also carries a standard Wiener process $\{W(t)\}_{t \geq 0}$, such that (1) holds with a rate function

$$r(t) = \mathrm{inv}\, H(t),$$

where $\mathrm{inv}\, H$ denotes the inverse function of H.

If we also assume that

(H) $$\liminf_{x \to \infty} H(\varepsilon x)/H(x) > 0 \quad \text{for all} \quad \varepsilon > 0,$$

then $O(\cdot)$ in (1) can even be replaced by $o(\cdot)$.

Hence, Theorems 1 and 2 cover parts of the results of Hanson and

Russo (1983 a, Section 5) and Steinebach (1986 a, Section 2).

EXAMPLE 2. In the situation of Example 1 assume that $EX_1 = \mu > 0$ and define

$$N(t) = \inf\{s : Z(s) > t\} \quad , \quad 0 \leq t < \infty \, ,$$

i.e. $\{N(t)\}_{t \geq o}$ is the usual generalized renewal process (or first-passage-time process) considered in the literature.

Csörgő, Horváth and Steinebach (1985, 1986) (cf. also Mason and van Zwet (1985) for a different approach) have shown, that one has a strong approximation

$$\sup_{o \leq s \leq t} \left| \frac{N(s)-s/\mu}{\sigma/\mu^{3/2}} - W(s) \right| = O(\text{inv } H(t)) \quad \text{a.s.} \, ,$$

with $\{W(t)\}_{t \geq o}$ being a Wiener process and $H(x)$, $x \geq 0$, satisfying the assumptions of Example 1. If (H) is also assumed, the $O(\cdot)$-term can again be replaced by $o(\cdot)$. To be in accordance with assertion (3.1o) on page 34 of Csörgő, Horváth and Steinebach (1985) note that $(\text{inv } H(t))/t$ is decreasing under the given assumptions.

Hence, Theorems 1 and 2 also extend the "if"-parts in Section 1 of Steinebach (1986 b). For the "only if"-parts see the discussion below.

EXAMPLE 3. Combining the partial sum and renewal process strong approximations discussed in Examples 1 and 2, Csörgő, Deheuvels and Horváth (1986) have recently obtained such invariance principles for stopped sums. Let $\{X_i\}_{i=1,2,\ldots}$ and $\{Y_i\}_{i=1,2,\ldots}$ be i.i.d. sequences, independent of each other, with $EX_1 = \mu$, $Var(X_1) = \sigma^2$, $EY_1 = \lambda > 0$, $Var(Y_1) = \tau^2$. For $t \geq 0$ set

$$S(t) = \sum_{i \leq t} X_i \quad ,$$

$$N(t) = \inf\{s \geq 0 : \sum_{i \leq s} Y_i > t\} \quad \text{and}$$

$$Z(t) = S(N(t)) \, .$$

If $EH(|X_1|)<\infty$ and $EH(|Y_1|)<\infty$, where either $H(x) = \exp(\alpha x)$ (for some $\alpha>0$) or $H(x) = x^r$ (for some $r>2$), then the following strong approximation holds:

$$\sup_{0\leq s\leq t} \left|\frac{Z(s)-(\mu/\lambda)s}{\theta} - W(s)\right| = O(\text{inv } H(t)) \quad \text{a.s.},$$

with $0<\theta^2 = (\sigma^2/\lambda) + (\mu^2\tau^2/\lambda^3)$. In case of $H(x) = x^r$, $O(\cdot)$ can be replaced by $o(\cdot)$. Thus the lag sums results of Hanson and Russo (1983 a) can also be extended to stopped lag sums results as indicated above.

A series of further examples could be added here, of course, covering e.g. dependent situations or nonlinear renewal processes (cf. Csörgő, Horváth and Steinebach (1985, Sections 3 and 4). We omit these examples, but instead we briefly discuss some methods to derive necessity of moment conditions in particular situations.

3. NECESSITY OF MOMENT CONDITIONS

Lai (1974, Theorem 2) has already shown that certain almost sure limit theorems for delayed sums correspond with appropriate moment conditions on the summands in both directions. Some further investigations, concerning partial sums as well as renewal processes, are to be found in Steinebach (1986 a,b) proving necessity and sufficiency of certain generalized moment conditions.

On the other hand, the results in Theorems 1 and 2 are immediate consequences of their analogues for the Wiener process via the strong approximation (1), and the latter approximation usually corresponds with certain moment assumptions on the approximated process (see e.g. Examples 1 - 3 above). Hence it seems only be natural to ask whether or not the limit relations of Theorems 1 and 2 are necessarily based upon finiteness of certain moments connected with the process $\{Z(t)\}$.

For a particular answer given below let us assume that

> there are random points $t_n = \tau_1 + \cdots + \tau_n$ such that, with $Z_i = Z(t_i) - Z(t_{i-1})$, $t_o = 0$, the random vectors (τ_1, Z_1), $(\tau_2, Z_2), \ldots$ are i.i.d., $\tau_i > 0$ a.s. and $(\tau_1 + \cdots + \tau_n)/n \to \alpha > 0$ a.s. (as $n \to \infty$).

Moreover, suppose that

> $r(t) > 0$ $(t > 0)$ is a continuous and strictly increasing (rate) function with $r^2(t)/\log t \to \infty$, $r(t) = O(t^p)$ (for some $p < 1/2$) as $t \to \infty$ and $r(t) \sim r(s)$ as $t \sim s$.

Put $H(x) = \text{inv } r(x)$ (for $x \geq x_o$ sufficiently large). Using the notations introduced above and in Section 2 we have the following result:

THEOREM 3. Suppose a_t is measurable, $0 < a_t \leq t$ $(t > 0)$, and satisfies

(A5) $\quad a_t \sim a\, r^2(t)/\log t \quad$ as $\quad t \to \infty$ (for some $a > 0$).

If, for some $c > 0$,

(7) $$\limsup_{t \to \infty} L^*(t, a_t)/d(t, a_t) \leq c \quad \text{a.s.},$$

and

(8) $$\lim_{n \to \infty} (t_n - t_{n-1})/a_{t_n} = 0 \quad \text{a.s.},$$

then

(M) $\quad EH(\varepsilon |Z_1 - \mu \tau_1|) < \infty \quad$ for some $\varepsilon > 0$.

In order to illustrate the applicability of Theorem 3, let us consider the situation of Example 3 again. Take $\{k_i\}_{i \geq 1}$ and $\{t_i\}_{i \geq 1}$ as the strictly ascending ladder indices and ladder heights, respectively, of the sequence $\{Y_i\}_{i \geq 1}$. By the arguments of Feller (1966, Chapter XII) and the independence of $\{X_i\}_{i \geq 1}$ and $\{Y_i\}_{i \geq 1}$, we have that $(\nu_i, \tau_i, Z_i)_{i \geq 1}$ is an i.i.d. sequence, where

$$\nu_i = k_i - k_{i-1} \quad , i = 1,2,\ldots, \quad k_o = 0 \quad ,$$
$$\tau_i = t_i - t_{i-1} \quad , i = 1,2,\ldots, \quad t_o = 0 \quad ,$$
$$Z_i = Z(t_i)-Z(t_{i-1}) \quad , i = 1,2,\ldots \quad .$$

For later use note that $\tau_i = T_{k_i} - T_{k_{i-1}}$, $i=1,2,\ldots$, where $T_o=0$, $T_n=Y_1+\cdots+Y_n (n=1,2,\ldots)$, denote the partial sums of the sequence $\{Y_i\}_{i\geq 1}$.

The following corollary of Theorem 3 underlines a remark of Csörgő, Deheuvels and Horváth (1986, p. 6o/61) on the optimality of their strong approximation rates for stopped sums under polynomial moments:

THEOREM 4. Let $\{Z(t)\}_{t\geq o}$ be as defined in Example 3. If

$$(1') \quad \sup_{0\leq s\leq t} \left| \frac{Z(s)-(\mu/\lambda)s}{\theta} - W(s) \right| = o(t^{1/r}) \quad \text{a.s.}$$

for some $r>2$, then

$$(M') \quad E|X_1|^r < \infty \text{ and } E(Y_1^+)^r < \infty .$$

In case of ordinary renewal processes $\{N(t)\}_{t\geq o}$, i.e. $Y_i\geq 0$ $(i=1,2,\ldots)$, this proves necessity of the moment condition assumed for the strong approximation with polynomial rate. In general, we get a condition on both tails of X_1, but only on the upper tail of Y_1. The latter fact, however, agrees with the one-sided definition of the stopping times $N(t)$ (first passage through t to the right), so that one cannot expect more than (M'). For a similar observation see also Csörgő, Horváth and Steinebach (1985, Theorems 3.3 and 3.4)

REFERENCES

Csörgő, M., Deheuvels, P., Horváth, L. (1986): An approximation of stopped sums with applications in queueing theory. In: Tech.Rep.Ser.Lab.Res.Stat.Probab., No. 77, 55-77, Carleton University, Ottawa.

Csörgő, M., Horváth, L., Steinebach, J. (1985): Invariance principles for renewal processes. In: Tech.Rep.Ser.Lab.Res.Stat. Probab., No. 59, 25-56, Carleton University, Ottawa.

Csörgő, M., Horváth, L., Steinebach,J. (1986): Strong approximations for renewal processes. C.R.Math.Rep.Acad.Sci.Canada 8, 151-154.

Csörgő, M., Révész, P. (1981): Strong Approximations in Probability and Statistics.Academic Press, New York.

Erdős, P., Rényi, A. (197o): On a new law of large numbers. J. Analyse Math. 23, 1o3-111.

Feller, W. (1966): An Introduction to Probability Theory and its Applications. Vol. II, Wiley, New York.

Hanson, D.L., Russo, R. (1981): On the law of large numbers. Ann. Probab. 9, 513-519.

Hanson, D.L., Russo, R. (1983 a): Some results on increments of the Wiener process with applications to lag sums of i.i.d. random variables. Ann. Probab. 11, 6o9-623.

Hanson, D.L., Russo, R. (1983 b): Some more results on increments of the Wiener process. Ann. Probab. 11, 1oo9-1o15.

Komlós, J., Major, P., Tusnády, G. (1975): An approximation of partial sums of independent r.v.'s and the sample d.f.I. Z. Wahrsch. Verw. Geb. 32, 111-131.

Komlós, J., Major, P., Tusnády, G. (1976): An approximation of partial sums of independent r.v.'s and the sample d.f.II. Z. Wahrsch. Verw. Geb. 34, 33-58.

Lai, T.L. (1974): Limit theorems for delayed sums. Ann. Probab. 2, 432-44o.

Major, P. (1976): The approximation of partial sums of independent r.v.'s. Z. Wahrsch. Verw. Geb. 35, 213-22o.

Mason, D., van Zwet, W. (1985): A note on the strong approximation to the renewal process. Preprint, University of Munich and University of Leiden.

Ortega, J., Wschebor, M. (1984): On the increments of the Wiener process. Z. Wahrsch. Verw. Geb. 65, 329-339.

Steinebach, J. (1986 a): Some remarks on strong approximations and the Hanson-Russo law of large numbers. Statist. & Decisions 4, 237-249.

Steinebach, J. (1986 b): Almost sure convergence of delayed renewal processes. Preprint, University of Marburg.

Strassen, V. (1964): An invariance principle for the law of the iterated logarithm. Z. Wahrsch. Verw. Geb. 3, 211-226.

Fachbereich Mathematik
Philipps-Universität
Hans-Meerwein-Straße
D-355o Marburg
Fed. Rep. Germany

RANDOM SETS IN $[0,1]$

František Straka, Josef Štěpán

Plzeň, Praha

Key words: Random set, random automorphism, inspection process of a random set, Brownian bridge

ABSTRACT

A random set is defined here as a r. v. X with values in the measure algebra $\mathbb{B}$ generated by Lebesgue measure λ on $[0,1]$. Theorems 1 and 2 provide equivalent definitions of such a random automorphism acting in the measure algebra $\mathbb{B}$. Theorem 3 and its corollaries investigate asymptotic properties of Lebesgue measure of random sets X_n employing an invariance principle technique for "inspection process" $t \longrightarrow \lambda(X_n \cap [0,t])$.

1. NOTATION AND DEFINITIONS

Let $\mathbb{B}$ be the algebra of Borel subsets of $I = [0,1]$ modulo sets of Lebesgue measure 0 and denote by λ the canonical extension of the measure to $\mathbb{B}$. Recall that $\mathbb{B}$ is a polish space w. r. t. λ - distance of elements A, B $\in \mathbb{B}$ defined by $\lambda(A \triangle B)$.

Denote by $\mathbb{A}$ the group of all automorphisms of the measure algebra $\mathbb{B}$. (Recall that T is an automorphism of $\mathbb{B}$ if it preserves the measure λ, countable unions and intersections. The weak topology of $\mathbb{A}$ is the coarsest topology which makes all maps $T \longrightarrow TB$ $(B \in \mathbb{B})$ to be continuous from $\mathbb{A}$ into $\mathbb{B}$. It is a known fact e. g. Halmos (1959), p. 87 - 89, that $\mathbb{A}$ endowed with the weak topology is a polish space).

Having a topological space, say $\mathbb{A}$, we shall denote its Borel σ- algebra by $\mathcal{B}(\mathbb{A})$.

Finally, consider a complete probability space $(\Omega, \mathcal{A}, P)$ and random variables

$$X : (\Omega,\mathcal{A}) \longrightarrow (\mathbb{B},\mathcal{B}(\mathbb{B})), \qquad T : (\Omega,\mathcal{A}) \longrightarrow (\mathbb{A}, \mathcal{B}(\mathbb{A}))$$

We shall say that X is a <u>random set</u> and T is a <u>random automorphism</u>. Having a random set X we define the random variable $p = \lambda(X)$ to be <u>the size of X</u> and the continuous stochastic process

$$X(t) = \lambda([0,t] \cap X), \qquad t \in I$$

<u>the inspection process</u> of X. Note that the distribution of an inspection process in $C(I)$ determines the distribution of the corresponding random set X in $\mathbb{B}$ and vice versa.

2. HOW TO PRODUCE A RANDOM SET

The most obvious way how to construct a random set is suggested by the following assertion.

Theorem 1. Let X be a random set. Then there exists a measurable stochastic process $Y = (Y(t), \ t \in I)$ such that

(1) $$Y(t) \text{ are } 0-1 \text{ variables},$$

and

(2) $$\{t : Y(t) = 1\} \in X \quad \text{a. s. } [P].$$

If Y, Z, are measurable processes satisfying (1) and (2) then $(P \times \lambda)\{(\omega,t) : Y(\omega,t) = Z(\omega,t)\} = 1$ and thus we may define correctly <u>probability of the inclusion</u> of $t_1, t_2, \ldots, t_n$ into a random set X, say $P[X \ni t_1,\ldots,t_n]$ as the equivalence class generated by the function

$$f(t_1, \ldots, t_n) = P[Y(t_1) = 1, \ldots, Y(t_n) = 1]$$

in $L_1(I^n, \mathcal{B}(I^n), \lambda^n)$ where Y is any measurable process satisfying (1) and (2).

Proof. First observe that for any $D \in \mathcal{A} \times \mathcal{B}(I)$ the map

$$\omega \longrightarrow D_\omega = \{t \in I : (\omega, t) \in D\}$$ generates a random set,

hence

$$\omega \longrightarrow \lambda(X(\omega) \cap D_\omega)$$ is a random variable

and finally

$$m(D) = \int_\Omega \lambda(X(\omega) \cap D_\omega) P(d\omega)$$

defines a finite measure on $\mathcal{A} \times \mathcal{B}(I)$ that is absolutely continuous w. r. t. $P \times \lambda$. Now, fix a measurable process

$$Z \in \frac{dm}{d(P \times \lambda)}$$ and observe that

$$\int_A \lambda(X \cap B)\, dP = \int_A \int_B Z(\omega, t)\, \lambda(dt)\, P(d\omega)$$

holds for all $A \in \mathcal{A}$ and $B \in \mathcal{B}(I)$. It follows from the separability of $(I, \mathcal{B}(I), \lambda)$ that

$$\lambda(X \cap B) = \int_B Z(\cdot, t)\, \lambda(dt), \quad B \in \mathcal{B}(I)$$

holds with probability one which immediately implies that

$$\{t : Z(\cdot, t) = 1\} \in X \qquad \text{a. s. } [P].$$

Finally, the process $Y = Z \cdot I_{\{0,1\}}(Z)$ is measurable and satisfies (1) and (2).

More sophisticated way how to produce a random set suggests the following procedure. Consider a random variable $0 \leq p \leq 1$ and a random automorphism T. Putting

$$(3) \qquad X(\omega) = T_\omega[0, p(\omega)], \quad \omega \in \Omega,$$

we recieve a random set X with the size p. Procedure (3) provides in fact a general representation for random sets.

Theorem 2. For any random set X with a size p there exists a random automorphism T such that (3) holds.

Proof. Putting

$$F(T,t) = T\left[0,t\right] \quad \text{for } (T,t) \in \mathbb{A} \times I$$

we can see that $F : \mathbb{A} \times I \to \mathbb{B}$ is a surjective Borel map (Halmos (1959), p. 104). As $\mathbb{A} \times I$ and $\mathbb{B}$ are polish spaces it follows from Von Neumann selection theorem (Schwartz (1973), p. 127) that there exists

(4) a universaly measurable map $S : \mathbb{B} \to \mathbb{A} \times I$

such that $F \circ S = \text{Identity}$.

Thus $S \circ X : \Omega \to \mathbb{A} \times I$ is $\mathcal{B}(\mathbb{A} \times I)$ $-\mathcal{A}$ measurable map ($(\Omega, \mathcal{A}, P)$ is a complete probability space!) and denoting

$$T = pr_{\mathbb{A}} \circ S \circ X, \text{ where } pr_{\mathbb{A}} : \mathbb{A} \times I \to \mathbb{A} \text{ is the projection,}$$

it follows from (4) that T is a random automorphism safisfying (3).

Example. Fix $n \in \mathbb{N}$ and $p \in (0,1)$ and consider a random automorphism T_n generated by the random uniform permutation of the intervals $I_{n,k} = \left[\frac{k-1}{n}, \frac{k}{n}\right]$, $k = 1, 2, \ldots, n$.

The random set X_n defined by $X_n = T_n\left[0,p\right]$ will be called the permutation random set of order n with the size p. Asymptotic properties of the sequence X_n will be studied in Section 3.

3. ASYMPTOTIC UNIFORMITY OF RANDOM SETS

The concept of the inspection process provides a nice tool when studying asymptotic properties of Lebesgue measure of the permutation random sets X_n.

Theorem 3. Consider a sequence of the permutation random sets X_n with a fixed size $p \in (0,1)$. Then

Applying suitable continuous functionals $H : C(I) \longrightarrow R$ to limiting procedure (5) we may get assertions as

Corollary 1. Let X_n be an asymptoticaly uniform sequence with a size $p \in (0,1)$. Then

$\lambda \{t \in I : X_n(t) > p \cdot t\} \longrightarrow Z$, where Z is uniformly distributed over the interval I.

See Billingsley (1968), p. 85 - 86.

Corollary 2. Let X_n be an asymptoticaly uniform sequence with a size $p \in (0,1)$. Then

$$(7) \quad D_n(f) = \sqrt{\frac{n}{p(1-p)}} \left(\int_{X_n} f(s)\, ds - p \int_0^1 f(s)\, ds \right) \xrightarrow{\mathcal{D}} \int_0^1 f(s) W^o(ds)$$

for all functions f which have the first derivation continuous on I.

We have denoted $\int f(s)\, W^o(ds) = \int f(s)\, W(ds) - W(1) \int f(s) ds$

where $\int f(s)\, W(ds)$ is Itó's stochastic integral w. r. t. the standart Wiener process W.

Proof. Denote

$$Z_n(s) = \sqrt{\frac{n}{p(1-p)}} (I_{X_n}(s) - p) , \quad s \in I$$

and integrate per partes to get

$$D_n(f) = \int_0^1 f(s)\, Z_n\,(s)\, ds = \int_0^1 f(s)\, U_n(ds) = -\int_0^1 U_n(s)\, f'(s)\, ds$$

which together with (5) implies (7) since $H(x) = -\int_0^1 x(s)\, f'(s) ds$, $x \in C(I)$ is a continuous functional on $C(I)$ and $H(W^o) =$ $= \int_0^1 f(s)\, W^o\,(ds)$.

Note that Corollary 2 suggests a procedure how to estimate

(5) $$U_n(t) = \sqrt{\frac{n}{p(1-p)}}\,(X_n(t) - p\cdot t) \xrightarrow{\mathcal{D}} W^o(t) \quad (\text{in } C(I)),$$

where $X_n = (X_n(t),\ t \in I)$ is the inspection process of the random set X_n and W^o is Brownian bridge over I.

Observing (5) is seems to be natural to say that a sequence of random sets X_n is asymptoticaly uniform with a size $p \in (0,1)$ if the corresponding inspection processes satisfy (5). For such a sequence we of course have

(6) $$\max_{t\in I} |X_n(t) - p\cdot t| \longrightarrow 0 \text{ in probability.}$$

To generalize Theorem 3 Straka (1986) has found sufficient conditions for a sequence X_n to be asymptoticaly uniform in terms of inclusion probabilities.

Proof. Denote

$$\xi_{n,i} = \sqrt{\frac{n}{p(1-p)}}\,\left(\lambda(X_n \cap I_{n,i}) - \frac{p}{n}\right),\quad 1 \leq i \leq n$$

and

$$Y_n(t) = \sum_{i=1}^{[nt]} \xi_{n,i}\,,\quad t \in I\,.$$

It is easy to verify that $\xi_{n,1}, \ldots, \xi_{n,n}$ are exchangable r. v.' s. such that

$$\sum_1^n \xi_{n,j} = 0,\quad \sum_1^n \xi_{n,j}^2 \longrightarrow 1,\quad \max_{1\leq j\leq n} |\xi_{n,j}| \longrightarrow 0 \text{ in probability,}$$

These facts directly imply (Billingsley (1968), p. 212) that $Y_n \xrightarrow{\mathcal{D}} W^o$ in Skorochod space D(I) which is equivalent to (5) as

$$\max_{t\in I} |U_n(t) - Y_n(t)| = \sigma(n^{-1/2}) \quad \text{a. s. } [P].$$

$\int_0^1 f(s)\, ds$ when observing the f on a random set X_n, only.

Corollary 3. Let X_n be an asymptoticaly uniform sequence of a size $p \in (0,1)$, f a function with the first derivation continuous on I such that

$$\sigma^2(f) = \int_0^1 f^2(s)\, ds - \left(\int_0^1 f(s)\, ds\right)^2 > 0 .$$

Then

$$(8) \quad p^{-1} \int_{X_n} f(s)\, ds \longrightarrow \int_0^1 f(s)\, ds \quad \text{in probability}$$

and

$$(9) \quad \frac{\sqrt{\frac{n}{1-p}}\, p \left(p^{-1} \int_{X_n} f(s)\, ds - \int_0^1 f(s)\, ds\right)}{\sqrt{\int_{X_n} f^2(s)\, ds - p^{-1}\left(\int_{X_n} f(s)\, ds\right)^2}} \xrightarrow{\mathcal{D}} N(0,1) .$$

It is easy to see that (8) is implied by (7). Thus

$$p^{-1} \int_{X_n} f^2(s)\, ds - p^{-2} \left(\int_{X_n} f(s)\, ds\right)^2 \longrightarrow \sigma^2(f) = \operatorname{Var} \int_0^1 f\, dW^o$$

in probability

and convergence (9) follows from (7).

4. REMARKS

The definitions and results in Section 1 and 2 may be easily formulated and proved replacing the measure space $(I, \mathcal{B}(I), \lambda)$ by any separable non atomic probability space in a view of the isomorphism theorem.

In connection with Theorem 2 a problem arises to characterize the geometry of the convex sets $D_X = \{\mathcal{L}(T,p),\ T,p \text{ satisfy } (3)\}$ where X is a fixed random set and to study "extremal representations" given by (3) where $\mathcal{L}(T,p)$ is an extremal point of D.

The techniques developed in Section 2 and 3 may be employed in some special cases to investigate the measure of a finite intersection of i. i. d. random sets. (See Hurt, Machek, Štěpán, Vorlíčková (1982) and Straka (1986)).

REFERENCES

Billingsley P. (1968): Convergence of Probability Measure. John Willey, New York.

Halmos P. (1959): Lecture on Ergodic Theory. IIL Moskva.

Hurt J., Machek J., Štěpán J., Vorlíčková D. (1982): The Intersection of Random Finite Sets. In: Math. Slovaca 32, No. 3, 229 - 237.

Schwartz L. (1973): Radon Measures. Oxford University Press.

Straka F. (1986): Náhodné množiny a jejich průniky. (Random Sets and their Intersections). PhD Theses, Plzeň (to appear in Czech).

Central Research Institute
ŠKODA Concern Enterprise
316 00 Plzeň - Bolevec

Department of Statistics
Faculty of Mathematics
and Physies
Charles University
Sokolovská 83
186 00 Praha 8
Czechoslovakia

OPTIMAL CHOICE PROBLEM WITH UNCERTAINTY OF SELECTION AND BACKWARD SOLICITATION

Krzysztof Szajowski

Wrocław

Key words: secretary problem, controlled Markov chain, backward induction, relative rank.

ABSTRACT

This paper deals with an optimal choice problem of the applicant with the absolute rank 1 or 2. It is allowed both recall of applicants and uncertainty of accepting the offer by a current applicant. Some properties of optimal selection procedures are derived in the general case. In the case with constant probability of successful solicitation an optimal procedure and the corresponding probability of obtaining the required applicant are obtained.

INTRODUCTION

The classical model of the secretary problem can be found in Gilbert and Mosteller (1966) and a survey of its extensions is given by Freeman (1983). The basic framework of the orginal problem can be stated as follows: n applicants appear before the decision-maker sequentially one by one in a random order. It means that all permutations are supposed to be equally likely. In early papers a restrictive assumption of the model was the impossibility of recalling previously rejected applicants. As each applicant appears, the decision-maker observes her rank relative to those preceding her and must decide whether to accept or to reject her and to interview futher applicant. The optimal stopping rule is one which maximizes the probability of choosing the best applicant.

For optimal choosing the best applicant the backward solicitation was considered by Yang (1974) and the uncertain employment was investigated by Smith (1975). Both, the uncertain employment and recall of observation in the best-choice problem has been solved independently by Tamaki (1979), Petruccelli (1981) and Choe and Bai (1983). In their formulation, when s-th applicant is viewed, the investigator can choose either to solicite any previously or actually interviewed applicant or to interview $s+1$-st one. If the offer is given to an applicant on a position $s-r$ then the applicant could accept it with known probability $p(r)$. Once unavailable applicant remains unavailable thereafter. Môri (1983) has dealt with the multi-choice problem when $p(0) \leq 1$ and $p(r) = 0$ for $r \neq 0$ and the aime is to choose the best applicant. Rose (1985) has looked at the problem differently. He has assumed that the investigator can accept, reject or hold applicant during some period with some cost. Szajowski (1982) has investigated the problem of optimal choice one of the two best applicants with backward solicitation. In the present paper the backward solicitation and uncertain employment is permitted.

NOTATIONS, ASSUMPTIONS AND FORMULAE

Let $Y_j(s)$ be *the current relative rank* of j-th examined applicant when s-th applicant is viewed $(j = 1,2,\ldots,s)$. Denote by t_s^l the relative (to s) position of the current l-th candidate i.e.

$$t_s^l = s - S_l(Y_1(s),\ldots,Y_s(s))$$

where $S_l(Y_1(s),\ldots,Y_s(s)) = j$ iff $Y_j(s) = l$. Let $t_s^l = \infty$ if an unsuccessful attemt has been made to procure the $s - t_s^l$ -th applicant. Our aime is to choose the best or the second best applicant with maximal probability so we can consider as candidates the relative 1 or 2 applicants only. Let us assume that if an attempt is made to procure the first (second) candidate, this attempt will be successful with the probability $p(t_s^1)$ $(p(t_s^2))$. We suppose that $p(\infty) = 0$. It implies that an applicant, which has been unavailable, remains unavailable thereafter, and we assume that $p(k)$ is non-increasing function.

The problem can be described as a *controlled Markov chain* with the state spaces $X_t = \{(s, t_s^1, t_s^2): s \le t, t_s^1, t_s^2 = 0,1,\dots,s-1,\infty\} \cup \{\delta_1(s),\delta_2(s)\}$ and the action spaces $A_t = \{f, b1, b2\}$, where the action f denotes: take the next applicant without solicitation of the current candidates: b1: solicit the first candidate and b2: solicit the second one. Assuming as in the classical secretary problem that all permutations are equally likely, the transition probabilities are given in Figure 1.

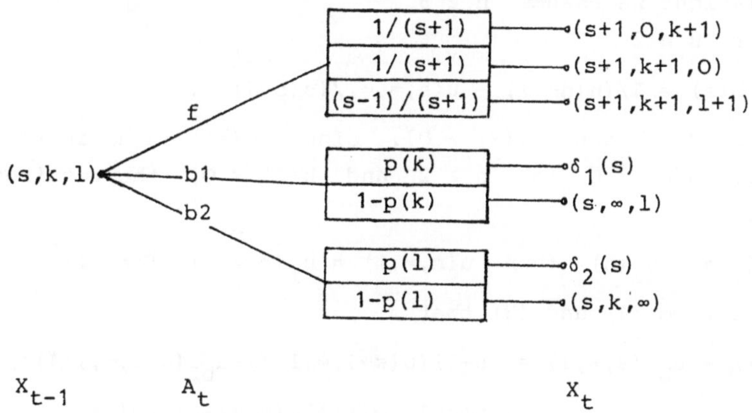

Figure 1.

Denote by $u_f(s,k,l)$, $u_{b1}(s,k,l)$, $u_{b2}(s,k,l)$ the probability of obtaining the required applicant when in the state (s,k,l) strategy f, b1, b2 is used, respectively , and the optimal one thereafter.

Let

$$u(s,k,l) = \max\{u_f(s,k,l),\ u_{b1}(s,k,l),\ u_{b2}(s,k,l)\}.$$

We obtain (see Figure 1) the following recursive formulae

(1) $$u_f(s,k,l) = (u(s+1,0,k+1) + u(s+1,k+1,0))/(s+1) + (s-1)u(s+1,k+1,l+1)/(s+1)$$

(2) $$u_{b1}(s,k,l) = j(2n-j-1)p(k)/(n(n-1)) + (1-p(k))u(s,\infty,l)$$

(3) $$u_{b2}(s,k,l) = j(j-1)p(l)/(n(n-1)) + (1-p(l))u(s,k,\infty)$$

$$u(n,k,l) = 1 - (1-p(l))(1-p(k)) .$$

We adopt the following strategy: if $u(s,k,l) = u_\gamma(s,k,l)$ then γ strategy in used. We prefer to strategy b type rather then f type. Solving the above system of the equations we obtain the best strategy, if the probabilities $p(k)$ are given for $k = 0,1,\ldots,n-1$.

GENERAL PRINCIPLES

In this section some properties analogues to those studied by Petruccelli (1981) and Szajowski (1982) are established in our more general formulation. We assume $n \geq 3$.

Denote for $s \geq 2$

$$b(s) = 1/(n(n-1)) - u(s,\infty,\infty)/(s(s-1)) .$$

Let $s_2(n) = \min\{2 \leq s \leq n:\ b(s) > 0\}$. Since $b(s)$ is an increasing function in s, $b(n) > 0$ for $n \geq 2$ and $b(2) < 0$, the required $s_2(n)$ exists.

Theorem 1. If $s < s_2(n)$ then $u(s,\infty,l) = u_f(s,\infty,l)$ for all l.

Proof. We have from (1) and (3) that

$$u_f(s,\infty,l) - u_{b2}(s,\infty,l) = (s-1)(u(s+1,\infty,l+1)-u_{b2}(s+1,\infty,l+1))/(s+1)+$$
$$+ s(s-1)[p(l+1)b(s+1)-p(l)b(s)].$$

From the definition of $s_2(n)$ we have that the right-hand side of above equation is non-negative for all l and $s < s_2(n)$.

Theorem 2. There exists an $s_1(n)$ such that if $s < s_1(n)$ then $u_f(s,k,l) > u_{b1}(s,k,l)$.

Theorem 3. If $u(s,k,l) \neq u_f(s,k,l)$ and $u(s,k,\infty) \neq u_f(s,k,\infty)$ then $u_{b2}(s,k,l) \leq u_{b1}(s,k,l)$.

Corollary. In the optimal procedure for the considered problem we should not solicit any applicant to some moment $s_1(n)$ and we should not solicit the second candidate to some moment $s_2(n) \geq s_1(n)$.

The proof of the Theorem 2,3 and Corollary follows from the basic equations (1)-(3) and are omitted here.

Theorem 4. In order that, under the optimal strategy, the offer to the second candidate is never made until all applicants have been interviewed, it is neccessary and sufficient that

$$(4)\qquad \frac{p(l+1)}{p(l)} \geq \frac{n-2-2p(0)}{n-2} \qquad \text{for all} \quad l < \infty .$$

Proof. From (1), (3) and (4) we have that $u_f(n-1,\infty,l) \geq u_{b2}(n-1,\infty,l)$ if and only if (4) holds.

Sufficiency is proved by induction. Suppose that (4) is fulfilled and for $s = n-1,\dots,t+1$ and all $l < \infty$, $u_f(s,\infty,l) \geq u_{b2}(s,\infty,l)$. Then

$$u_f(s,\infty,l) = u(s,\infty,\infty) + t(t-1)p(l+n-t)/(n(n-1))$$

and we have

$$u_f(s,\infty,l) - u_{b2}(s,\infty,l) = p(l)u(t,\infty,\infty)+t(t-1)[p(l+n-t)-p(l)]/(n(n-1)).$$

Write

$$f(t) = (u_f(t,\infty,l) - u_{b2}(t,\infty,l))n(n-1)/(p(l)t(t-1)).$$

$f(t)$ is a decreasing function in t and $f(n-1) \geq 0$. Thus Theorem 4 is proved.

Write $c(n) = 0$,

$$c(s) = 2p(0)[1 + \sum_{k=s+1}^{n-1} \prod_{j=s+1}^{k} (1 - 2p(0)/j)] =$$

$$= 2p(0)[\prod_{k=s}^{n-1} (1 + (1 - 2p(0))/k) - 1]/(1 - 2p(0))$$

for $s = n-1, n-2, \dots, 1$. Let $\varphi(s) = 1 - (n-1)c(s)/(s-1)$ and $s_o = \min\{s : \varphi(s+1) > 0\}$.

Theorem 5. Let $p(k) > 0$ for each finite k. If there exists a τ such that τ is the smallest $s > \max(s_o, s_2(n))$ satisfying

$$(5)\qquad \frac{p(k+1)}{p(k)} \leq \frac{\varphi(s)}{\varphi(s+1)} \qquad \text{for all} \quad k < \infty ,$$

then the solicitation to the first candidate should be made, if she refuses, then solicitation to the second one should be made whenever the procedure does not stop before τ.

Proof. Since right-hand side of (5) is an increasiong function for $s > s_o$, inequality (5) is fulfilled for $s \geq \tau$. Clearly, the statement of the theorem holds for $j = n-1$. Assume it is true for $j = n-1, n-2, \ldots, t+1$ and all finite k.

Write

$$u_b(s,k,l) = p(k)\frac{s(2n-s-1)}{n(n-1)} + (1 - p(k))p(l)\frac{s(s-1)}{n(n-1)} + \\ + (1 - p(k))(1 - p(l))u(s,\infty,\infty) .$$

Then

$$u_f(t,k,l) - u_b(t,k,l) \leq \left(\frac{2t(n-t-p(0))}{n(n-1)} - \frac{2(1-p(0))}{t+1}u(t+1,\infty,\infty)\right)p(k+1) - \\ - \frac{2t(n-t)}{n(n-1)}p(k) - (1-p(k))\left(\frac{2p(0)}{n} + \frac{2(1-p(0))}{t+1}u(t+1,\infty,\infty)\right) + \\ + (1-p(k))\frac{t(t-1)}{n(n-1)}(p(l+1)\varphi(t+1) - p(l)\varphi(t)).$$

For $t > \max(s_o, s_2(n))$ we have

$$\left(\frac{2t(n-t-p(0))}{n(n-1)} - \frac{2(1-p(0))}{t+1}u(t+1,\infty,\infty)\right)\varphi(t)/\varphi(t+1) - \frac{2t(n-t)}{n(n-1)} + \\ + 2(1-p(0))u(t+1,\infty,\infty)/(t+1) \leq -2p(0)/n .$$

because it can be verified that the left-hand side of above inequality is an increasing function in t and for $t = n-1$ the inequality is fulfilled. Therefore, by the condition (5) we have $u_f(t,k,l) - u_b(t,k,l) \leq 0$. The induction stops at $t = \tau$.

CONSTANT PROBABILITY OF SUCCESSFUL SOLICITATION

We deal with the case $p(0) = p \geq p(k) = q$ for $k = 1,2,\ldots,n-1$.

Theorem 6. Under the above assumption the optimal strategy is to proceed to stage $s_1(n) - 1$ without solicitation; solicit the first candidate as she appears among the remaining $n - s_1(n) - 1$ applicants; beginning with the $s_2(n)$-th applicant as the second candidate appears, solicit the first one; if she refuses, choose the second one. If the procedure does not stop to stage $n-1$, solicit the first candidate. The probability of obtaining the required applicant $u(1,1,\infty) = w(s_1) = w(s_1,s_2,k,l)$ (i.e. at s_1 the state is (s_1,k,l) and $s_2 \geq s_1$)

can be calculated as follows:

$$(6)\quad w(s_1,s_2,k,l) = \sum_{j=s_1+1}^{s_2} \frac{s_1(s_1-1)}{j(j-1)(j-2)}[p\,\frac{j(2n-j-1)}{n(n-1)} + (1-p)q\,\frac{j(j-1)}{n(n-1)} +$$
$$+ (1-p)w(j,s_2,\infty,\infty) + w(j,s_2,k+s_1-j,0)] +$$
$$+ \frac{s_1(s_1-1)}{(n-1)(n-2)}\, w(n-1,n-1,k,l) +$$
$$+ \sum_{j=s_2}^{n-1} \frac{s_1(s_1-1)}{j(j-1)(j-2)}[(p+q)\,\frac{j(2n-j-1)}{n(n-1)} +$$
$$+ ((1-p)q + (1-q)p)\,\frac{j(j-1)}{n(n-1)} +$$
$$+ (1-p)(2-q)w(j,j,\infty,\infty)].$$

The proof of this theorem is similar to the proof of Theorem 6 in Szajowski (1982) and follows from the basic equations (1)-(3) .

For the asymptotic solution let us denote

$$v_1(x,y) = \lim_{n\to\infty} w([xn],[yn],\infty,\infty)$$

and

$$v(x,y) = \lim_{n\to\infty} w([xn],[yn],\, k,l)$$

for $0 < x \le y \le 1$.

We have from (6) that

$$v_1(y,y) = y^2 \int_y^1 [(2pt + 2(1-p)v_1(t,t))/t^3]dt$$

with condition $v_1(1,1) = 0$ and

$$v_1(x,y) = x^2 \int_x^y [(2pt + (q-p)t^2 + (2-p)v_1(t,y))/t^3]dt +$$
$$+ x^2 v_1(y,y)/y^2 .$$

(6) follows that

$$v(y,y) = y^2 \int_y^1 [((p+q)t(2-t) + (p+q-2pq)t^2 + (1-p)(2-q)v_1(t,t))/t^3]dt$$
$$+ y^2[1 - (1-q)^2] .$$

Since $w(r,s,k,0) = w(r,s,k,l)$ for $r < s$, therefore

$$v(x,y) = x^2 v(y,y)/y^2 + x^2 \int_x^y (v(t,y)/t^3)dt +$$

$$+ x^2 \int_x^y [(pt(2-t) + (1-p)qt^2 + (1-p)v_1(t,y))/t^3]dt .$$

Consequently

$$v(y,y) = [2(p+q) - (2p(1-p)(2-q)/(1-2p))]y + (1-(1-q)^2-q(2+p))y^2 +$$

$$+ 2pqy^2 \ln y + p(2-q)y^{2p}/(1-2p)$$

and

$$v(x,y) = ax^2 + b(y)x^p + C(y)x$$

where

$$a = (1-p)(q-p)/(2-p) - q + p + pq$$

$$b(y) = (2p/(1-p) - 2p/(1-2p))y^{1-p} + ((q-p)/(2-p))y^{2-p} + (2p/(1-2p))y^p$$

$$C(y) = -ay - b(y)y^{p-1} + v(y,y)/y.$$

The asymptotically optimal solution of the problem can be obtained taking $s_1(n) = [\alpha n]$ and $s_2(n) = [\beta n]$, where α and β are the solution of the equations (unique such that $0 < \alpha \le \beta \le 1$):

$$v(\beta,\beta) = q\beta(2-\beta) + (1-q)p\beta^2 + (1-p)(1-q)v_1(\beta,\beta)$$

and

$$2a\alpha + pb(\beta)\alpha^{p-1} + C(\beta) = 0 .$$

For $p = 0,5$ and $p = 1$ the asymptotic solution is the limit of the above solution when p tends to 0,5 and 1, respectively.
The nmerical examples are given in TABLE 1.

TABLE 1

The Optimal Decision Numbers $s_1(n)$, $s_2(n)$ and Probability of Choosing the Best or Second Best Applicant when $p(0) = p$, $p(k) = \rho p$.

ρ	n	p = 0,25: $s_1(n)$	p = 0,25: $s_2(n)$	p = 0,25: $u(1,1,\infty)$	p = 0,5: $s_1(n)$	p = 0,5: $s_2(n)$	p = 0,5: $u(1,1,\infty)$	p = 0,75: $s_1(n)$	p = 0,75: $s_2(n)$	p = 0,75: $u(1,1,\infty)$
	5	2	3	0,3361	2	2	0,5526	3	4	0,6891
	10	2	6	0,3186	4	7	0,5101	4	8	0,6456
0,25	20	4	11	0,3095	7	14	0,4935	8	15	0,6263
	50	10	28	0,3038	16	33	0,4850	20	36	0,6146
	100	20	54	0,3019	31	65	0,4808	40	71	0,6106
	∞	0,196n	0,536n	0,3001	0,308n	0,640n	0,4778	0,390n	0,698n	0,6067
	5	2	4	0,3600	2	4	0,5958	3	4	0,7735
	10	3	7	0,3443	5	8	0,5738	6	8	0,7426
0,5	20	6	13	0,3371	9	15	0,5627	11	16	0,7300
	50	14	32	0,3329	21	37	0,5562	26	40	0,7222
	100	28	63	0,3316	42	73	0,5540	52	79	0,7197
	∞	0,273n	0,624n	0,3302	0,412n	0,727n	0,5520	0,512n	0,780n	0,7172
	5	2	4	0,3871	3	4	0,6672	4	4	0,8583
	10	5	8	0,3792	6	9	0,6533	7	9	0,8448
0,75	20	9	15	0,3756	12	17	0,6472	14	18	0,8383
	50	21	38	0,3735	29	42	0,6438	34	44	0,8348
	100	42	75	0,3728	58	84	0,6427	68	88	0,8337
	∞	0,420n	0,747n	0,3721	0,575n	0,834n	0,6419	0,679n	0,876n	0,8326

REFERENCES

Choe K.I. and Bai D.S. (1983): A secretary problem with backward solicitation and uncertain employment. J. Appl. Prob. 20, 891-896.

Freeman P.R. (1983): The secretary problem and its extensions: a review. Int. Statist. Rev. 51, 189-206.

Gilbert J.P. and Mosteller F. (1966): Recognizing the maximum of a sequence. J. Amer. Statist. Assoc. 73, 775-780.

Mόri T.F. (1983): The secretary problem with hesitating candidates. Proc. 4th. Pann.Symp. Math. Statist. Sept. 1983, Bad Tatzmannsdorf.

Petruccelli J.D. (1981): Best-choice problems involving uncertainty of selection and recall of observations. J. Appl. Prob. 18, 415-425.

Rose J.S. (1985): Optimal sequential selection based on relative ranks with renewable call options. J. Amer. Statist. Assoc. 79, 430-435.

Smith M.H. (1975): A secretary problem with uncertain employment. J. Appl. Prob. 12, 620-624.

Szajowski K. (1982): Optimal choice problem with backward solicitation. Zastos. Matem. 17, 455-471.

Tamaki M. (1979): Secretary problem with uncertain employment when backward solicitation is permitted. Math. Japonica 24, 439-450.

Yang M.C.K. (1974): Recognizing the maximum of a sequence based on relative rank with backward solicitation. J.Appl. Prob. 11, 504-512.

Institute of Mathematics
Technical University of Wrocław
Wybrzeże Wyspiańskiego 27
50-370 Wrocław
Poland

INVESTIGATION OF THE INDEPENDENCE

OF BINARY VARIABLES

László Telegdi

Budapest

Key words: binary variables, independence

ABSTRACT

The independence of binary variables indicating the presence or absence of dichotomous characters is investigated by the estimation of terms characterizing the distribution of size of the character combinations.

The paper deals with the investigation of the structure of binary (random) variables indicating the presence or absence of dichotomous characters. Let us denote by $\underline{n}$ their number, by $\underline{W}_1$, $\underline{W}_2$, ... $\underline{W}_n$ the variables, by $\underline{A}_1$, $\underline{A}_2$, ... $\underline{A}_n$ the characters, by $\underline{G}_g$ ($\underline{g}$ = 1, 2, ... 2^n-1) their possible combinations, and by $\underline{M}$ the number of objects $\underline{y}_g$ where the variables are observed. For arbitrary

$$G_g = \{A_{i_1}, A_{i_2}, \dots A_{i_k}\}, \quad 1 \leq k \leq n, \quad 1 \leq g \leq 2^n-1,$$

let us denote by $\underline{P}_T(\underline{i}_1,\underline{i}_2,\dots\underline{i}_k)$ the probability that an object has the characters belonging to $\underline{G}_g$ and possibly others too, and by $\underline{\sigma}_T(\underline{i}_1,\underline{i}_2,\dots\underline{i}_k)$ the number of such objects. The question whether the variables are independent can be settled theoretically by the χ^2 test belonging to the null hypothesis according to which for each $\underline{G}_g$

$$P_T(i_1, i_2, \ldots i_k) = \prod_{j=1}^{k} P_T(i_j),$$

on the basis of the corresponding 2^n-fold contingency table. The paper deals with the case when it is illusory to perform the test, because 2^n is too large and/or the probabilities $\underline{P}_T(\underline{i})$ are too small.

Let us denote by $\underline{P}^{(k)}$ the probability that an object has exactly $\underline{k}$ characters, and by $\underline{\sigma}^{(k)}$ the number of such objects, furthermore let

$$P = P^{(0)}.$$

Let us introduce the following notations too:

$$q_i = \frac{P_T(i)}{1-P_T(i)}, \quad i = 1, 2, \ldots n,$$

$$S_0 = 1, \quad S_k = \sum_{1 \leq i_1 < i_2 < \ldots < i_k \leq n} \prod_{j=1}^{k} q_{i_j}, \quad k = 1, 2, \ldots n.$$

Let us suppose that the variables are independent. The definitions of $\underline{P}^{(k)}$ and $\underline{S}_k$ imply that

$$P^{(k)} = PS_k, \quad k = 0, 1, \ldots n,$$

i.e. for producing the probabilities $\underline{P}^{(k)}$ corresponding to the relative frequencies $[\underline{\sigma}^{(k)} / \underline{M}]$, it is sufficient to calculate - besides $\underline{P}$ - the terms $\underline{S}_k$, which depend - through the terms $\underline{q}_i$ - only on the probabilities $\underline{P}_T(\underline{i})$. Let

$$T_j = \sum_{i=1}^{n} q_i^j, \quad j = 1, 2, \ldots n.$$

The terms $\underline{S}_k$ can be determined recursively, namely according to Newton's formula

$$S_k = \frac{1}{k} \sum_{j=1}^{k} (-1)^{j-1} T_j S_{k-j}, \quad k = 1, 2, \ldots n.$$

For arbitrary $\underline{A}_i$ let us denote by $\underline{P}(\underline{i})$ the probability that an object has $\underline{A}_i$, but no other characters, and by $\underline{\sigma}(\underline{i})$ the number of such objects. Let $\underline{e}_{gi}$ be the value (1 or 0) of the variable $\underline{W}_i$ observed on the object $\underline{y}_g$. Because of the independence of the variables

$$P(e_{gj}=0,\ j\neq i \mid e_{gi}=1) = P(e_{gj}=0,\ j\neq i \mid e_{gi}=0),$$

thus

$$\frac{P(i)}{P_T(i)} = \frac{P}{1-P_T(i)},$$

whence

$$P_T(i) = \frac{P(i)}{P+P(i)}. \tag{1}$$

Let

$$\sigma = \sigma^{(0)},$$

and let the probabilities $\underline{P}_T(\underline{i})$ be estimated in such a way that

$$\hat{P} = \frac{\sigma}{M}$$

and

$$\hat{P}(i) = \frac{\sigma(i)}{M}, \quad i = 1, 2, \ldots n,$$

should hold. Then

$$\hat{P}_T(i) = \frac{\sigma(i)}{\sigma+\sigma(i)}, \quad i = 1, 2, \ldots n. \tag{2}$$

Because of the independence

$$(3) \qquad P = \prod_{i=1}^{n} [1-P_T(i)].$$

(1) and (3) imply

$$P = \prod_{i=1}^{n} [1 - \frac{P(i)}{P+P(i)}] = \prod_{i=1}^{n} \frac{P}{P+P(i)},$$

therefore

$$(4) \qquad \sigma = M \prod_{i=1}^{n} \frac{\sigma}{\sigma+\sigma(i)}$$

should hold. Testing Equation (4) may be the first step of the investigation of independence.

Estimates (2) are not efficient estimates of the probabilities $\underline{P}_T(\underline{i})$. Their efficient estimates are the relative frequencies $[\underline{\sigma}_T(\underline{i}) / \underline{M}]$. Under them

$$(5) \qquad \sigma = M \prod_{i=1}^{n} \frac{M-\sigma_T(i)}{M}$$

should hold in case of independence. However, the frequencies $\underline{\sigma}_T(\underline{i})$ - in contrast with the frequencies $\underline{\sigma}(\underline{i})$ - are usually not directly given data. This implies that the right side of Equation (5) - in contrast with that of Equation (4) - usually can't be calculated directly.

In the recent treatment the decisive moment of investigating the independence of the variables is the comparison of the series $\{\underline{\sigma}^{(k)}\}$ and $\{\underline{M}\hat{\underline{P}}^{(k)}\}$, where the estimates $\hat{\underline{P}}^{(k)}$ are produced in the above-mentioned way. About this comparison the following can be told. If the probabilities $\underline{P}^{(k)}$ were known, the expression

$$\sum_{k=0}^{n} \frac{[\sigma^{(k)} - MP^{(k)}]^2}{MP^{(k)}}$$

would asymptotically have χ^2 distribution with degree of freedom n [see, e.g., Lehmann (1959)]. Now they are not known, but estimated. In such cases the degree of freedom must be decreased by the number of the estimated parameters. Our suggestion is that the above described estimation of probabilities can be traced back to the estimation of two parameters, thus the expression

$$\sum_{k=0}^{n} \frac{[\sigma^{(k)} - MP^{(k)}]^2}{MP^{(k)}}$$

asymptotically has χ^2 distribution with degree of freedom (n-2). This suggestion is supported by the result of Monte Carlo trials.

From conception until birth various defects may develop in the embryo and fetus. Such a defect present at birth is called congenital abnormality (CA). The concurrence of more different CAs in the same person is called multiple CA (MCA). On the basis of the material of the Hungarian Congenital Malformation Register the occurrences of MCAs were statistically analysed [Czeizel et al. (1987)]. Within this framework the independence of CAs was investigated in the aforesaid way.

REFERENCES

Czeizel A., Telegdi L., Tusnády G. (1987) Multiple Congenital Abnormalities. Publishing House of the Hungarian Academy of Sciences, Budapest.

Lehmann E. L. (1959) Testing Statistical Hypotheses. Wiley, New York.

Computer and Automation Institute
Hungarian Academy of Sciences
H-1502 Budapest - P.O. Box 63
Hungary

CONTRIBUTION TO THE GENERAL TREATMENT OF RANDOM PROCESSES USED IN CHEMICAL REACTION KINETICS

János Tóth

Budapest

Key words: stochastic kinetics, Markovian pure jump process, embedded Markov chain, classification of states, irreducibility, aperiodicity

ABSTRACT

The usual stochastic model of complex chemical reactions delineates a class of continuous time Markovian jump processes with the state space $\mathbb{N}_0^M$, where M is the number of chemical components or species. As a continuation of previous investigations structure of the state space (including irreducibility and aperiodicity) is fully studied in the special cases where either the number of chemical components or the number of elementary reactions making up the complex chemical reaction equals one.

INTRODUCTION

Let us consider a vessel containing a finite number of chemical components or species and let us suppose that among these components elementary reactions take place.

As a result of these elementary reactions the quantities of the chemical components change. The usual stochastic model for this phenomenon is a continuous time discrete state Markovian jump process.

It is only a minor branch of the literature that strives to state general results on all the reactions or on great classes of reactions simultaneously. The most important papers in this spirit are those written or initiated by T. G. Kurtz, see e. g. Kurtz (1981) or Chapter 5 of Érdi and Tóth (1987). These works almost exclusively dealt with similarities and differences between the stochastic and the usual deterministic model.

The present paper, similarly to Tóth (1986), is devoted to investigate the structure of the state space from the point of view of connectedness and aperiodicity. In the disconnected cases structure of all the connected components of the graph is characterized.

PRELIMINARIES

Here we present necessary definitions and statements mainly from Tóth (1986).

DEFINITION 1. A (complex chemical) reaction is a quadruple $\langle\mathcal{M},\mathcal{R},(\alpha,\beta),w\rangle$ consisting of a mechanism $\langle\mathcal{M},\mathcal{R},(\alpha,\beta)\rangle$ and a kinetics w where

(i) $\mathcal{M}$ and $\mathcal{R}$ are finite sets of M and R elements respectively ($M,R \in \mathbb{N}$);

(ii) α and β are MxR matrices of nonnegative integers;

(iii) for all $r \in \mathcal{R}$ w_r is a real valued function on $\mathbb{R}^M$ or on $\mathbb{R}^{\mathcal{M}}$ (this kind of identification will be made here and below) i. e. $w:\mathcal{R} \to (\mathbb{R}_0^+)^{(\mathbb{R}^{\mathcal{M}})}$.

A usual way of displaying the reaction $\langle\mathcal{M},\mathcal{R},(\alpha,\beta),w\rangle$ is the following:

$$(1) \qquad \sum_{m=1}^{M} \alpha(m,r)X(m) \xrightarrow{w_r} \sum_{m=1}^{M} \beta(m,r)X(m) \quad (r \in \mathcal{R} \text{ or } r=1,2,\ldots,R)$$

and here (1) is called the rth elementary reaction, $X(m) \in \mathcal{M}$ is the mth chemical component or species, the nonnegative integers $\alpha(m,r)$ and $\beta(m,r)$ are the stoichiometric coefficients or molecularities expressing the number of X(m) needed to and produced in the rth elementary reaction. The matrix $\gamma := \beta - \alpha$ and its columns $\gamma(.,r)$ - the elementary reaction vectors - will often be used below.

DEFINITION 2. The stochastic model of the reaction $\langle \mathcal{M}, \mathcal{R}, (\alpha,\beta), w\rangle$ is a continuous time Markovian jump process with the state space $\mathbb{Z}^M$ and with the infinitesimal transition probabilities

$$a_{np} := \sum_{r:\ \gamma(\cdot,r)=p-n} w_r(n) \qquad (n,p \in \mathbb{Z}^M).$$

From now on it will be supposed that the kinetics w fulfils

$$w_r(n) > 0 \iff n \geq \alpha(\cdot,r) \quad (r \in \mathcal{R},\ n \in \mathbb{Z}^M);$$

as the most often used form for w is:

$$w_r(n) = k(r) \prod_{m=1}^{M} \prod_{j_m=0}^{\alpha(m,r)-1} (n_m - j_m) \quad (r \in \mathcal{R},\ k(r) \in \mathbb{R}^+,\ n \in \mathbb{Z}^M).$$

Here we are interested in the best possible description of the directed graph corresponding to the embedded Markov chain of the stochastic model of complex chemical reactions. This graph (or synonymously, relation) can obviously be defined in the following equivalent way:

$$\Gamma := \left\{ (n,p) \in \mathbb{Z}^M \times \mathbb{Z}^M;\ \exists r \in \mathcal{R} (p-n = \gamma(\cdot,r)) \wedge (w_r(n) > 0) \right\}.$$

For an arbitrary set $X \subset \mathbb{Z}^M$ let us denote by $X_{\rightarrow}$ the union of the vertices of X and those vertices of Γ that can be reached from the vertices of X by a directed route.

As $(\mathbb{N}_0^M)_{\rightarrow} = (\mathbb{N}_0^M)$, the domain of kinetics can be restricted to $\mathbb{N}_0^M$. Γ will also be used in this naturally restricted sense.

DEFINITION 3. The mechanism $\langle \mathcal{M}, \mathcal{R}, (\alpha,\beta)\rangle$ is said to be reversible, if for all $r \in \mathcal{R}$ there exists an $r' \in \mathcal{R}$ such that $\alpha(\cdot,r') = \beta(\cdot,r)$ and $\beta(\cdot,r') = \alpha(\cdot,r)$.

Being a natural assumption that $\alpha(\cdot,r) = \alpha(\cdot,r')$, $\beta(\cdot,r) = \beta(\cdot,r')$ implies $r=r'$ we take it for granted. Therefore uniqueness of r' in the definition is a consequence. Another consequence is that if the mechanism is reversible then Γ is symmetric.

DEFINITION 4. The mechanism $\langle \mathcal{M}, \mathcal{R}, (\alpha, \beta)\rangle$ is said to be mass conservative if there exists a $\varrho \in (\mathbb{R}^+)^M$ such that $\varrho^T\alpha = \varrho^T\beta$.

RESULTS

Let us introduce the notation

$$K(\alpha) := \{\alpha + n;\ n \in \mathbb{N}_0^M\} \qquad (\alpha \in \mathbb{N}_0^M)$$

for the closed convex cone with vertex α obtained from the first orthant of $\mathbb{Z}^M$ through a shift by α. It is obvious that

$$K(\beta) \setminus K(\alpha) = \{n \in \mathbb{N}_0^M;\ \exists m\ \alpha_m > n_m \wedge \forall\ m\ n_m \geqq \beta_m\}$$

therefore this set is empty if and only if $\beta \geqq \alpha$.

Let us consider the reaction (1). In this general case the following statement holds.

STATEMENT 1. The points of $K(0) \setminus \bigcup_{r \in \mathcal{R}} (K(\alpha(\cdot,r)) \cup K(\beta(\cdot,r)))$ represent closed classes with a single absorbing state: they are isolated absorbing states.

The case R=1

At first we restrict ourselves to the detailed investigation of the reaction:

$$\sum_{m=1}^{M} \alpha(m)X(m) \longrightarrow \sum_{m=1}^{M} \beta(m)X(m),$$

and introduce the notation $\alpha := (\alpha(1), \ldots, \alpha(M))$, $\beta := (\beta(1), \ldots, \beta(M))$.

STATEMENT 2. The points of $K(\beta) \setminus K(\alpha)$ are nonisolated absorbing states. No other absorbing states exist then these and the isolated absorbing states of $K(0) \setminus (K(\alpha) \cup K(\beta))$.

Turning to the connected subgraphs of Γ (those that have edges too) one finds three qualitatively different cases.

THEOREM 1.
(i) If $\alpha \leqq \beta$ then these subgraphs are

$$\{(k, k+\gamma), (k+\gamma, k+2\gamma), \ldots\}$$

where $k \in K(\alpha) \setminus K(\beta)$ is arbitrary. As $K(\beta) \setminus K(\alpha) = \emptyset$, there are no nonisolated absorhing states. All states are transient.

(ii) If $\alpha \geq \beta$ then these subgraphs are

$\{(k-\gamma, k), (k-2\gamma, k-\gamma), \ldots\}$

where $k \in K(\beta) \setminus K(\alpha)$ is arbitrary and it is the single absorbing state in this subgraph, all the other states are transient.

(iii) If $\gamma := \beta - \alpha$ has positive and negative coordinates too (what is equivalent to saying that the reaction is mass conservative) then these subgraphs are finite and they are

$\{(k-\gamma, k), (k-2\gamma, k-\gamma), \ldots, (k-\ell\gamma, k-(\ell-1)\gamma)\}$

where $k \in K(\beta) \setminus K(\alpha)$ is arbitrary and it is the single absorbing state in this subgraph, all the other states are transient. The number $\ell \in \mathbb{N}$ is such that $k-\ell\gamma \in K(\alpha)$ while $k-(\ell+1)\gamma \notin K(\alpha)$.

The time evolution of the process is deterministic under a deterministic initial condition: it either remains in an isolated absorbing state or wanders through a part of one of the directed routes above depending on α, β and the initial condition.

The reversible case R=2

It is relatively easy to extend the arguments for the case

$$\sum_{m=1}^{M} \alpha(m)X(m) \rightleftharpoons \sum_{m=1}^{M} \beta(m)X(m).$$

There are no nonisolated absorbing states in this case. One has two qualitatively different cases regarding the connected subgraphs. (As Γ is symmetric because the mechanism is reversible Γ, the edges will be considered undirected and they will only be given once below.)

THEOREM 2.

(i) If all the coordinates of γ are of the same sign (in the wide sense) then let $\alpha' := \min\{\alpha, \beta\}$, $\beta' := \max\{\alpha, \beta\}$. With these notations the connected subgraphs of Γ (those that have edges too) are

$\{(k, k+\gamma), (k+\gamma, k+2\gamma), \ldots\}$

where $k \in K(\alpha') \setminus K(\beta')$ is arbitrary.

(ii) If γ has positive and negative coordinates too (what is again equivalent to saying that the reaction is mass conservative) then these subgraphs are finite and they are

$\{(k, k+\gamma), (k+\gamma, k+2\gamma), \ldots, (k+(\ell-1)\gamma, k+\ell\gamma)\}$

where $k \in K(\alpha) \setminus K(\beta)$ is arbitrary an the number $\ell \in \mathbb{N}$ is such that $k+\ell\gamma \in K(\beta)$ and $k+(\ell+1)\gamma \notin K(\beta)$. (The same subgraphs are obtained if the role of α and β are interchanged.)

In accordance with Theorem 1 of Tóth (1986) and with the remark after it Γ may only be connected in the case R=1 and in the reversible case R=2 if M=1. But the statements to come below show that if M=1 then no restriction on R is needed in order to obtain similar results.

The case M=1

Here we are concerned with the complex chemical reaction consisting of R elementary reactions

$$\alpha(r)X \longrightarrow \beta(r)X \qquad (r=1,2,\ldots,R).$$

STATEMENT 3. The number of connected subgraphs of Γ (those that have edges too) is the greatest common divisor of $\gamma(1),\ldots,\gamma(R)$ denoted here by $((\gamma(1),\ldots,\gamma(R)))$ where $\gamma(r):=\beta(r)-\alpha(r)$ $(r \in \mathcal{R})$.

A consequence of this statement and of Statement 1 is that Γ is connected if and only if the greatest common divisor of $\gamma(1),\ldots,\gamma(R)$ is 1 and at least one of the stoichiometric coefficients is zero.

The next statement concerns aperiodicity and is related to reversible reactions. Let us decompose the numbers $\gamma(r)$ in the following way:

$$\gamma(r)=2^{g(r)}h(r) \qquad (2\nmid h(r);\ r=1,2,\ldots,R).$$

With these notations the following statement holds.

THEOREM 3. If a reversible complex chemical reaction contains a single chemical component then the embedded Markov chain of its stochastic model is of period 2 if and only if $g(1)=\ldots=g(R)$ and otherwise it is aperiodic.

As a consequence of Statement 3 and Theorem 3 the full classification of the possible structures of Γ for the reversible case when M=1 looks like this.

In all the cases below one may have the following isolated absorbing states:

$$\{n;\ 0 \le n \le \min\{\alpha(r), \beta(r);\ r=1,2,\ldots,R\}\}$$

Disregarding these states what one obtains is denoted by Γ_0.

CORROLARY 1.

(i) $R=1$, $\gamma(1)=1$

Γ_0 is irreducible and aperiodic.

(ii) $R>1$, $((\gamma(1),\ldots,\gamma(R)))=1$, $g(1)=\ldots=g(R)$

Γ_0 is irreducible and of period 2.

(iii) $R>1$, $((\gamma(1),\ldots,\gamma(R)))=1$, not all $g(r)$s are equal

Γ_0 is irreducible and aperiodic.

(iv) $R=1$, $\gamma(1)=d>1$

Γ_0 consists of d connected subgraphs all of which are of period 2.

(v) $R>1$, $((\gamma(1),\ldots,\gamma(R)))=d>1$, $g(1)=\ldots=g(R)$

Γ_0 consists of d connected subgraphs all of which are of period 2.

(vi) $R>1$, $((\gamma(1),\ldots,\gamma(R)))=d>1$, not all $g(r)$s are equal

Γ_0 consists of d connected aperiodic subgraphs.

Detailed proofs of the statements will be published later.

REFERENCES

Érdi P., Tóth J. (1987): Mathematical models of chemical reactions. Manchester University Press (in preparation).

Érdi P., Tóth J., Hárs V. (1981): Some kinds of exotic phenomena in chemical systems. In: Qualitative Theory of Differential Equations, 1979, János Bolyai Math. Soc. and North Holland Publ. Co., Amsterdam-London, 205-229.

Kurtz T. G. (1981): Approximation of Population Processes, SIAM, Philadelphia, PA.

Niven I., Zuckerman H. S. (1970): An Introduction to the Theory of Numbers (Third ed.), J. Wiley and Sons Inc., New York-London-Sydney-Toronto.

Tóth J. (1986): Structure of the state space in stochastic kinetics. In: Proc. 5th Pann. Symp. Math. Stat., 1985 (in press).

Hungarian Academy of Sciences
Computer and Automation Institute
P. O. B. 63
H-1502 Budapest
Hungary

f-PROJECTION OF σ-FINITE MEASURES

AND ITS INFORMATION-THEORETIC AND STATISTICAL APPLICATIONS

Igor Vajda

Prague

Key words: f-divergence, f-projection, σ-finite measures, probability measures, spectral measures, minimum distance estimation.

ABSTRACT

f-divergences of finite and σ-finite measures are considered. A finite measure minimizing on a given set the f-divergence with a fixed σ-finite measure is called an f-projection of the σ-finite measure on the given set. Applications of f-projections of this type in source coding, in particular speech coding, and in minimum distance estimation are considered. Existence and characterization theorems for f-projections are presented.

PRELIMINARIES

Let (Ω, S) be a measurable space, $\mathbb{P}(S) \subset \mathbb{S}(S)$ correspondingly the sets of all probability and σ-finite measures on S, and $\mathbb{F}(S)$ the set of all finite signed measures on S. For every $\mu \in \mathbb{S}(S)$, $\mathbb{F}(\mu)$ denotes the set of all measures from $\mathbb{F}(S)$ dominated by μ.

Let $|\nu|$ be the total variation of a signed measure $\nu \in \mathbb{F}(S)$ and

$$\|\nu\| = |\nu|(\Omega).$$

It is well known (cf. III.7 in Dunford and Schwartz (1958)) that

$\|.\|$ is a norm making a Banach space out of $\mathbb{F}(S)$. The linear subsets $\mathbb{F}(\mu)$, $\mu \in \mathbb{S}(S)$, of $F(S)$ are Banach subspaces. To this end it suffices to take into account that the mapping sending $\nu \in \mathbb{F}(\mu)$ to the Radon-Nikodym density $d\nu/d\mu$ is an isometrical isomorphism between $\mathbb{F}(\mu)$ and the Banach space $L_1(\mu)$ of S-measurable and absolutely μ-integrable real valued functions of variable $\omega \in \Omega$.

In the sequel we consider convex sets in Banach spaces and convex functions of variables from Banach spaces in the sense of Barbu and Procupanu (1978). Let $\mathbb{R}$ be the real line and

$$^*\mathbb{R} = \mathbb{R} \cup \{\infty\}, \qquad \mathbb{R}^* = {}^*\mathbb{R} \cup \{-\infty\}$$

and let us assume the usual arithmetic and topology in $^*\mathbb{R}$ and $\mathbb{R}^*$ (cf. the Prerequisities section in Halmos (1964)). We denote by $K[0,\infty)$ the set of all continuous convex functions $f: [0,\infty) \mapsto {}^*\mathbb{R}$ with

$$0 \le f(x) < \infty \quad \text{for } x \in (0,\infty), \qquad f(1) = 0.$$

Obviously, every function f from $K[0,\infty)$ is uniquely given by its values $f(x)$ for $x \in (0,\infty)$ and the values $f(x)$ are nonnegative for all $x \in [0,\infty)$. As examples one may consider the functions χ^α, $\alpha \in [1,\infty)$, or I_α, $\alpha \in \mathbb{R}$, defined for $x \in (0,\infty)$ by

$$\chi^\alpha(x) = |x-1|^\alpha$$

or

$$I_\alpha(x) = \frac{x^\alpha - \alpha x + \alpha - 1}{\alpha(\alpha-1)} \quad \text{if } \alpha \ne 0,\ \alpha \ne 1.$$

The values $I_0(x)$, $I_1(x)$ are assumed to be defined for every $x \in (0,\infty)$ by the continuous extension of the above considered expression $I_\alpha(x)$, i. e.

$$I_0(x) = -\ln x + x - 1, \quad I_1(x) = x \ln x - x + 1$$

(cf. Sec. 2 in Liese and Vajda (1986)).

For every $f \in K[0,\infty)$ and every $\mu \in \mathbb{S}(S)$, $\nu \in \mathbb{S}(S) \cap \mathbb{F}(\mu)$ we consider the f-divergence $f(\nu\|\mu)$ of ν and μ defined by

$$(1) \qquad f(\nu\|\mu) = \int f\left(\frac{d\nu}{d\mu}\right) d\mu.$$

It follows from (A.28) in Liese and Vajda (1986) that

$$(2) \qquad 0 \le f(\nu\|\mu) \le f(0) + f(\infty)/\!\!/\infty$$

where

$$f(\infty)/\!\!/\infty = \lim_{x\to\infty} f(x)/x.$$

Further (cf. ibid.), if $f(x)$ is strictly convex at the point $x = 1$, then $f(\nu\|\mu) = 0$ iff $\nu = \mu$ and the f-divergence upper bound $f(0) + f(\infty)/\!\!/\infty$ may be attained on ν, μ under consideration only if $f(0) + f(\infty)/\!\!/\infty = \infty$.

f-PROJECTION AND ITS APPLICATIONS

Let $\mu \in \mathbb{S}(S)$ and let us consider a nonvoid $\mathbb{A} \subset \mathbb{S}(S) \cap \mathbb{F}(\mu)$. An f-projection of μ on $\mathbb{A}$ denotes a measure $\nu_* \in \mathbb{A}$ for which

$$f(\nu_*\|\mu) = \inf_{\nu\in\mathbb{A}} f(\nu\|\mu).$$

f-projections P_* of probability measures $Q \in \mathbb{P}(S)$ on sets $\mathbb{A} \subset \mathbb{P}(S) \cap \cap \mathbb{F}(\mu)$, with different restrictions on f and $\mathbb{A}$, have been considered by many authors (e. g. Kullback (1959), Perez (1967), Csiszár (1975), Liese (1975), Rüschendorf (1984)). Next we consider several information-theoretic applications of f-projections of σ-finite measures on sets of finite measures, in addition to the applications considered by the cited authors.

Let us consider two Gaussian stationary time series $\xi = (\xi_1, \xi_2, \ldots)$, $\eta = (\eta_1, \eta_2, \ldots)$ with Borel measurable power spectral densities $p(\omega)$, $q(\omega)$ on $\Omega = [-\pi, \pi]$. Define spectral measures μ, ν on the σ-algebra S of Borel subsets of Ω by

$$\mu(E) = \frac{\pi}{4}\int_E p(\omega)d\omega, \qquad \nu(E) = \frac{\pi}{4}\int_E q(\omega)d\omega.$$

According to Gray, Neuhoff and Shields (1975), the generalized Ornstein distance of the series ξ and η equals

$$\frac{1}{2\pi}\int_{-\pi}^{\pi}(\sqrt{q(\omega)}-\sqrt{p(\omega)})^2 d\omega = I_{\frac{1}{2}}(\nu\|\mu).$$

The $I_{\frac{1}{2}}$-divergence is obviously a metric in the set of all spectral measures of above considered type. A source coding minimizing the generalized Ornstein distance thus leads to the problem of $I_{\frac{1}{2}}$-projection of finite measures.

A non-metric distance essentially equal the I_0-divergence of above considered spectral measures ν, μ has been successfully tested in coding of speech signals (cf. Nedoma, Feistauerová, Pacovský and Vajda (1985)). Evaluation of a linear model minimizing $I_0(\nu\|\mu)$ for a signal with spectral measure μ leads to nothing but the I_0-projection of μ on the set $\mathbb{A}$ of all spectral measures of linear stationary models of given order.

The last application considered in the present paper is based on the frequently used representation of the empirical evidence of mathematical statistics (the sample vector) by a measure, in particular by a probability measure $Q \in \mathbb{P}(S)$. Some representations make realistic the assumption that the parametrized family of theoretical models $\mathbb{A} = \{P_\theta : \theta \in \Theta\} \subset \mathbb{P}(S)$ is dominated by (or rather equivalent with) Q. This in particular applies to the kernel representation of Beran (1977). If the assumption holds then the minimum f-divergence estimation reduces to the f-projection of Q on $\mathbb{A}$.

f-PROJECTIONS ON PARAMETRIZED FAMILIES

Let us consider the problem of f-projection described in the preceding paragraph with the additional assumption $\Theta \subset \mathbb{R}^n$. This is a good model to start with since it reduces the f-projection problem to a minimization of function $\phi(\theta) = f(P_\theta \| Q)$ on a set Θ from the relatively simple Euclidean space $\mathbb{R}^n$.

The existence of f-projection and chances to its evaluation depend heavily on whether the function $\phi(\theta)$ is continuous on Θ. For I_α-divergences, $0 < \alpha < 1$, this problem has been solved by Víšek (1981). There is however a more general solution under weaker assumptions on the family $\mathbb{A}$. For the proof we refer to Sec. 10 of Liese and Vajda (1986), where it is shown that our result holds in fact for arbitrary topological space Θ and for arbitrary probability measure $Q \in \mathbb{P}(S)$ (i. e. not only of $Q \gg \mathbb{A}$ considered in the present paper).

Theorem 1. Let $f(0) + f(\infty)/\!\!/\infty < \infty$ and let $\mathbb{A}$ be dominated by $\mu \in \mathbf{S}(S)$. If the densities $dP_\theta/d\mu$ are μ-a. e. continuous on Θ, then $f(P_\theta \| Q)$ is continuous on Θ.

Obviously, if Θ is convex and $f(dP_\theta/dQ)$ is convex on Θ Q-a.s. then $f(P_\theta \| Q)$ is convex on Θ. This further simplifies the question of existence and uniqueness, as well as the evaluation, of the f-projection of Q on $\mathbb{A}$.

GENERAL EXISTENCE THEOREM

Let us consider $\mu \in \mathbb{S}(\mathcal{S})$ and a convex nonvoid $\mathbb{A} \subset \mathbb{S}(\mathcal{S}) \cap \mathbb{F}(\mu)$. In addition to the total-variation topology we consider on $\mathbb{F}(\mu)$ the weak topology sent from $L_1(\mu)$ via the above considered isometric isomorphism. The following theorem generalizes the main result of Liese (1975) or Proposition 8.5 in Liese and Vajda (1986).

Theorem 2. If $\mathbb{A}$ is a closed set in the Banach space $\mathbb{F}(\mu)$ and $f(\infty)/\!/\infty = \infty$ then the f-projection of μ on $\mathbb{A}$ exists.

Proof. We shall follow the idea of the proof of the above cited Proposition 8.5. Let $0 \le v \le \infty$ be the infimum of $f(\nu\|\mu)$ on $\mathbb{A}$. If $v = \infty$ then it is nothing to prove. Let us assume $0 \le v < \infty$ and let us consider the nonvoid

$$\mathbb{B} = \{\nu \in \mathbb{A}: f(\nu\|\mu) \le 2v\}.$$

By what is presented on pp. 102, 103 in Barbu and Precupanu (1978), $f(\nu\|\mu)$ is convex and lower semicontinuous on $\mathbb{A}$. The assumption that $\mathbb{A}$ is closed and convex thus implies that $\mathbb{B}$ is closed and convex too. By a Mazur's theorem (Theorem 2 in V.1 of Yosida (1965)), each convex closed set in a Banach space is weakly closed. This implies that $\mathbb{B}$ is weakly closed and that $f(\nu\|\mu)$ is weakly lower semicontinuous on $\mathbb{B}$. It suffices to prove that $\mathbb{B}$ is weakly sequentially compact because the Weierstrass theorem (Theorem 1.1 in Barbu and Precupanu (1978)) then implies that the minimum of $f(\nu\|\mu)$ on $\mathbb{B}$ exists. By IV.9.1 in Dunford and Schwartz (1958), it suffices to prove that for every $\nu \in \mathbb{B}$ with $p = d\nu/d\mu$ and every sequence $E_1 \supset E_2 \supset \ldots$ of sets from $\mathcal{S}$ with empty intersection it holds

$$\lim_{n\to\infty} \int_{E_n} p\,d\mu = 0$$

and that this convergence is uniform for all $\nu \in \mathbb{B}$. If $\mu(E_n) = 0$ for some n then it is nothing to prove. In the opposite case define a sequence $c_1 \le c_2 \le \ldots$ tending to ∞ as follows

$$c_n = \frac{1}{\sqrt{\mu(E_n)}}.$$

If $c_n > 1$ then it holds for every $\nu \in \mathbb{B}$

$$\int_{E_n} p\,d\mu \le c_n\mu(E_n) + \int_{\{p>c_n\}} \frac{p-1}{f(p)} f(p)\,d\mu + \mu(E_n)$$

$$\le \frac{1}{c_n} + \frac{c_n-1}{f(c_n)} f(\nu\|\mu) + \frac{1}{c_n}$$

$$\le \frac{2}{c_n} + \frac{c_n-1}{f(c_n)} 2v.$$

The desired result now follows from the assumption $f(\infty)/\infty = \infty$.

GENERAL CHARACTERIZATION THEOREM

Let us consider the same model as before with the following additional assumptions:

(1) $f(x)$ is differentiable at $x \in (0,\infty)$ with the derivative $f'(x)$; we put $f'(0)$ to be equal the right-hand derivative of $f(x)$ at $x=0$;

(2) there exists $c > 1$ such that

$$\limsup_{x\to\infty} \frac{f(cx)}{f(x)} < \infty;$$

(3) $f(\nu\|\mu) < \infty$ for every $\nu \in \mathbb{A}$.

The following generalization of Theorem (8.10) of Liese and Vajda (1978) follows from the obvious modification of the argument presented in Sec. 8 ibid.

Theorem 3. $\nu_* \in \mathbb{A}$ is an f-projection of μ on $\mathbb{A}$ iff $f'(d\nu_*\|d\mu) \in L_1(\nu)$ for all $\nu \in \mathbb{A}$ and

$$\int f'\left(\frac{d\nu_*}{d\mu}\right) d\nu_* = \inf_{\nu\in\mathbb{A}} \int f'\left(\frac{d\nu_*}{d\mu}\right) d\nu.$$

This result sharpens part a) of Theorem 5 of Rüschendorf (1984). For $f = I_1$ Theorem 3 implies the following result.

Corollary. $\nu_* \in \mathbb{A}$ is an I -projection of a finite measure μ on $\mathbb{A}$ iff for all $\nu \in \mathbb{A}$ it holds $\nu \ll \nu_*$ and

$$I_1(\nu_*\|\mu) + I_1(\nu\|\nu_*) \le I_1(\nu\|\mu).$$

If μ and $\mathbb{A}$ satisfy the additional condition $\{\mu\} \cup \mathbb{A} \subset \mathbb{P}(S)$ then Corollary reduces to the Csiszár (1975) necessary and sufficient condition for I_1-projection of a probability measure μ on a set of probability measures.

ACKNOWLEDGEMENT

The author would like to express his thanks to F. Liese and J. Michálek for valuable discussions and suggestions concerning the subject of present paper.

REFERENCES

Dunford N, Schwartz J. T. (1958): Linear Operators. Part I: General Theory. Interscience Publishers, New York.

Barbu V., Precupanu T. (1978): Convexity and Optimization in Banach Spaces. Editura Academiei, Bucharest.

Liese F., Vajda I. (1986) Convex Statistical Distances. Teubner Verlag, Leipzig.

Kullback S. (1959): Information Theory and Statistics. J. Wiley, New York.

Perez A. (1967): Information-theoretic risk estimates in statistical decision. Kybernetika 3, 1-21.

Csiszár I. (1975). I-divergence geometry of probability distributions and minimization problems. Ann. Probab. 3, 146-158.

Liese F. (1975): On the existence of f-projections. Colloq. Math. Soc. J. Bolyai 16, 431-446.

Rüschendorf L. (1984): On the minimum discrimination information theorem. Recent Results in Estimation Theory. Ed. E. J. Dudewicz, D. Plachky, P. K. Sen. Oldenbourg, München.

Gray R. M., Neuhoff D. L., Shields P. C. (1975): A generalization of Ornstein's $\bar{d}$-distance with applications to information theory. Ann. Probab. 3, 315-328.

Nedoma J., Feistauerová J., Pacovský J., Vajda I. (1985): Analysis of improved methods of speech signal compression (in Czech). Res. rep. N. 1341, Institute of Information Theory and Automation, Czechoslov. Acad. Sci., Prague.

Beran R. (1977): Minimum Hellinger distance estimates for parametric models. Ann. Statist. 5, 445-463.

Víšek J. Á. (1981): On the continuity of minimal α-entropy. Kybernetika 17, 32-44.

Yosida K. (1965): Functional Analysis. Springer Verlag, Berlin.

Czechoslovak Academy of Sciences
Institute of Information Theory and Automation
182 08 Prague 8
Pod vodárenskou věží 4
Czechoslovakia

ON THE INFORMATION OF EXPERIMENTS

WHEN OBSERVING TIME SERIES

Sándor Veres

Budapest

Key words: statistical information, ARMA processes, Monte Carlo study

ABSTRACT

This paper is devoted to estimation problems of ARMA processes in view of statistical information theory. First Lindley's definition of the information content of a statistical experiment is extended to the information on the model structure. It is shown that the whole information supplied by the experiment can be decomposed into the information on the model structure and the information on the parameters of the model. For nearly nonstationary and seasonal models the properties of estimators can not be evaluated using classical information measures. The arising problems are illustrated by a special example, where the distribution of estimators are studied by the Monte Carlo method.

INTRODUCTION

Measures of information contained in an experiment has been introduced by Lindley (1956), Mallows (1959), Rényi (1961), De Groot (1962) and a deep study was made by Nemetz (1972). The most natural approach is Bayesian. If we have one prior, Fisher's parametric measure and its generalizations can be used, which are defined as follows.

Let $P = \{ P_\Theta, \Theta \in \underline{\Theta} \}$ be a parametric family of probability measures on a measurable space $(X,\mathcal{B})$ dominated by a σ-finite measure μ. Let $f(x|\Theta) = dP_\Theta/d\mu$, $\Theta \in \underline{\Theta}$. With the notations $\ell(x,\Theta) = \log f(x|\Theta)$, $\dot\ell = \partial\ell/\partial\Theta$, $\ddot\ell = \partial^2\ell/\partial\Theta\partial\Theta^T$ the most important parametric measures of information were defined by

- Fisher (1925): $I_F(\Theta) = E_\Theta \dot\ell \dot\ell^T$
- Vajda (1973) : $I_V(\Theta) = E_\Theta|\dot\ell|^\alpha$, $\alpha \geq 1$
- Mathai (1967): $I_{Mat}(\Theta) = [E_\Theta|\dot\ell|^\alpha]^{1/\alpha}$, $\alpha \geq 1$
- Boekee (1977): $I_{Bo}(\Theta) = [E_\Theta|\dot\ell|^{s/s-1}]^{s-1}$.

The main nonparametric measures of information express the amount of information supplied by the data for discriminating in favour of a distribution f_1 against another f_2. These are the measures of

- Kullback-Leibler (1951): $I_{KL}(f_1,f_2) = E_{f_1}\log(f_1/f_2)$
- Rényi (1961): for $\alpha>0, \alpha\neq 1$ $I_R(f_1,f_2) = (\alpha-1)^{-1}\log E_{f_2}(f_1/f_2)^\alpha$
- Csiszár (1963): $I_C(f_1,f_2) = E_{f_2}(\phi(f_1/f_2))$ for ϕ convex function
- Vincze (1982): $I_V(f_1,f_2) = \int_X \frac{(f_1-f_2)^2}{f_1+f_2}\, d\mu$

In this paper we are mainly interested in the information content of experiments when observing time series. For stationary processes a practically useful method is to compute the estimated Fisher information matrix, i.e. the Hesse matrix, and to approximate by its inverse the covariance matrix of the maximum likelihood (ML) estimator.

Let us consider ARMA processes $\{y_t\}$ satisfying the equation

$$y_t + a_1 y_{t-1} + \ldots + a_p y_{t-p} = e_t + b_1 e_{t-1} + \ldots + b_q e_{t-q} \qquad (1)$$

where e_t is a Gaussian white noise sequence with $Ee_te_s = \sigma^2\delta_s^t$. Assume that the characteristic polinomials $a(z) = 1+a_1z+\ldots+a_pz^p$ and $b(z) = 1+b_1z+\ldots+b_qz^q$ have all their zeros outside the complex unit disc. Seasonal time series satisfy equations of the type

$$(1 - z^c)\, a(z)\, y_t = (1 - z^d)\, b(z)\, e_t$$

with the backward shift operator $z \cdot y_t = y_{t-1}$, $t \in Z$, and $c \neq d$

are positive integers.

For small sample sizes and especially if the zeros of $a(z)$ or $b(z)$ are close to the unite circle the Fisher information matrix may give a poor characterization of the ML estimator. To demonstrate the necessity of a study on the behaviour of estimators for small sample sizes let us consider the simplest AR(1) process $\{y_t\}$ satisfying the equation

$$y_t = q y_{t-1} + e_t , \qquad t = 1,\dots,T$$

where for the sake of simplicity $e_t \sim N(0,1)$ is assumed. It is known that the asymptotic distribution of the estimator

$$\hat{q} = \Big(\sum_{t=1}^{T} y_t y_{t-1} \Big) \Big/ \sum_{t=1}^{T} y_t^2$$

is as $T \to \infty$

$$\sqrt{T}(\hat{q} - q) \xrightarrow{\mathcal{D}} N(0,d), \ d = (1-q^2)^{-1} \text{ for } |q| < 1 \tag{2}$$

$$\sqrt{T}(\hat{q} - q) \xrightarrow{\mathcal{D}} \Big(\int_0^1 W_t dW_t\Big)\Big(\int_0^1 W_t^2 dt\Big)^{-1} , \text{ where } W_t,\ t \geq 0, \text{ is a standard Wiener process, for } q = \pm 1 \tag{3}$$

$$q^T . \frac{(\hat{q} - q)}{(q^2-1)^{1/2}} \xrightarrow{\mathcal{D}} z , \text{ for } q > 1, \text{ where } P(z < x) = \int_{-\infty}^{x} \frac{\pi^{-1}}{1+t^2} dt . \tag{4}$$

The proofs of limit relations (2)-(4) can be found in Rao (1984) and in Veres (1985). These asymptotic relations suggest that for increasing q and fixed sample size $\hat{q}$ gives better and better estimator of q . Since the asymptotic distributions in (2)-(4) change discontinuously, therefore there are necessarily sample sizes T and parameters q close to 1 such that the asymptotic distributions give very bad approximations of the true distribution of $\hat{q}$. In general, if a zero of $a(z)$ is close to the unit circle, it would be important to find methods for the correct evaluation of ML estimators.

In the case of AR(1) processes the first derivation of the distribution of small sample estimators was given by Arató (1982), where the exact distributions were computed in the continuous time case and approximations were used for discrete time processes.

STRUCTURAL INFORMATION

By Lindley's definition (1956) an experiment is a quadruple $E = (X, \mathcal{B}, \underline{\Theta}, P)$, where P is an a priori probability measure on the composed set of parameters

$$\underline{\Theta} = \bigcup_{i \in S} \Theta_i \quad .$$

Assume that Θ_i, $i \in S$, are open subsets of R^{d_i}, and Θ_i is the set of the possible parameter vectors of the model M_i. Furthermore let the densities $f(x|\Theta)$ satisfy all the customary regularity conditions referring to their differentiability with respect to Θ . S is an index set of the possible models. Then $p_i = P(\Theta \in \Theta_i)$, $i \in S$, is the a priori distribution of the model structure. Denote by $p(M_i|x)$, $i \in S$, the posteriori distribution of the model M_i after observing $x \in X$.

DEFINITION. The information content of an experiment on the model structure is

$$Q(x,E) = -\sum_{i \in S} p_i \log p_i + \sum_{i \in S} p(M_i|x) \log p(M_i|x)$$

and the average information content is defined as

$$Q(E) = \int_X Q(x,E) f(x) d\mu(x),$$

where $f(x)$ is the density of the marginal distribution of x,

$$f(x) = \sum_{i \in S} \int_{\Theta_i} f(x|\Theta) dP(\Theta),$$

which we also call the density of the "averaged statistical model". Let the definition of information on the parameter vector be given with respect to the different models M_i, $i \in S$, as Lindley's (1956) definition

$$\mathcal{J}_{\Theta}(E|M_i) = \int_X \mathcal{J}(x,E|M_i) f(x) d\mu(x)$$

where

$$\mathcal{J}(x,E|M_i) = \left[-\int_{\Theta_i} \log p(\Theta) dP(\Theta) + \int_{\Theta_i} p(\Theta|x) \log p(\Theta|x) dP(\Theta) \right] / p_i$$

The whole information content of the experiment is defined as

$$\mathcal{J}(E) = I((i,\Theta),x),$$

where $I(\cdot,\cdot)$ denotes the mutual information of two random variables (for its definition see e.g. Pinsker (1961) or its generalization in Veres (1982)). Then the following proposition holds.

THEOREM. The information of an experiment splits into two parts: the structural information and the sum of parametric informations weighted by the a priori distribution of the models, i.e.

$$\mathcal{J}(E) = Q(E) + \sum_{i\in S} p_i \mathcal{J}_\Theta(E|M_i)$$

The proof of this theorem can be obtained by some straighforward computations.

Furthermore, the parametric information $\mathcal{J}_\Theta(E|M_i)$ of the experiment is the average of the Kullback-Leibler information in favour of the different models against the "averaged stochastic model". Denote the Kullback-Leibler information numbers by

$$I_{KL}(\Theta,A) = \int_X f(x|\Theta) \log \frac{f(x|\Theta)}{f(x)} d\mu(x) .$$

Then

$$\mathcal{J}_\Theta(E|M_i) = p_i^{-1} \int_{\Theta_i} I_{KL}(\Theta,A) dP(\Theta) .$$

We note that the above measures of structural and parametric information are interesting not only in those cases, where an a priori distribution is known. These definitions make possible comparative studies of different structure and parameter estimation procedures. A discussion of this approach will be given in a forthcoming paper (cf. Akaike (1979)).

PARAMETRIC INFORMATION

Besides Lindley's measure of parametric information the practically most useful measures are those, which give some characterization of the estimators or describe the power of discrimination of models at a parameter point. Kullback-Leibler measure of information is used to define the divergence between two distributions

$$D(f_1,f_2) = I_{KL}(f_1,f_2) + I_{KL}(f_2,f_1),$$

which is used as the measure of discrimination information. For two close parameters Θ and $\Theta + \Delta\Theta$

$$D(f(x|\Theta),f(x|\Theta+\Delta\Theta)) = \Delta\Theta' I_F(\Theta)\Delta\Theta + o(\|\Delta\Theta\|^3)$$

holds (see in Kullback (1959)) and therefore the local discrimination information essentially leads to the computation of the Fisher information.

For illustrating the problems in connection with the use of information measures in time series analysis we analyse a simple nonstationary model. Let y_t be a linear process satisfying the equation

$$y_t - q\, y_{t-1} - y_{t-6} + q\, y_{t-7} = \sigma e_t , \qquad E e_t e_s = \delta_s^t \quad (6)$$

which is a seasonal time series with period 6. We are interested in values of q close to 1 and in small sample sizes. For the sake of simplicity assume that the initial values are $y_t = 0,\ t \le 0$. In the following Figures both the results of Monte Carlo studies and the theoretical Fisher information are illustrated.

The simulations were made for the conditional model (6) with $\sigma^2 = 1$ and for number of observations T= 50 with 100 independent repetitions. The source noise $\{e_t\}$ was generated by a normal random generator verified and tested in many respects in Dunham (1983) . In Fig. 1 the average (for 100 independent repetitions) bias of the ML estimator of q are shown. Observe that the bias are small for small and large values of the true q . The largest bias were obtained approximetly at q = .8 .

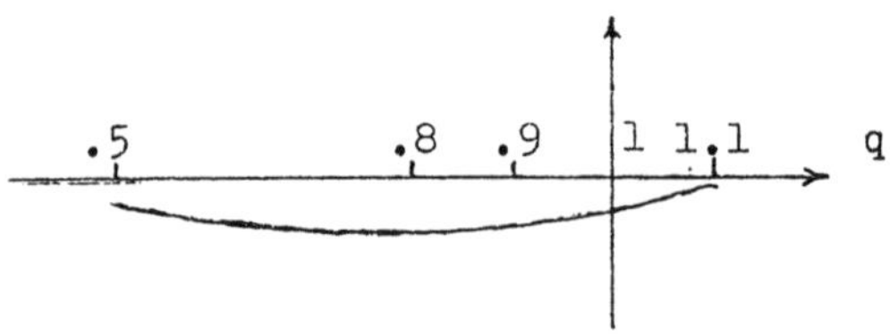

Fig. 1

Fig. 2 shows four main characteristics of parametric statistical information:

1: The average (for 100 independent repetitions) of $1/\hat{I}_F(q)$, where $\hat{I}_F(q)$ denotes the estimated Fisher information.
2: The exact values of $1/I_F(q)$.(Its computation was made by a recursive procedure on a computer.)
3: The estimated variance of the ML estimator
4: The exact values of $1/\tilde{I}_F(q)$, where $\tilde{I}_F(q)$ denotes the Fisher information for the (unconditional) stationary model

5: The average value of $I_F(q)^{-1}(1+\gamma^2(q))$, where $\gamma^2(q)$ denotes the statistical curvature defined in Efron (1975).

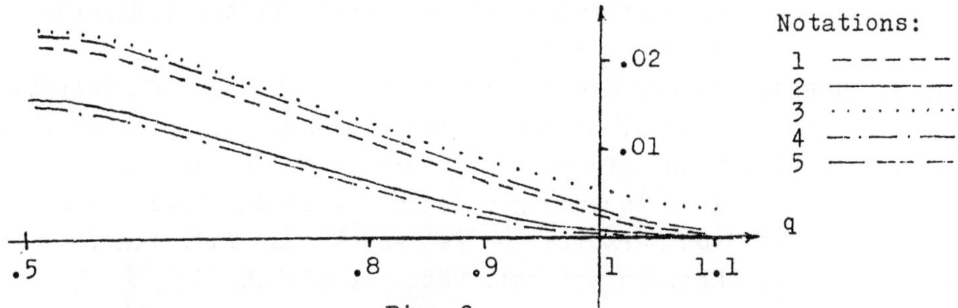

Fig. 2

The following observations throw some light on the nature of changes of the parametric information content for nearly non-stationary and seasonal models.

- $1/I_F(q)$ significantly underestimates the true variance of the ML estimator (which was here estimated by repeated simulations).
- for stable values of q $(q \ll 1)$ $1/\hat{I}_F(q)$ approximates the true variance of the estimator. For q in the neighbourhood of 1 the approximation becomes extremely bad.
- there is a general tendency that the estimators are continuously improving when approaching from stationarity towards nonstationarity of the observed process. It is seen that the parameters of seasonal and explosive processes can be better approximated (or at least as well) as for stationary processes.

Considering the above observations it seems to be an important problem to find methods for the correct evaluation of the variance of the estimators for seasonal and nearly nonstatinary processes. One approach would be to use Efron's statistical curvature for the modification of the estimated variance of the ML estimator as

$$\mathrm{Var}(\hat{q}) \simeq I_F(q)^{-1}(1 + \gamma_q^2), \qquad (7)$$

where γ_q^2 denotes the statistical curvature of an experiment defined in Efron (1975). In our example, this approximation has given better results only for the stationary values of q . In fact, the closer q is to 1 the larger number of observations is needed for the correct application of the approximation (7).

REFERENCES

Akaike, H. (1979): A Bayesian extension of the minimum AIC procedure of autoregressive model fitting, Biometrika 66, 237-242.

Arató, M. (1982): Linear Stochastic Systems with Constant Coefficients, Lecture Notes in Cont. 45, Springer Vlg.

Csiszár, I.(1963): Eine informationstheoretische Ungleichung und ihre Anwendung auf den Beweis der Ergodizitat von Markoffschen Ketten, Magyar Tud. Akad. Mat. Kutató Int. Közl. 8, 85-108.

DeGroot, M.(1972): Uncertainty, information and sequential experiments. Ann. Math. Statist. 33, 404-419.

Efron, B. (1975) : Defining the Curvature of a Statistical Problem. Ann. Stat. 3, 1189-1242.

Kullback,S.(1959): Information Theory and Statistics, John Wiley,N.Y.

Lindley, D.(1956): On a measure of the information provided by an experiment. Ann. Math. Statist. 27, 986-1005.

Mallows, C.(1959): The information in an experiment. J. Roy. Stat. Soc. Ser. B 21, 79-86.

Nemetz, T. (1972): Information type measures and their application in statistics. Lecture Notes, Carleton Univ., Ottawa

Pinsker, M.(1960): Information and Informational Stability of Stochastic Variables and Processes. (in Russian) Akademia Nauk SSSR, Moscow.

Rényi, A. (1961): On measures of entropy and information. In: Proc. 4th Berk. Symp. Statist., Univ. California Press, Vol. I, 547-561.

Rao, M. M.(1984): Probability Theory and Its Application, Academic Press, New York - London.

Veres, S. (1983): Common generalization of information measures, Trans. 9th Prague Conf. on Inf. Th., 263-368.

Vincze, I. (1982): A measure of information and its role in uncertainty relations in mathematical statistics, In: Exchangeability in Prob. Statist. (eds: G. Koch and F. Spizzichino), North Holland P. C.

CSM Institute for Computer Technique
H-1751 Budapest, P.O. Box 65
Hungary

ON REASONS FOR INTRODUCING

MEANS FOR APPROXIMATELY OPTIMAL CHOICE

OF ROBUST PROCEDURE.

Jan Ámos Víšek

Prague

Key words: Robust testing, efficiency rate, local deficiency

ABSTRACT

Reasons for introduction of helping means for approximately optimal choice of robust statistics are explained and supported by an example. A proposal of a characteristic (of statistical procedure) which together with Hampel's influence curve or size and power dependency (see Víšek (1986b)) may serve as such helping means is given and illustrated again by an example.

INTRODUCTION

Huber's solution of a robust estimation and testing problem given in 1964 and 1965 (Huber (1964), (1965)), respectively, has started a new period in development of statistics. Namely it brought to our disposal whole families of statistics in which every statistic corresponded to given value of "contamination level". It implies that selection of one of them (and pretending that this one is the best possible e. g. for estimation of location parameter) was more or less connected - explicitly or implicitly - with expressing our guess about level of contamination. Since it was actually more guess than any statistically based estimation this selection was questionable relying only on an experience or authority. In such a situation it was natural that there was felt a lack of a tool for studying and classifying estimators or tests and their behaviour namely under a change of contamination level. Hence when in 1968 F. R. Hampel has proposed to use for this purpose the Fréchet derivative (or directional derivative - influence curve) it met with the great interest (Hampel (1968)). What was however much more important the influence curve has proved to be in the region of point estimation really powerful tool for these purposes (namely for asymptotic representation and hence for studying asymptotic behaviour of estimators). It was the reason why the idea of influence curve which can't be directly used for testing was modified for it and supported by adequate theory (see Víšek (1983b), Jurečková, Víšek (1984), Víšek (1986a,b). Nevertheless it occurred that for finite sample size situation we would have needed to know a little more about behaviour of robust procedures to be able to provide an advice what statistic was to be chosen. It inspired performance of large simulation studies. Let us recall as an example the Princeton one. Results of these studies as well as the results obtained by means of appropriate approximations of the distributions of test statistics for finite sample size (by means of Ch. Field's and F. R. Hampel's small sample asymptotics (Field, Hampel (1982)) hinted that although the difference between the assumed and the actual underlying probabilistic model can't imply a catastrophic change of behaviour of corresponding robust statistic it may cause (in the case when the level of contamination was guessed wrongly - namely too low) changes of values of statistical characteristics of procedures which are not negligable (Víšek (1983a), Michálek-Vajda-Víšek (1984) - see also the below given example). It was necessary to cope with this problem. One of possible way was to find estimators of contamination level for corresponding models of contaminacy. It was done for Huber's model and for model with general neighbourhoods (Rieder's model) in the form of (strongly) consistent estimators (Víšek (1985), (1986c)). But since the estimators of contamination level include free parameters (e. g. width of window, quantile processes) they are convenient only for special situation when we meet with processing data which we obtain repeatedly and which have in some sense stable character.

It was the source of endeavour to provide for the statisticians the helping means giving them an orientation in problem of selecting procedure. Such means should offer them at first a hint about the rate of changes of statistical characteristics of procedures when the underlying probabilistic model changes. At the second these means should indicate what loss they suffer when choosing the procedure (from given family) which does not correspond to actual contamination level but to some value which is only near to the actual one.

The influence curve (Hampel (1968)) and the size and power of dependency (Víšek (1983b)) may fulfil the first demand. They say statisticians whether the actual values of statistical characteristics (e. g. size and power of test) would be considerably different from the assumed ones when the actual contamination level is different from the assumed one, i. e. they characterize the distance between the assumed curve and the actual one in below given picture. Then, if statistician needs to know precisely the values of them (and it is an almost typical case for testing statistical hypotheses) he has to find what is the actual contamination level.

However sometimes statisticians don't need to learn the value(s) of statistical characteristic(s) of chosen procedures (e. g. value of variance of a point estimator) but they would like to know whether, having known the precise value of contamination level, they could gain a lot (in the sense of more adequate selecting statistical procedure). It is they would appreciate any idea about distance between the actual and attainable curve in below drawn picture. And that is the goal of this paper to give a proposal of means for numerical characterization of mentioned distance.

NOTATION

Let (X,A) be a measurable space and let M denote the set of all probability measures defined on it. Moreover for any subset G of a topological space T denote by G^0 the interior of G.

EXAMPLE

We shall use Huber's model of contaminacy. Let P_0, $P_1 \in M$, $P_0 \neq P_1$ and put for $\varepsilon \in [0,1]$

$$P_i(\varepsilon) = \{Q \in M: Q = (1-\varepsilon)P_i + \varepsilon H;\ H \in M\}.$$

(For the sake of simplicity we assume $\varepsilon_0 = \varepsilon_1 = \varepsilon$.) Let us restrict ourselves on an interval $(0,a) \subset (0,1)$ such that for any $\varepsilon \in (0,a)$ $P_0(\varepsilon) \cap P_1(\varepsilon) = \emptyset$. Put $F(n,\varepsilon,\alpha) = \{\Psi: X^n \to [0,1],\ \sup_{P \in P_0(\varepsilon)} E_P\Psi \leq \alpha\}$. Due to Huber (1965) we may find for any $\varepsilon \in (0,a)$ a family of the most powerful tests (of size α) $\{\Psi(n,\varepsilon,\alpha,\underline{x})\}_{n \in N, \varepsilon \in (0,a), \alpha \in (0,1)}$ ($\Psi(n,\varepsilon,\alpha,\underline{x}): X^n \to [0,1]$ for testing $H(\varepsilon): P^{(n)} \in P_0^{(n)}(\varepsilon)$ against $A(\varepsilon): P^{(n)} \in P_1^{(n)}(\varepsilon)$, i. e. $\sup_{P \in P_0(\varepsilon)} E_P\Psi(n,\varepsilon,\alpha,\underline{x}) \leq \alpha$ and $\inf_{P \in P_1(\varepsilon)} E_P\Psi(n,\varepsilon,\alpha,\underline{x}) = \sup_{\Psi \in F(n,\tau,\alpha)} \inf_{P \in P_1(\varepsilon)} E_P\Psi$. Let us recall briefly construction of these tests. Let p_0 and p_1 are densities of P_0 and P_1, respectively. Having found out the solution $c_0(\varepsilon)$ and $c_1(\varepsilon)$ of the equations $(1-\varepsilon)\{P_1(p_1/p_0 > c_0(\varepsilon)) + c_0(\varepsilon)P_0(p_1/p_0 \leq c_0(\varepsilon))\} = 1$, and $(1-\varepsilon)\{P_0(p_1/p_0 < c_1(\varepsilon)) + c_1^{-1}(\varepsilon)P_1(p_1/p_0 \geq c_1(\varepsilon))\} = 1$, and putting $w_\varepsilon(x) = \log\{\max[c_0(\varepsilon), \min[p_1(x)/p_0(x), c_1(\varepsilon)]]\}$ we may define $T(n,\varepsilon,\underline{x}) = \sum_{i=1}^n w_\varepsilon(x_i)$ and finally establish the test $\Psi(n,\varepsilon,\alpha,\underline{x})$ assigning an appropriate critial value.

In below given graph by the curve going through "+" $\sup_{P \in P_1(\varepsilon)} (1-E_P\Psi(n,\varepsilon,\alpha,\underline{x}))$ (as a function of α) is shown for $P_0 = N(0,1)$, $P_1 = N(1,1)$, $n = 40$ and $\varepsilon = .03$. In another words, this curve displayes the *assumed* error probabilities of the tests $\{\Psi(40,.03,\alpha,\underline{x})\}_{\alpha \in (0,1)}$, i. e. error probabilities of these tests with respect to $P_0(.03)$ and $P_1(.03)$, respectively. The curve running through "*" offers the *actual* error probabilities of the tests $\{\Psi(40,.03,\alpha,\underline{x})\}_{\alpha \in (0,1)}$ if the *actual* contamination level is $\tilde{\varepsilon} = .05$, i. e. this curve represents the set of pairs

$$\{\sup_{P \in P_0(.05)} E_P\Psi(40,.03,\alpha,\underline{x}),\ \sup_{P \in P_1(.05)} (1-E_P\Psi(40,.03,\alpha,\underline{x}))\}_{\alpha \in (0,1)}.$$

Finally, the third curve drawn through "→" gives the *attainable* values of error probabilities for contamination level $\tilde{\varepsilon} = .05$, i.e.

it collects error probabilities of the tests $\{\Psi(40,.05,\alpha,\underline{x})\}_{\alpha\in(0,1)}$ with respect to $P_o(.05)$ and $P_o(.05)$, respectively.

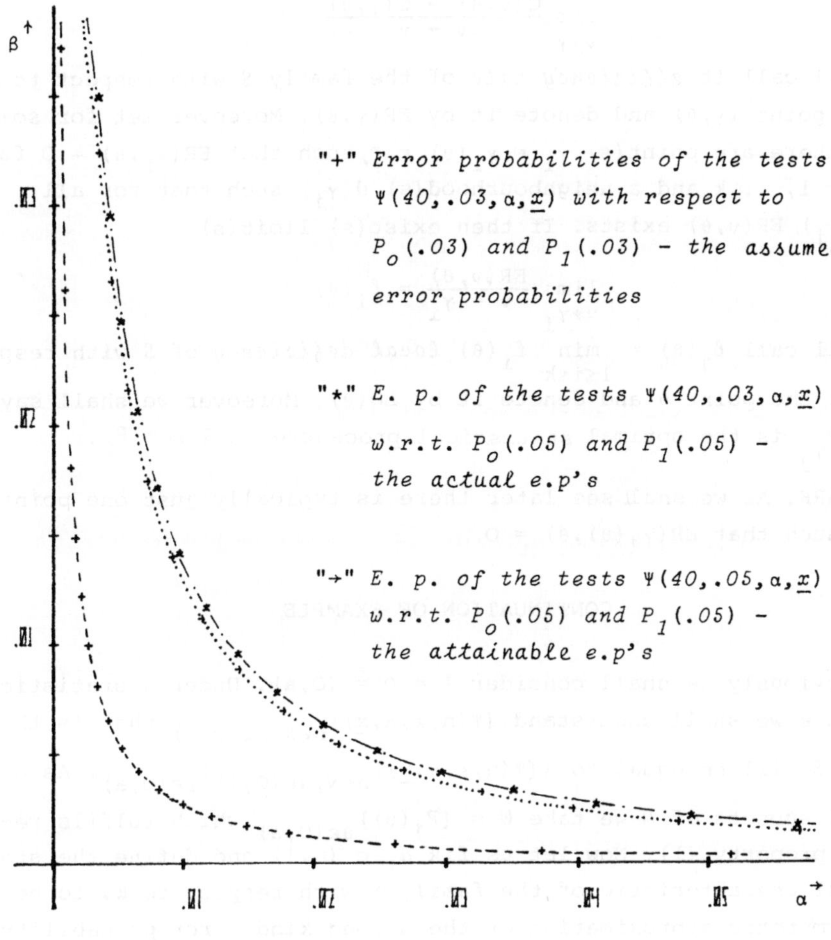

DEFINITION OF EFFICIENCY RATE AND LOCAL DEFICIENCY

Let Γ and Θ be linearly ordered topological spaces and $S = \{SP_\gamma\}_{\gamma\in\Gamma}$ and $W = \{P_\theta\}_{\theta\in\Theta}$ a family of statistical procedures and a family of subsets of M, respectively, such that for any $\theta_o \in \Theta^o$ we have

$$(1) \qquad P_{\theta_o} = \bigcup_{\theta\le\theta_o} P_\theta = \bigcap_{\theta\ge\theta_o} P_\theta .$$

Moreover let $C(\gamma,\theta): \Gamma \times \Theta \to R$ be a statistical characteristic of the statistical procedure SP_γ with respect to a class P_θ. Then if for some $(\gamma,\theta) \in \Gamma \times \Theta$ exists a limit

$$\lim_{\nu\to\gamma} \frac{C(\nu,\theta) - C(\gamma,\theta)}{\nu - \gamma}$$

we shall call it *efficiency rate* of the family S with respect to W at the point (γ,θ) and denote it by $ER(\gamma,\theta)$. Moreover let for some $\theta \in \Theta$ there are point(s) $\gamma_i = \gamma_i(\theta) \in \Gamma$, such that $ER(\gamma_i,\theta) = 0$ for all $i = 1,\dots,k$, and a neighbourhood(s) $O(\gamma_i)$ such that for all $\nu \in O(\gamma_i)$ $ER(\nu,\theta)$ exists. If then exist(s) limit(s)

$$\lim_{\nu\to\gamma_i} \frac{ER(\nu,\theta)}{\nu - \gamma_i} = \ell_i(\theta)$$

we shall call $\ell_j(\theta) = \min_{1\le i\le k} \ell_i(\theta)$ *local deficiency* of S with respect to W at the point θ and denote it by $LD(\theta)$. Moreover we shall say that SP_{γ_j} is the optimal statistical procedure in S for P_θ.

REMARK. As we shall see later there is typically just one point $\gamma_1(\theta)$ such that $ER(\gamma_1(\theta),\theta) = 0$.

CONTINUATION OF EXAMPLE

Obviously we shall consider $\Gamma = \Theta = (0,a)$. Under a statistical procedure we shall understand $\{\Psi(n,\varepsilon,\alpha,\underline{x})\}_{n\in N,\alpha\in(0,1)}$ that is the family S will be equal to $\{\{\Psi(n,\varepsilon,\alpha,\underline{x})\}_{n\in N,\alpha\in(0,1)}\}_{\varepsilon\in(0,a)}$. As a family of subsets of M we take $W = \{P_1(\theta)\}_{\theta\in(0,a)}$ which fulfils required property (1). Now let us fix $\alpha_0 \in (0,1)$ and define the statistical characteristic of the family S with respect to W, to be the asymptotic approximation of the second kind error probability of the test $\Psi(n,\varepsilon,\alpha_0(\varepsilon,\theta),\underline{x})$ with respect to $P_1(\theta)$, where $\alpha_0(\varepsilon,\theta) = \sup\{\alpha: E_{Q_{0\theta}}\Psi_{n,\alpha,\varepsilon} \le \alpha_0\}$, i. e. $C(\varepsilon,\theta) = \sup_{P\in P_1(\theta)} \Phi(a(\varepsilon,\theta,P))$ where

$$a(\varepsilon,\theta,P) = \frac{u_{\alpha_0}\sqrt{\operatorname{var}_{Q_{0\theta}} w_\varepsilon(x)} + n^{1/2}\left[E_{Q_{0\theta}} w_\varepsilon(x) - E_P w_\varepsilon(x)\right]}{\sqrt{\operatorname{var}_P w_\varepsilon(x)}},$$

Φ is the standard normal distribution, u_{α_0} its α_0-upper quantile and $Q_{0\theta}$ the first member of the least favourable pair $(Q_{0\theta},Q_{1\theta})$ of the pair $(P_0(\theta),P_1(\theta))$ (see Huber 1965). (For reasons to use this statistical characteristic see Víšek (1987a).) It is easy to see that the main role for attaining $\sup_{P \in \mathcal{P}_1(\theta)} \Phi(a(\varepsilon,\theta,P))$ will play (for n sufficiently large) the difference $E_{Q_{0\theta}} w_\varepsilon(x) - E_P w_\varepsilon(x)$. Hence, it implies that P should be taken to be equal to $R(\varepsilon,\theta) = (1-\theta)P_1 + \theta \cdot \delta\{p_1/p_0 < c_0(\varepsilon)\}$, where $\delta\{p_1/p_0 < c_0(\varepsilon)\}$ is any probability measure having all mass distributed only over the set $\{p_1/p_0 < c_0(\varepsilon)\}$.

THEOREM 1. In the above given framework the efficiency rate $ER(\varepsilon,\theta)$ exists for any $\varepsilon \in (0,a)$ and $\theta \in (0,a)$ and is equal to $\phi(a(\varepsilon,\theta)) \cdot b(\varepsilon,\theta)$, where ϕ is the density of the standard normal distribution, $a(\varepsilon,\theta) = a(\varepsilon,\theta,R(\varepsilon,\theta))$,

$$b(\varepsilon,\theta) = [\mathrm{var}_{R(\varepsilon,\theta)} w_\varepsilon(x)]^{-1}\Big[\{u_{\alpha_0} \cdot \tfrac{1}{2}[\mathrm{var}_{Q_{0\theta}} w_\varepsilon(x)]^{-\frac{1}{2}} \cdot D_0(\varepsilon,\theta) + n^{\frac{1}{2}}[G_0(\varepsilon,\theta) - G_1(\varepsilon,\theta)]\} \cdot [\mathrm{var}_{R(\varepsilon,\theta)} w_\varepsilon(x)]^{\frac{1}{2}} - \tfrac{1}{2}\,\mathrm{var}^{-\frac{1}{2}}_{R(\varepsilon,\theta)} w_\varepsilon(x)\; D_1(\varepsilon,\theta)\,\{u_{\alpha_0}[\mathrm{var}_{Q_{0\theta}} w_\varepsilon(x)]^{\frac{1}{2}} + n^{\frac{1}{2}}[E_{Q_{0\theta}} w_\varepsilon(x) - E_{R(\varepsilon,\theta)} w_\varepsilon(x)]\}\Big],$$

where again for i = 0,1

$$G_i(\varepsilon,\theta) = \frac{d \log c_0(\varepsilon)}{d\varepsilon}\{(1-\theta)P_i(\frac{p_1}{p_0} < c_0(\varepsilon)) + i.\theta\} + \frac{d \log c_1(\varepsilon)}{d\varepsilon}\{(1-\theta)P_i(\frac{p_1}{p_0} > c_1(\varepsilon)) + (1-i).\theta\}$$

$$\tfrac{1}{2} D_i(\varepsilon,\theta) = \frac{d \log c_0(\varepsilon)}{d\varepsilon} \cdot \log c_0(\varepsilon)\ (1-\theta)P_i(\frac{p_1}{p_0} < c_0(\varepsilon)) + i.\theta\} + \frac{d \log c_1(\varepsilon)}{d\varepsilon} \cdot \log c_1(\varepsilon)\{(1-\theta)P_i(\frac{p_1}{p_0} > c_1(\varepsilon)) + (1-i)\theta\} - E_{M(i)} w_\varepsilon(x) \cdot G_i(\varepsilon,\theta)$$

(with $M(0) = Q_{0\theta}$ and $M(1) = R(\varepsilon,\theta)$) and

$$\frac{dc_0(\varepsilon)}{d\varepsilon} = [(1-\varepsilon)^2 P_0(\frac{p_1}{p_0} \leq c_0(\varepsilon))]^{-1}$$

and

$$\frac{dc_1(\varepsilon)}{d\varepsilon} = -c_1^2(\varepsilon)[(1-\varepsilon)^2 P_1(\frac{p_1}{p_0} > c_1(\varepsilon)]^{-1}.$$

The proofs of the Theorem 1 and Theorem 2 are similar to proofs of analogous results in Víšek (1987a) and hence will be omitted.

LEMMA 1. For every $\varepsilon \in (0,a)$ there is $\tilde{\varepsilon}$ such that $ER(\tilde{\varepsilon},\varepsilon) = 0$.

REMARK. Since the family *S* consists of the most powerful tests (built up for the considered Huber's model of contaminacy) and since the normal approximations of the error probabilities are relatively good we have $ER(\varepsilon,\varepsilon) \stackrel{\sim}{=} 0$. The following table is a hint for building up an idea what is represented by "$\stackrel{\sim}{=}$". It may be also of interest to have an idea about corresponding second type error probabilities, hence they are displayed, too (denoted by $\beta(\varepsilon)$). The table was evaluated for $P_0 = N(0,1)$, $P_1 = N(1,1)$, $n = 40$ and $\alpha_0 = .05$.

ε	.03	.04	.05	.06	.07	.08	.09
$\beta(\varepsilon).10^2$	.015	.046	.123	.287	.616	1.23	2.24
$ER(\varepsilon,\varepsilon).10^3$	.4316	.7608	.1232	.6163	2.478	3.246	3.887

THEOREM 2. Local deficiency of our system *S* with respect to *W* exists and is approximately equal to $\phi(a(\varepsilon,\varepsilon))\{d(\varepsilon)-a(\varepsilon,\varepsilon).b(\varepsilon,\varepsilon)\}$ $\stackrel{\sim}{=} \phi(a(\varepsilon,\varepsilon)).d(\varepsilon)$ where $d(\varepsilon) = \left[\frac{ds(\tau,\theta)}{d\tau}\right]_{\tau=\theta=\varepsilon}$.

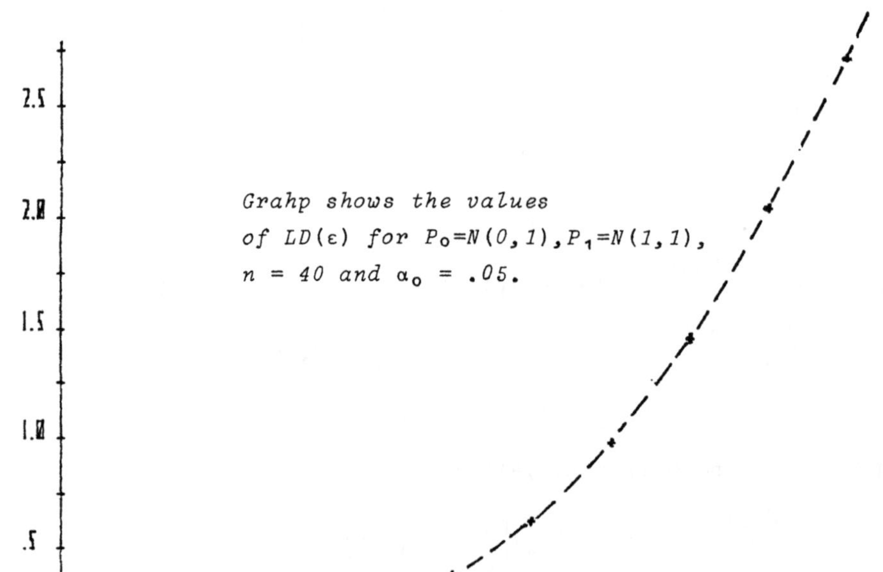

Grahp shows the values of $LD(\varepsilon)$ for $P_0=N(0,1)$, $P_1=N(1,1)$, $n = 40$ and $\alpha_0 = .05$.

CONCLUDING REMARKS

REMARK 1. Having in our disposal values of $LD(\varepsilon)$ for $\varepsilon \in (0,a)$ we may easy create an idea about behaviour of the power of tests from a neighbourhood of any $\varepsilon \in (0,a)$. We have for $\varepsilon^* \in \mathcal{O}(\varepsilon)$

$$2.[C(\varepsilon^*,\varepsilon) - C(\varepsilon,\varepsilon)] \cong LD(\varepsilon) \; . \; (\varepsilon^*-\varepsilon)^2,$$

i. e. the supremum over $\mathcal{P}_1(\varepsilon)$ of the power of the test $\Psi(n,\alpha_0,\bar{\varepsilon},\underline{x})$ will differ from the greatest possible power (which corresponds to the most powerful test $\Psi(n,\alpha_0,\varepsilon,\underline{x})$) approximately about $\frac{1}{2}LD(\varepsilon).(\varepsilon^*-\varepsilon)^2$.

REMARK 2. For another examples of usage $LD(\varepsilon)$ (and especially of $ER(\varepsilon,\theta)$) consult Víšek (1987b). There is presented example illustrating possibility of finding an optimal statistical procedure (from given family $\mathcal{S}$) with respect to given $\mathcal{P}(\varepsilon)$ from a family $\mathcal{W}$ by means of $ER(\varepsilon,\theta)$.

REFERENCES

Field Ch. A., Hampel F. R. (1982): Small sample asymptotic distribution of M-estimates of location, Biometrika 69, 29-46.

Hampel F. R. (1968): Contributions to the theory of robust estimation. Unpublished dissertation. Univ. of California, Berlekey.

Huber P. J. (1964): Robust estimation of a location parameter, Ann. Math. Statist. 35, 73-101.

Huber P. J. (1965): A robust version of the probability ratio test, Ann. Math. Statist. 36, 1753-1758.

Jurečková J., Víšek J. Á. (1984): Sensitivity of Chow-Robins procedure to the Contamination, Commun. Statist. - Sequential Analysis 3(2), 175-190.

Michálek J., Vajda I., Víšek J. A. (1984): New topics in robust statistics with applications, COMPSTAT84, Phisica-Verlag, Wien, 73-83.

Víšek J. Á. (1983a): On second order efficiency of a robust test and approximations of its error probabilities, Kybernetika 9(5), 387-407.

Víšek J. Á. (1983b): Sensitivity of the test risk with respect to contamination, Commun. Statist. - Sequential Analysis 2(3), 243-258.

Víšek J. Á. (1985): Estimating contamination level, to appear in Transactions of the Fifth Pannonian Symposium, Visegrad.

Víšek J. Á. (1986a): Weighted risk of test: Sensitivity to contamination, to appear in Decision and Statistics.

Víšek J. Á. (1986b): Sensitivity of the test error probabilities with respect to the level of contamination in general model of contaminacy, to appear in J. Statist. Plann. Inference.

Víšek J. Á. (1986c): Estimation of contamination level in model of contaminacy with general neighbourhoods, in preparation.

Víšek J. Á. (1987a): Efficiency rate and local deficiency of the most powerful tests in the model of contaminacy with general neighbourhoods, in preparation.

Víšek J. Á. (1987b): Efficiency rate and local deficiency of Huber's estimators of location and of α-estimators, in preparation.

Czechoslovak Academy of Sciences
Institute of Information
Theory and Automation
Pod vodárenskou věží 4
182 08 Prague 8
Czechoslovakia

COMPARISON OF SEVERAL HOMOGENEITY TESTS

FOR SAMPLES WITH CENSORED DATA

Petr Volf

Prague

Key words: Censoring, homogeneity tests, Monte Carlo method, power of test.

ABSTRACT

The contribution studies the power of several nonparametric homogeneity tests, provided the data are randomly censored. The modified Kolmogorov-Smirnov and Cramér-von Mises statistics are considered, as well as Cox's test and Peto&Peto test. The results are obtained by Monte Carlo method. They are compared and summarized in tables.

INTRODUCTION

The problem of comparison of two samples has been already studied even for the case with censored observations. Several nonparametric homogeneity tests have been proposed. Some of them are modifications of common procedures. Mostly, the asymptotic properties of test statistics are known. But their behavior for small and moderate sample size is the subject of studies, mainly by Monte Carlo method.

We intend to present the modified Kolmogorov-Smirnov (KS) and Cramér-von Mises (CM) procedures (Hall, Wellner (1980); Pettitt, Stephens (1976)). The results, as to the power of the tests, are encouraging. Their use is not complicated, moreover we are able to test the homogeneity on various intervals. Two variants of mentioned procedures are examined, with survival function and cumulative hazard function estimators. For the comparison, we investigate two frequently met tests, Cox's one and Peto&Peto Wilcoxon-type procedure (Lee et al. (1975)).

If random variable Y is censored from the right side by random variable V, we observe

$$Z = \min(Y,V), \quad \delta = I[Y \le V].$$

In two-sample problem we deal with M+N independent copies of two random variables $Y^{(1)}$, $Y^{(2)}$. Let $F^{(1)}$, $F^{(2)}$ be their distribution functions, we wish to test the hypothesis $F^{(1)} = F^{(2)}$. Censoring variables V_i, $i = 1,\dots,M+N$ are independent mutually and of Y-s. They are the copies of random variable V distributed according to some distribution function G. Throughout the paper we assume the continuity of $F^{(1)}$, $F^{(2)}$, G. Generally, the asymptotic results hold even when the censoring variables differ.

TEST PROCEDURES

Let M, N be the numbers of observations in two samples, denote $Z_1 \le Z_2 \le \dots \le Z_{M+N}$ all observations after ordering. Further denote, for $i = 1,\dots,M+N$:

δ_i - the indicator of censoring: $\delta_i = 1$ if Z_i is noncensored, $\delta_i = 0$ otherwise,

$a_i = 1$ if Z_i is from the first sample, $a_i = 0$ for the second sample,

A_i - the number of observations from the first sample not less than Z_i,

B_i - the number of observations from the second sample not less than Z_i,

$K_i = M + N - i + 1 = A_i + B_i$.

We shall remind the test statistics of Kolmogorov-Smirnov type and their asymptotic properties. Let us construct the product-limit estimators of survival functions from both samples:

$$P_M^{(1)}(t) = \prod_{i=1}^{M+N} \left(\frac{A_i - 1}{A_i}\right)^{\delta_i a_i I[Z_i \le t]}, \quad P_N^{(2)}(t) = \prod_{i=1}^{M+N} \left(\frac{B_i - 1}{B_i}\right)^{\delta_i (1-a_i) I[Z_i \le t]}$$

The estimators of the cumulative hazard functions are

$$L_M^{(1)}(t) = \sum_{i=1}^{M+N} \frac{\delta_i a_i}{A_i} I[Z_i \le t], \quad L_N^{(2)}(t) = \sum_{i=1}^{M+N} \frac{\delta_i (1-a_i)}{B_i} I[Z_i \le t].$$

Let us remind also the estimators of their asymptotic variance, according to Breslow, Crowley (1974):

$$C_M^{(1)}(t) = \sum_{i=1}^{M+N} \frac{M\delta_i a_i}{A_i^2} I[Z_i \le t], \quad C_N^{(2)}(t) = \sum_{i=1}^{M+N} \frac{N\delta_i (1-a_i)}{B_i^2} I[Z_i \le t].$$

All estimators are constructed for $t \in (-\infty, T]$, T such that $(1-F^{(1)}(T)).(1-F^{(2)}(T)).(1-G(T)) > 0$. There the estimators are strongly consistent. For the asymptotic conclusions suppose that M/N has positive and finite limit, as M, N $\to \infty$. Now introduce, again for $t \in (-\infty, T]$:

$$V_{M,N}(t) = \sqrt{\frac{M\ N}{M+N}} \left(\frac{P_M^{(1)}(t)}{P_N^{(2)}(t)} - 1 \right), \quad W_{M,N}(t) = \sqrt{\frac{M\ N}{M+N}} \left(L_M^{(1)}(t) - L_N^{(2)}(t) \right),$$

$$A_{M,N}(t) = \frac{1}{M+N} \{N\ C_M^{(1)}(t) + M\ C_N^{(2)}(t)\}$$

$$K_{M,N}(t) = \frac{A_{M,N}(t)}{1+A_{M,N}(t)} \qquad K(t) = \lim_{M,N\to\infty} K_{M,N}(t) \quad \text{a.s.}$$

Suppose, that our hypothesis H_0 of homogeneity $Y^{(1)}$ and $Y^{(2)}$ holds on $(-\infty, T]$. Then, as the consequence of the main result of Breslow, Crowley (1974), the asymptotic distribution of $V_{M,N}(t)/(1+A_{M,N}(t))$ on $(-\infty, T]$ is the distribution of $\mathcal{B}(K(t))$. There $\mathcal{B}(u)$ denotes the standard Brownian bridge process. Its distribution is well known, for $c \ge 0$ and $U \in (0,1)$:

$$\Pr\{\sup \mathcal{B}(u) > c,\ u \in (0,U]\} = \Pr\{\inf \mathcal{B}(u) < -c,\ u \in (0,U]\} =$$

$$= \bar{\Phi}(D.c) + \exp(-2c^2)\ .\ \bar{\Phi}(D(1-2U).c),$$

where $\bar{\Phi}$ denotes 1-distribution function of $\mathcal{N}(0,1)$, D stands for $[U(1-U)]^{-\frac{1}{2}}$. Denote this probability as $P^+(U,c)$, the following probability as $P(U,c)$:

$$\Pr\{\sup|\mathcal{B}(u)|>c,\ u \in (0,U]\} = 2\bar{\Phi}(Dc) - 2.\sum_{\ell=1}^{\infty} (-1)^{\ell} \exp(-2c^2\ell^2)\ .$$

$$.\ \{\bar{\Phi}(Dc[2\ell(1-U)-1]) - \bar{\Phi}(Dc[2\ell(1-U)+1])\}.$$

We shall test the hypothesis H_0: $F^{(1)} = F^{(2)}$ on $(-\infty, T]$, on the asymptotic level $\alpha \in (0,1)$. It can be accomplished by following way:

Find $c_1 = \max V_{M,N}(Z_i)/(1+K_{M,N}(Z_i))$, $i=1,\dots,M+N$, $Z_i \le T$.

If $P^+(K_{M,N}(T), c_1) < \alpha$, we reject H_0 in favour of the alternative A_1: $F^{(1)} < F^{(2)}$. By $c_2 = \min V_{M,N}(Z_i)/(1+K_{M,N}(Z_i))$ we test H_0 against A_2: $F^{(1)} > F^{(2)}$. The alternative $F^{(1)} \neq F^{(2)}$ is accepted and H_0 rejected, when $P(K_{M,N}(T), c) < \alpha$, where $c = \max(c_1, |c_2|)$.

If we wish to use the estimators of cumulative hazard functions, we replace $V_{M,N}$ by $W_{M,N}$.

Cramér-von Mises criterion has been investigated by Pettitt, Stephens (1976). In their paper we can find the table of quantiles of $\omega(\tau) = \int_0^\tau B^2(u)du$, for various $\tau \in (0,1]$. Test statistics

$$\sum_{i=1}^{M+N} \delta_i \left(\frac{V_{M,N}(Z_i)}{1+K_{M,N}(Z_i)}\right)^2 (K_{M,N}(Z_i)-K_{M,N}(Z_{i-1})).I[Z_i \le T], \text{ with } K_{M,N}(Z_0)=0$$

is asymptotically distributed as $\omega(K(T))$. In practice, the value of the test statistics is compared with quantiles of $\omega(K_{M,N}(T))$.

Notice, that statistics $V_{M,N}$ is related to $P_M^{(1)}/P_N^{(2)}-1$. Let us imagine that $P^{(2)} > P^{(1)}$. Then $|P^{(1)}/P^{(2)}-1| < |P^{(2)}/P^{(1)}-1|$, the difference can be considerable. Obviously, for small samples the test based on the second term is far more powerful. When one-sided alternative is expected, one has to choose the proper version of test. Under H_0, the asymptotic probabilities of the first kind error are equal.

Let us now recapitulate the forms of two well known test statistics:

Cox's test: Let

$$U = \sum_{i=1}^{M+N} \delta_i (a_i - A_i/K_i), \quad I = \sum_{i=1}^{M+N} \delta_i A_i B_i / K_i^2,$$

then $U/\sqrt{I}$ is asymptotically distributed as $N(0,1)$, provided H_0 holds.

Peto&Peto test: Let us construct the product limit estimator of survival function from all data:

$$\hat{P}(t) = \prod_{i=1}^{M+N} \left(\frac{K_{i-1}}{K_i}\right)^{\delta_i . I[Z_i \le t]},$$

define the scores $v_i = \hat{P}(Z_i) + \hat{P}(Z_{i-1}) - 1$ for $\delta_i = 1$, $v_i = \hat{P}(Z_i) - 1$ for $\delta_i = 0$. Then the statistics

$$W = \sum_{i=1}^{M+N} a_i v_i \cdot \left\{ \frac{MN \sum_{i=1}^{M+N} v_i^2}{(M+N)(M+N-1)} \right\}^{-\frac{1}{2}}$$

asymptotically follows the distribution of $N(0,1)$, provided H_0 holds.

For both tests, small value of test statistics indicates, that the first sample tends to be greater than the second sample.

RESULTS

Several Monte Carlo studies on the two-sample problem have been published in recent years. Lee at al. (1975) in their large study have summarized some conclusions about Cox's, Peto&Peto and other tests, but not about KS or CM procedures.

We simulated 100 times two samples of sizes M, N from 20 to 80. Some cases were checked by 500 times simulation, but the results did not change significantly. The samples followed exponential, Weibull or normal distribution, with accuracy of pseudo-random numbers generator. We used following notation:
$Exp(\lambda)$: $F(t) = 1 - \exp(-\lambda t)$, $W(a,b)$: $F(t) = 1 - \exp(-at^b)$, $N(0,\sigma^2)$. Values of parameters were $\lambda = 1/L$, $a = 1/\sqrt{L}$, $\sigma = L$, $L = 1,\dots,5$, $b = \frac{1}{2}$. Both samples were censored by realizations of appropriety chosen random variable. Approximately 50 % censoring was reached.

Asymptotic levels $\alpha = 0.05$ and $\alpha = 0.1$ were chosen. The hypothesis of homogeneity was accepted if the value of test statistics lied between $\frac{\alpha}{2}$ and $1 - \frac{\alpha}{2}$ quantile of asymptotic distribution, except of CM statistics, where $1 - \alpha$ quantile was the criterion. When comparing some Weibull or exponential samples with Exp(1), in fact we were testing H_0 against one-sided alternative on asymptotic level $\frac{\alpha}{2}$. The KS and CM tests were computed throughout interval $[Z_1, T)$, where $T = \min\{\max(Z_i: \delta_i=1, a_i=1), \max(Z_i: \delta_i=1, a_i=0)\}$. It corresponded approximately to 90 % quantile of the distribution of Y-s.

As we expected, the KS and CM tests were the most powerful, especially the variants with survival function estimators. But real frequencies of the first kind error were higher for these tests. This effect and its dependence on the extent of samples were studied, too, the result are displayed by Table 5.

As it was said, interchange of samples could lead to considerable decrease of power of KS and CM tests constructed from the survival function estimators. One has to be aware of the fact, the sample with expected larger survival should be treated as the first one. This effect disapears with increasing sample sizes (Table 2). The other procedures are stable, their statistics are symmetrical with respect to interchange of samples. Dependence of the tests' behavior on the upper boundary T was not investigated.

The following Tables 1 - 4 display the frequencies of rejections of H_0, from 100 cases. KS(P) means modified Kolmogorov-Smirnov test based on survival functions, KS(L) is based on cumulative hazard functions. CM is test of Cramér-von Mises, based on survival functions. By "*" we denote the columns with the results after interchange of samples.

TABLE 1

The first sample has Exp(1/L) distribution, M = 20, the second sample - Exp(1), N = 30, censoring - Exp (½)

L \ α	Cox		Peto, Peto		KS(P)				KS(P)*		KS(L)		CM		
	0.05	0.1	0.05	0.1	0.01	0.025	0.05	0.1	0.05	0.1	0.05	0.1	0.025	0.05	0.1
1	3	8	2	8	7	12	14	21	12	20	1	3	5	13	17
2	41	48	36	45	60	66	75	79	1	8	20	35	50	56	65
3	75	82	72	78	85	89	89	90	0	0	41	63	85	90	92
4	91	95	90	93	95	98	98	99	0	0	56	72	93	95	96
5	96	98	95	97	98	98	99	100	0	0	69	87	100	100	100

TABLE 2

The distributions are the same as in Table 1, M = 80, N = 50

L \ α	Cox		Peto, Peto		KS(P)		KS(P)*		KS(L)	
	0.05	0.1	0.05	0.1	0.05	0.1	0.05	0.1	0.05	0.1
1	5	11	7	12	8	14	8	15	2	6
2	85	90	77	85	84	88	22	39	70	78
3	99	100	99	99	99	99	72	89	93	98
4	99	100	99	100	100	100	91	97	99	99
5	100	100	100	100	100	100	98	99	99	100

TABLE 3

$W(1/\sqrt{L},\frac{1}{2})$, M = 50 against Exp(1), N = 50, censored by V=2(Exp(1)+0.3)

L \ α	Cox		Peto, Peto		KS(P)		KS(P)*		KS(L)		CM	
	0.05	0.1	0.05	0.1	0.05	0.1	0.05	0.1	0.05	0.1	0.05	0.1
1	7	12	16	23	44	59	51	60	27	41	51	74
2	17	37	19	28	80	88	12	19	32	48	62	78
3	39	60	32	40	94	96	4	10	39	58	76	92
4	90	93	62	70	95	98	1	8	49	70	98	99
5	98	99	87	90	100	100			61	77	99	99

TABLE 4

$N(0,L^2)$, M = 50 against $N(0,1)$, N = 30, censored by $N(0.3,1)$

L \ α	Cox		Peto, Peto		KS(P)		KS(L)		CM	
	0.05	0.1	0.05	0.1	0.05	0.1	0.05	0.1	0.05	0.1
1	6	8	6	11	7	12	1	2	7	10
2	5	10	9	15	24	35	6	24	25	37
3	6	10	16	24	40	58	35	53	49	70
4	5	9	20	29	65	82	56	75	70	86
5	5	11	22	35	70	87	71	86	78	93

TABLE 5

Relative frequencies of the first kind error (i. e. rejection of H_0), when both samples follow Exp(1) distribution, censored by Exp(½), from 500 cases.

α:	KS(P)				KS(L)			
M=N:	0.01	0.025	0.05	0.1	0.01	0.025	0.05	0.1
20	0.082	0.118	0.146	0.182	0.002	0.004	0.014	0.022
40	0.048	0.078	0.114	0.146	∅	0.01	0.014	0.036
60	0.032	0.06	0.082	0.122	0.002	0.008	0.016	0.044
80	0.02	0.042	0.076	0.11	0.006	0.01	0.026	0.050
100	0.014	0.028	0.054	0.090	0.004	0.012	0.022	0.060

REFERENCES

Breslow N., Crowley J. (1974): A large sample study of the life table and product limit estimates under random censorship. Ann.Statist. 2, 437-453.

Hall W. J., Wellner J. A. (1980): Confidence bands for a survival curve from censored data, Biometrika 67, 133-143.

Lee E. T., Desu M. M., Gehan E. A. (1975): A Monte Carlo study of the power of some two-sample tests, Biometrika 62, 425-432.

Pettitt A. N., Stephens M. A. (1976): Modified Cramér - von Mises statistics for censored data, Biometrika 63, 291-298.

Czechoslovak Academy of Sciences
Institute of Information Theory and Automation
182 08 Prague 8
Pod vodárenskou věží 4
Czechoslovakia

ADAPTIVE ROBUST D-ESTIMATOR

Miloslav Vošvrda

Prague

Key words: Partial adaptation, D-estimator, robust estimation, selector statistics.

ABSTRACT

The paper suggests the use of D-estimator for a simultaneous estimation of parameters of a location and scale. A family of generalized exponential power distributions is used for estimating as a projection family. The suggested D-estimator is compared with the ML-estimator by mean square errors in a Monte Carlo analysis of different populations.

1. INTRODUCTION

A simultaneous estimating of a location and scale is considered in case that underlying distribution is unknown. We shall use a partial adaptation method (cf. Hogg & Lenth (1984)) for estimating parameters. That method is based on using of a selector statistic for choosing of a density from a family of densities (cf. Hogg (1967)). Our selector statistic will be of a different form. The estimated type will be used as a projection probability measure in a minimum divergence procedure. Monte Carlo study will demonstrate the behaviour of the resulting estimators in term of mean square errors (MSE).

2. PRELIMINARIES

Let $\Theta = \mathbb{R} \times \mathbb{R}^+$ be the parameter space of a location and scale and let $\theta = (\mu, \sigma) \in \Theta$. Let $(\mathbb{R}, \mathcal{B})$ be the usual

Borel measurable sample space. Let us consider a nonempty interval of positive real numbers $M = \langle 1, m \rangle$ and let us denote

(1) $$\eta = (\theta, t) = (\mu, \sigma, t) \in \Theta \times M$$

We shall consider the family of probability measures

(2) $$\mathcal{P} = \{ P_\eta \mid \eta \in \Theta \times M \}$$

on the sample space $(\mathbb{R}, \mathcal{B})$ defined by distribution functions

(3) $$F_\eta(x) = \frac{t\, y(t)}{2\Gamma(t^{-1})} \int_{-\infty}^{x} \exp\left\{ -\left[\frac{y(t)\,|z-\mu|}{\sigma} \right]^t \right\} dz$$

where $\Gamma(.)$ is the gamma function and

(4) $$y(t) = \left[\Gamma(3t^{-1}) / \Gamma(t^{-1}) \right]^{1/2}.$$

Hereafter, this family will be considered as the projection family.

In addition to the family $\mathcal{P}$, we shall consider a family of probability measures

(5) $$\mathcal{Q} = \{ Q_\theta \mid \theta \in \Theta \}$$

on $(\mathbb{R}, \mathcal{B})$. This family describes parameter-depending probability law governing random sample. The random sample vector $\underline{X} = (X_1, \ldots, X_n)$ of any sample size n is assumed to be generated in accordance with the probability measure Q_θ^n on the product space $(\mathbb{R}, \mathcal{B})$.

For every sample vector $\underline{x} = (x_1, \ldots, x_n) \in \mathbb{R}^n$, we consider the empirical probability measure Q_n on $(\mathbb{R}, \mathcal{B})$ defined by

(6) $$Q_n(B) = n^{-1} \sum_{i=1}^{n} \chi_B(x_i)$$

where $\chi_B(.)$ is the characteristic function of $B \in \mathcal{B}$.

For every $u \in (0,1)$ and for every distribution function F on $\mathbb{R}$, we consider the usual quantile function

(7) $$F^{-1}(u) = \inf \{ x \mid F(x) \geq u \}$$

with the obvious identity

(8) $$F^{-1}(u) = \inf \{ x \mid F(x) = u \}$$

in the case that F is continous.

3. AD-ESTIMATOR

We define AD-estimator of a location and scale as a sequence of mappings $T_n : R^n \to \Theta$ for $n = 1,2,\ldots$ by the following two steps.

I) Estimation of type $t \in M$. Let n be an arbitrary positive integer and $\underline{x} = (x_1,\ldots,x_n)$ an arbitrary sample vector from R^n. The estimator of type $\hat{t} : R^n \to M$ is defined as follows

(9) $$\hat{t}^{-1}(\underline{x}) = \ln \frac{|x_{([n\alpha']+1)}|}{|x_{([n\alpha'']+1)}|}$$

where $x_{(i)}$ denotes the i-th order statistics of the sample $\underline{x}$ for $i = 1,\ldots,n$, $[x]$ is the integer part of $x \in R$, and $\alpha' = 0.0025$, $\alpha'' = 0.05$. This estimator can be explained as follows. Filliben (1969) has proposed a measure $\tau(\alpha',\alpha'')$ of "tail length" of an arbitrary distribution function which applied to the family of distributions $\{F_\eta \mid \eta \in \Theta \times M\}$ and to the above given α', α'' yields

(10) $$\tau^{-1}(0.0025, 0.05) = \ln \frac{F^{-1}_{(0,1,t)}(0.0025)}{F^{-1}_{(0,1,t)}(0.05)}$$

Approximately, it holds $\tau \sim c \cdot t$ for

(11) $$c^{-1} = \ln \frac{\ln \dfrac{t\, y(t)}{0.0125\, \Gamma(t^{-1})}}{\dfrac{t\, y(t)}{0.1\, \Gamma(t^{-1})}}$$

Since for typical values $(t \geq 5)$ the coefficient c is approximately equal 1, the above considered estimator $\hat{t}(\underline{x})$ of the Filliben's measure in fact estimates the unknown type t as well.

II) The projection into the family $\{P_{\theta,\hat{t}} \mid \theta \in \Theta\}$ for the given type $\hat{t} = \hat{t}(\underline{x})$. Let $\underline{x}$ be the same arbitrary sample as in the step I) above. Let us consider a continuous convex function $f : (0,\infty) \to R$, strictly convex at the point 1 with $f(1) = 0$. Each f under consideration defines an f-divergence $D_f(Q_n ; P_{\theta,\hat{t}})$ between the empirical probability measure Q_n corresponding to the sample $\underline{x}$ and the probability measure $P_{\theta,\hat{t}}$, the type of which is determined by the sample $\underline{x}$. In accordance with Vajda (1984), we consider projections $P_{T_n,\hat{\varepsilon}}$ of empirical probability measures Q_n into the projection family $\{P_{\theta,\hat{\varepsilon}} \mid \theta \in \Theta\}$ defined by

$$T_n = T_n(\underline{x}) = \arg\min_{\theta \in \Theta} D_f(Q_n; P_{\theta,\hat{\varepsilon}}) . \tag{12}$$

The estimator $\{T_n, n=1,2,\ldots\}$ defined in the step II, with the help of the step I, is called A$\mathcal{D}$-estimator. Obviously, it differs from estimators considered in Vajda (1984) by that it is partially adaptive (adaptive D-estimator).

4. ADAPTIVE D-ESTIMATOR

As an example of the above determined A$\mathcal{D}$-estimator, we may consider Aα-estimator defined by the following function

$$\begin{aligned} f(x) &= \alpha^{-1}(1-x^{\alpha}) \quad \text{for} \quad \alpha \in (0,1), \quad x \in R \\ &= -\log x \quad \text{for} \quad \alpha = 0, \quad x \in R . \end{aligned} \tag{13}$$

Let us note that a minimum divergence estimator with $\alpha = 0$ coincides with the maximum likelihood estimator (cf. Vajda (1984)). Because

$$D_f(Q_n; P_{\theta,\hat{\varepsilon}}) = E_{Q_n}[f(p(\cdot;\theta,\hat{\varepsilon}))], \tag{14}$$

where $p(\cdot;\theta,\hat{\varepsilon})$ is a density of probability measure $P_{\theta,\hat{\varepsilon}}$ it holds

$$\begin{aligned} D_f(Q_n; P_{\theta,\hat{\varepsilon}}) &= E_{Q_n}[1-p^{\alpha}(\cdot;\theta,\hat{\varepsilon})] \quad \text{for } \alpha \in (0,1) \\ &= E_{Q_n}[-\log p(\cdot;\theta,\hat{\varepsilon})] \quad \text{for } \alpha = 0 . \end{aligned} \tag{15}$$

By definition (12), the A$\mathcal{D}$-estimator $\{T_n\}_{n=1}^{\infty}$ is obtained as follows

$$\begin{aligned} T_n = T_n(\underline{x}) &= \arg\min_{\theta \in \Theta} E_{Q_n}[1-p^{\alpha}(\cdot;\theta,\hat{\varepsilon}(\underline{x}))], \quad \alpha \in (0,1) \\ &= \arg\min_{\theta \in \Theta} E_{Q_n}[-\log p(\cdot;\theta,\hat{\varepsilon}(\underline{x}))], \quad \alpha = 0 . \end{aligned} \tag{16}$$

Asymptotical properties of A$\mathcal{D}$-estimators (strong consistency, asymptotic efficiency, second order efficiency in Rao's sense, adaptivity in Bickel's sense) are demonstrated in VOŠVRDA (1986). For Monte Carlo analysis, we are putting $\alpha = 0.1$. This value has been picked out in accordance with the robustness of the resulting Aα-estimator (cf. VOŠVRDA (1984)). For minimalization, it has been used an adaptive nongradient algorithm of Rosenbrock's type with the adjustement which introduced in the paper of Špitálský (1985).

5. MONTE CARLO ANALYSIS

The $A\alpha$-estimator, with $\alpha = 0.1$, was examined in the following situations:

Samples had been generated in the size 4/4/20 (in the standard notation of Monte Carlo studies) with double exponential, normal or uniform underlying distribution. For each combination sample size-underlying distribution, MSE of the location and scale estimates of the procedure being tested were computed by the formula

$$MSE_j = L^{-1} \sum_{i=1}^{L} (\theta_j - \hat{\theta}_{ij})^2 \quad , j = 1,2 \quad , \tag{17}$$

where L is a number of experiments and $\hat{\theta}_{ij}$ is the j-th estimate of the i-th experiment. For each combination sample size-underlying distribution, the relative efficiencies RE of the location and scale estimators of the procedure being tested were calculated by

$$RE = MSE_{ML} / MSE_I \tag{18}$$

where MSE_{ML} is the MSE of the maximum likelihood estimator with known underlying distribution and MSE_I is the MSE of our estimator of location or scale (cf. Table 1). The robustness of the $A\alpha$-estimator is demonstrated by the Table 2, where the contamination of the following form has been considered

$$Q_{(0,1)} = (1-\varkappa)\, \mathcal{N}(0,1) + \varkappa\, \delta_{10} \tag{19}$$

where δ_x is the ptobability measure concentrating all mass at the point x (in our case $x = 10$) and $\varkappa \in \langle 0, 1 \rangle$

6. CONCLUSIONS

The Table 1 shows that the $A\alpha$-estimator provides good estimates of location and scale when the probability law Q_θ for the data is unknown. The Table 2 shows very expressively that the $A\alpha$-estimator is outfitted with a strong robust property.

7. ACKNOWLEDGEMENTS

The autor would like to thank to I. Vajda for valuable comments.

Table 1. RE of the location and scale estimators, L = 1000

n		4	8	12	16	20
underlying distribution						
normal	location	0.65823	0.65451	0.62049	0.74894	0.69433
	scale	0.69580	0.78913	0.90371	0.93816	0.82919
uniform	location	0.51729	0.36368	0.31737	0.26528	0.28709
	scale	0.32639	0.48176	0.53654	0.53057	0.38317
double	location	0.68600	0.63730	0.82279	0.92034	0.90383
exponential	scale	0.61583	0.23544	0.70666	0.91091	0.87220

Table 2. Values and MSE of the location and scale estimators with the contamination type $/1-\varkappa/\mathcal{N}/0,1/+\varkappa\delta_{10}$, n=20, L = 1000

$\varkappa$	0	MSE	0.05	MSE	0.1	MSE	0.2	MSE
ADE	0.03555	0.08413	0.10356	0.08109	0.17642	0.07642	0.67816	0.40412
	0.96754	0.02737	0.40040	0.17898	2.11200	0.31303	3.50310	0.13827
MLE	-0.03720	0.05608	0.46830	0.06208	0 99998	0.04321	2.02070	0.03231
	0.99462	0.08930	6.00710	0.17710	10.37200	0.27635	17.53100	0.59974

ADE - AD-estimator MLE - Maximum likelihood estimator

REFERENCES

Bickel P.J. (1982): On adaptive estimation. Ann. Statist. 3, 647 - 671.

Filliben J.J. (1969): Simple and robust estimation of the location parameter of a symmetric distribution. Ph. D. dissertation, Princeton University, Princeton, NJ.

Hogg R.V. (1967): Some observations on robust estimation. JASA, 62, 1179 - 1186.

Hogg R.V. (1974): Adaptive robust procedures: A partial review and some suggestions for future applications and theory (with comments). JASA, 69, 900 - 927.

Hogg R.V., Lenth R.V. (1984): A review of some adaptive statistical techniques. Commun. Statist. 13, 1551 - 1579.

Parzen E. (1979): Nonparametric Statistical Data Modelling. JASA, 74, 105 121.

Špitálský J. (1985): Efficient adaptive nongradient minimization algorithm (in Czech), Report ÚTIA, 1320.

Vajda I. (1984): Minimum divergence principle in statistical estimation, Statistics and Decisions, Supplement Issue No 1, Recent Results in Estimation Theory 239 - 261.

Vošvrda M. (1984): Statistical Experiments with Efficiency, Robustness and Adaptability of Modified Directed D-estimators. Compstat'84, Poster information.

Vošvrda M. (1986): On adaptive robust D-estimators (in a preparation).

Institution of Information Theory
and Automation,
Czechoslovak Academy of Science
Pod vodárenskou věží 4
182 08 Prague 8
Czechoslovakia

PROBABILISTIC MODEL OF A RANDOM MANUFACTURING PROCESS

Barbara Wójcik

WARSZAWA

Key words: probabilistic model, control operation, tolerance interval, dynamic programming.

ABSTRACT

A probabilistic model of a manufacturing process from a products' quality point of view is considered. This model is to be used in minimization of production cost.

CHARACTERISTIC OF A MANUFACTURING PROCESS

In this paper a random manufacturing process is shown as a series structure in which N production stands $\{1,2,\ldots,n,\ldots,N\}$ and N+1 control operations $\{0,1,\ldots,N\}$ exist aligned one after the other (starting from the zero control operation and ending to the N+1-th). Such the structure is met in practice very often, especially in the electronic and machine industry. Every production stand $n, n \in \langle 1,2\ldots,N\rangle$ processes a set I_n of input goods (raw materials, semi-products) into a set I_n of output goods. To every good one quality parameter considered to be the most representative of the product state has been assigned. An alternative classification applied to every automatic control operation separates examined products into good and bad ones depending upon a quality criterion assumed. In the simpliest case products are checked on the grounds od one, particular feature. In general, it is possible to consider a few features as random parameter.

The features of a single n-th production stand can be defined by the following:

$f_n(x)$ - probabilistic characteristic (a pdf) of a continuous measurable feature X describing a raw- material,

$g_n(y)$ - a pdf of a continuous measurable feature Y decribing a semi-product,

$g_n(y|x)$ - a conditional pdf describing a quality of n-th production process,

u_n - quota of input goods corresponding to the unit of an output product,

b_n - an expected value of good usage rate in the defined time interval,

a_n - an expected value of good production rate by the n-th device in the same time interval.

It is obvious that transformation of the input product into the output one is of a probabilistic nature. Thus, it particular operation on a given detail is carried out at lathe stand, then dimensions of the details machined are not identical even if the operation in executed on highly identical details. These dimensions make a random sequence of values and some of then may be even recognized as rejects.

If therefore x_i denotes the i-th detail dimensions at the input of the n-th production stand, Y_j - random dimensions of the j-th product (obtained after the operation on the detail i is executed)

$$Y_j = Y_j(x) \qquad j \in J_n$$

where $\quad x=x_i \qquad i \in I_n$

Obviously the dimensions of the detail X_i machined are also random variables X_i in the production practice.If $\bar{x}_i$ determines the expected value of the variable X_i then the dependence y_j on the value X_i may be obtained from the formula.

$$(1) \qquad Y_j = \phi_j(\bar{x}) + \sum_{i \in I_n} h_{ij} \, (X_i - \bar{x}_i) + \delta_j \,, \qquad i \in I_n;\ j \in J_n$$

where

Y_j - random variable of the value of the parameter determining the j-th product at the n-th process output,

$\phi_j(x)$ - the quality parameter values of the j-th good produced in the n-th process, provided that the process is executed in ideal conditions and the input characteristic of products are identical and equal to $\bar{x}$,

h_{ij} - coefficients determining the influence of the input good parameters on the values of quality parameters of products produced in the n-th process,

$X_i-\bar{x}_i$ - a random variable being the difference between the real value of the quality parameter of the i-th product, transmitted to the input of the n-th device, and its expected value $\bar{x}_i$,

δ_j - a random variable related the n-th process quality determinimg process disturbances caused by random changes of the process execution.

Consequently, the n-th process is formally a transformation of the I_n group of random variables X_i (where $i \in I_n$) into the I_n group of random variables Y_j ($j \in J_n$). This transformation may be determined by the set of conditional probability density functions

$$\{g_n (y|x)\}$$

Note, that for the "lathe operation" type process the set of input products I_n has only one element and size of the I_n set (a cardinal number) equals one:

$$\text{card } I_n = \text{card } J_n = 1$$

In case of the known density function $f_n(x)$ the density functions of variable Y can be easily computed by using a linear approximation (1) .

In this case, the additional dispersion of the y_j random variable values introducted by the n-th operation is characterized by the random variable δ_j , which is interpreted as an "accuracy" characteristic of a stand executing the n-th process.

For the given value of the input good quality parameter x_i $i \in I_n$, the value y_j of the parameter of output good $j \in J_n$ is a random variable. We deal with N succesive manufacturing processes, each of them introduces a dispersion of output goods parameters.Hence, the appropriate control operations should be applied in order to eliminate bad products in due time. In the simpliest case of an alternative classification, the products, outside the range of tolerance assumed $d = |y_{min}, y_{max}|$ are rejected. It has been a priori assumed, that control error is negligible.

The final product quality is defined by the density function $g_N(y)$, which depends on the raw material quality $f_o(x)$ and on the probabilistic characteristics of quality of product obtained in succesive processes of production cycle $g_1(y)$, $g_2(y)$ ..., $g_{N-1}(y)$ depending on the tolerance intervals in successive control operations.

To determine the quality probabilistic characteristics of products obtained in each production stand we must know the pdf of the good transmitted to the n-th device input.When card $I_n=1$ for the given density function $f_n(x)$ we have

(2)
$$g_n(y) = \int_{x_{min}}^{x_{max}} g_n(y|x)\ f_n(x)dx$$

where (x_{min}, x_{max}) - the tolerance interval for the product investiqated in (n-1)-th control operation, $g_n\ (y|x)$ - the conditional pdf /probabilistic characteristic/ of the n-th process quality.

The quality of the raw material for the n-th production stand can be found from the relation:

(3)
$$f_n(x) = \begin{cases} \dfrac{g_{n-1}(y)}{G_{n-1}(y_{max}) - G_{n-1}\ (y_{min})} & \text{for } y_{min} < y < y_{max} \\ 0 \text{ for the other values of } y & \end{cases}$$

where $G_n(y)$ - distribution function,

(y_{min}, y_{max}) - tolerance interval for the product y.

The pdf $f_n(x)$ obtained from (2) is normalized.

A PARTICULAR CASE OF AN IDEAL PROCESS QUALITY

By $\delta_j = 0$ we mean an ideal quality of the n-th process in which one input good and one output good take part i.e. $I_n = J_n = \{1\}$, n. Thus, the equation (1)takes the form:

(4)
$$Y = \phi(\bar{x}) + h\ (X-\bar{x})$$

Assuming that

(5) $a=h$

$b=\phi(\bar{x})-h\bar{X}$

we obtain from (4) the following expression

(6) $Y = aX + b$

where Y,X are random variables.

Then the $G_n(y)$ distribution functions is

(7) $G_n(y) = P_n\,(aX+b<y) = P_n\,(X<\frac{y-b}{a}) = F_n\,(\frac{y-b}{a})$

and the density function $g_n(y)$ takes the form:

(8) $g_n\,(y) = \frac{d}{dy}\,G_n(y) = \frac{1}{a}\,f\,(\frac{y-b}{a})$

where f denotes a normal density function.

Applying the symbols used in (5) and (8) we get:

(9) $g_n(y) = \frac{1}{h}\,f\,(\frac{y-\phi(\bar{x})}{h}+\bar{x})$

In case of the control operation executed after the n-th process, the quality probabilistic characteristic is

(10) $$f_{n+1} = \frac{g_n(x)}{G_n(y_{max}) - G_n(y_{min})} = \frac{1}{h}\,\frac{f(\frac{x-\phi^n(\bar{x})}{h})}{|G_n(y_{max})-G_n(y_{min})|}$$

where $y\in(y_{min}, y_{max})$

RELATIONS BETWEEN EXPECTED VALUES OF USAGE AND PRODUCTION OF PRODUCTS

A control operation executed after the n-th production process enables to separate rejections from the set of goods produced in this process with the following probability

(11) $P_n = 1-G_n(y_{max}) + G_n(y_{min})$

where $G_n(y_{max})$, $G_n(y_{min})$ - distribution functions. New let us try to follow an expected value of the good usage rate in successive stages of manufacturing process

(12) $(1-P_{n-1})\,a_{n-1} = b_n$

where

a_{n-1} an expected value of good production rate by the n-1th device in some time interval,

b_n - an expected value of good usage rate in defined time interval.

The expected value of the production rate a_n for the n-th device depends on the expected value of good usage rate b_n and on the technical featúres of the device.

(13) $$u_n = \frac{a_n}{b_n}$$

where

u_n - quota of output goods corresponding to the unit of an input good.

After the subsequent control operation is executed we have

(14) $$a_n(1-P_n) = b_{n+1}$$

considering the relations (12)and (13) we obtain .

(15) $$u_n b_n (1-P_n) = b_{n+1}$$

considering the subsequent processes upto the n-th we obtain

(16) $$b_n = b_o \prod_{i=0}^{n-1} u_i(1-P_i)$$

where $u_o = 1$

For the final product the relation (16) has the form

(17) $$b_{N+1} = b_o \prod_{i=0}^{N} u_i(1-P_i) = a_{N+1}$$

where $u_{N+1} = 1$

The expected value of the input raw material used in the manufacturing process has the form

(18) $$b_o = \frac{a_{N+1}}{\prod_{i=0}^{N} u_i(1-P_i)}$$

And the expected value of the good wear rate for the n-th process may be described as follows:

(19a) $$b_n = a_{N+1} \frac{\prod_{i=0}^{n-1} u_i(1-P_i)}{\prod_{i=0}^{N} u_i(1-P_i)}$$

and the expected values of the good production rate is

(19b) $$a_n = (1-P_{n-1})a_{n-1} u_n$$

OPTIMIZATION PROBLEM

Described above probabilistic model could be treated as a Markov process. Then to solve the problem of minimizing the production costs connected with bad products rejection dynamic programming can be utilized. This method is based on the property of Markov's process which proclaims that independently from the initial state and decisions defining approach to a specified state, remaining decisions must create the optimal policy which defines the optimal transition from this state to the last point.
The production costs consist of a constant part containing costs of the execution of production process and of a second part connected with products rejection in the succesive control operations.
The decision variables in the discussed model are the tolerance intervals for the raw material in the input of the first production stand /investigated at the "zero" control operation/ and for the products obtained in successive production stands /through semi - products to final ones/.

For the n-th control operation a set of possible and admissible tolerance intervals at our disposal is as follows

(20) $\{D\}_n = \{d_1, d_2, \ldots, d_k\}n$

For the whole production process there is a set $\{D\}$ being a family of sets in the all control operations

(21) $\{D\} = \{\{D\}_o, \{D\}_1, \ldots, \{D\}_N\}$

We are looking for the optimal set $\{D\}_{opt}$ whose elements form the optimal ranges of tolerances in the succeding control operations

(22) $\{D\}_{opt} = \{d_o, d_1, \ldots, d_N\}_{opt}$

where $d_{o\,opt} \in \{D\}_o$, $d_{1opt} \in \{D\}_1$, ..., $d_{Nopt} \in \{D\}_N$.

$\{D\}_o$ denotes a set of tolerance intervals under investigation in the control operation preceding the first technological process, $\{D\}_1$ - a set of tolerances for production investigated after the first technological operation, $\{D\}_N$ denotes a set of tolerances for the final product. It was assumed that the set $\{D\}_N$ contains one element-i.e. there exists fixed tolerance limit $\{D\}_N = \{d_1\}_N$

The goal function assumed /the expected value of costs related to bad products rejection/ is in the following form:

(23) $E(C) = \sum_{n=0}^{N} a_n P_n K_n$

where

- $E(C)$ - the expected value of costs of defects rejection,
- a_n - the expected value of products obtained at the n-th production stand,
- P_n - the probability of rejection of product by the n-th control operation,
- K_n - the cost per piece of the product by the n-th control operation.

General relation utilized in the optimization process is in the form:

(24) $h_n(s) = \min_{(s,j)} |E(C_{s,j}) + h_{n-1}(j)|$ for $n=0,1,\ldots,N$

where

- $h_n(s)$ - the minimum cost of rejection of bad products in the state s, when to the end of the process remains n stages,
- $E(C_{sj})$- the expected value of costs related to defective products in the state s, when in $|N-n+1|$ -th control operation the optimal is the state j,
- $h_{n-1}(j)$-the minimum cost of defective products rejection in the state j for $|N-n+1|$th control operation, when to the end of the process remains n-1 stages.

The state notion is the quality of a product defined by the given tolerance interval. $D_n(s)$ denotes decision which enables to reach $h_n(s)$. For the initial state n=0 the costs of bad products rejection can be found as follows

(25) $E(C_o)= P_o a_o K_o$

To find P_o we must utilize $f_o(x)$ at equation (11)having the set $\{D\}_o = \{d_1,d_2,\ldots,d_m\}_o$ at disposal, s_o , for n=0 we have the set $\{E(C_o)\}$ such that

(29) $\forall_{i\in\{1,\ldots,m\}} d_{oi} : E(C_{oi}) \in \{E(C_o)\} = P_{oi} a_o K_o$

the minimal value of the set $\{E(C_o)\}$is defined as $h_o(s)$ and the adequate tolerance interval as $d_o(s)$.

Utilizing the relations describing the quality of products obtained in the successive production stands we are able to find the optimal decision $d_N(s)$. This decision determines the optimal tolerance interval for the (N-1)-th control operation , when in the N-th control we are in the state described by the first tolerance interval. If we know the optimal state in the (N- 1)-th control operation, it is possible to find the optimal decision $d_{N-1}(s)$ and this reasoning can be repeated to obtain the decision $d_o(s)$, In such the way we will obtain the set of the optimal tolerance intervals $\{D\}_{opt}$:

(27) $$\{D\}_{opt} = \{d_{oi}, d_{1m}, \ldots, d_{N1}\}_{opt}$$

where $d_{oi} \in \{D\}_o$, $d_{1m} \in \{D\}_1$, ..., $d_{N1} \in \{D\}_N$.

Thus $\{i, m, \ldots, 1\}$ denotes the set of optimal indices in the separate control operations. This set determines the optimal amounts of semi products manufactured in the successive technological stands.

REFERENCES

Piasecki S., Wójcik B.: A Probabilistic Model of Manufacturing Process from a Quality of Products Point of View, with Reqard to Interoperation Controls. Systems Sciences , vol.4 ,pp.281-289.

Polish Academy of Sciences
Systems Research Institute
str. Newelska 6
01-447 Warsaw
POLAND

decision $d_N(s)$. This decision determines the optimal tolerance interval for the (N-1)-th control operation, when in the N-th control we are in the state described by the first tolerance interval. If we know the optimal state in the (N-1)-th control operation, it is possible to find the optimal decision $d_{N-1}(s)$ and this reasoning can be repeated to obtain the decision $d_0(s)$. In such the way we will obtain the set of the optimal tolerance intervals $\{B\}_{opt}$:

(27) $\{D\}_{opt} = \{d_{oi}, d_{1m}, \ldots, d_{Nl}\}_{opt}$

where $d_{oi} \in \{D\}_o$, $d_{1m} \in \{D\}_1$, ..., $d_{Nl} \in \{D\}_N$.

Thus {i,m,...,l} defines the set of optimal indices in the separate control operations. This set determines the optimal amounts of semi-products manufactured in the successive technological stands.

References

[illegible] Manufacturing Process [illegible] Productivity [illegible] pp. 271-282.

Polish Academy of Sciences
Systems Research Institute
ul. Newelska 6
01-447 Warszawa
Poland

NOTE ON A SOLUTION OF THE DEGUÉ'S PROBLEM

Anna Wolińska - Welcz

Lublin

ABSTRACT

A solution of Dugue s problem of finding characteristic functions φ_1 , φ_2 such that $\frac{\varphi_1(t)+\varphi_2(t)}{2}=\varphi_1(t)\varphi_2(t)$ is presented . Some results in this direction are given .

One of the most powerful tools used for the investigation in the probability theory and its many applications are the characteristic functions. There are still open problems for characteristic functions in the arithmetic of probability distributions .

Let us consider the following problem due to Daniel Dugué (1957) : For which couples φ_1 , φ_2 of characteristic functions the condition

$$(D) \qquad \frac{\varphi_1(t)+\varphi_2(t)}{2} = \varphi_1(t)\,\varphi_2(t) , \qquad -\infty<t<\infty$$

is satisfied ? Dugué gave one solution of this equation , namely

$$\varphi_1(t) = \frac{1}{1+it} \qquad \varphi_2(t) = \frac{1}{1-it}$$

and asked about others .

The special case of this equation with $\varphi_2(t) = \varphi_1(-t)$ was discussed by Vallander, Ibragimov and Lindtrop (1969) , Puri and Rubin (1970) . Some results connected with this problem were given by Rossberg (1972) and Kubik (1969).

In this paper we intend to give the solution of the problem (D) for characteristic functions of distributions with supports on semi-axes . Furthermore, a method for the construction of other solutions of the discussed problem is presented .

THEOREM 1 . Let φ_1 and φ_2 be two characteristic functions of one-sided distribution functions F_1 and F_2 , respectively, concentrated on semi-axes . Then the condition (D) is satisfied only by the following characteristic functions :

(i) $$\varphi_1(t) = 1 , \qquad \varphi_2(t) = 1 ;$$

(ii) $$\varphi_1(t) = e^{itb} , \qquad \varphi_2(t) = \frac{1}{2 - e^{-itb}} , \qquad b \neq 0 , b \in R;$$

(iii) $$\varphi_1(t) = \frac{1}{2} + \frac{1}{2} e^{itb} , \qquad \varphi_2(t) = \frac{1}{2} + \frac{1}{2} e^{-itb} , \qquad b \neq 0 , b \in R;$$

(iv) $$\varphi_1(t) = p + (1-p)e^{itb} , \qquad \varphi_2(t) = \frac{1}{2} + \frac{1}{2} e^{-itb} \frac{1}{2(1-p)-(1-2p)e^{-itb}} ,$$
$$0 < p < 1/2 , \qquad b \neq 0 , b \in R;$$

(v) $$\varphi_1(t) = \frac{r_1 - 1}{r_1 - e^{itb}} , \qquad \varphi_2(t) = \frac{(1 - r_2)e^{-itb}}{1 - r_2 e^{-itb}} ,$$
$$1 < r_1 < 2 , \quad r_1 + r_2 = 2 , \quad b \neq 0 , b \in R;$$

(vi) $$\varphi_1(t) = \left[p + (1-p)e^{itb}\right]\frac{r_1 - 1}{r_1 - e^{itb}} , \qquad \varphi_2(t) = \left[1 - p + pe^{-itb}\right] \frac{1 - r_2}{1 - r_2 e^{-itb}} ,$$
$$0 < p < 1 , \quad 1 < r_1 < 2 , \quad r_2 = 1 - (r_1 - 1)/[1 + 2(1-p)(r_1 - 1)] , \qquad b \neq 0 , b \in R;$$

(vii) $$\varphi_1(t) = \frac{\alpha}{\alpha - it} , \qquad \varphi_2(t) = \frac{\alpha}{\alpha + it} , \qquad \alpha \neq 0 , \alpha \in R .$$

This solution of the problem (D) for characteristic functions of distributions with supports on different semi-axes was obtained recently by applying analytical methods in probability considerations (see Wolińska-Welcz (1986)) .

We note that characteristic functions of such one-sided distributions are analytic in half planes and can be represented there by Fourier-Stieltjes transformations of their distribution functions. We construct meromorphic extensions $\tilde{g}_1$, $\tilde{g}_2$ of these functions to the whole complex plane, such that :

$$(D_z) \qquad \tilde{g}_1(z) + \tilde{g}_2(z) = 2\tilde{g}_1(z)\tilde{g}_2(z) , \qquad z \in \mathbb{C}$$

putting

$$\tilde{g}_1(z)=\begin{cases} g_1(z) & , \text{Im } z \geqslant 0 \\ \dfrac{g_2(z)}{2g_2(z)-1} & , \text{Im } z < 0 \end{cases} \qquad \tilde{g}_2(z) = \begin{cases} \dfrac{g_1(z)}{2g_1(z)-1} & , \text{Im } z > 0 \\ g_2(z) & , \text{Im } z \leqslant 0 \end{cases}$$

where $g_1(z)= \int_0^\infty e^{izx}\, dF_1(x)$, $\qquad g_2(z)= \int_{-\infty}^0 e^{izx}\, dF_2(x)$.

The function $\tilde{g}_1$ is analytic in a horizontal strip $\text{Im } z > -\tau_1$, $\tau_1 > 0$, while the function $\tilde{g}_2$ is analytic for $\text{Im } z < \tau_2$, $\tau_2 > 0$, and we have

$$\tilde{g}_1(t) = \varphi_1(t) , \qquad \tilde{g}_2(t) = \varphi_2(t) , \qquad t \in \mathbb{R} .$$

This leads the Dugué problem(D) to finding the solution of (D_z) on the real axis . So we take our considerations to the field of complex analysis .

Remark, that under above assumption the Dugué condition is satisfied by characteristic functions of both lattice distributions or by characteristic fun ctions of both no-lattice distributions.

Let us assume first for example that F_1 and F_2 defined no-lattice distributions. We ought to consider equations :

$$(1) \qquad 2\, g_1(is) - 1 = 0 , \qquad s \geqslant 0$$
$$(2) \qquad 2\, g_2(is) - 1 = 0 , \qquad s \leqslant 0 .$$

It is easy to see that each of these equations has at most one real solution. If $s = \alpha , \alpha > 0$ is a solution of (1) then one can show that $z = i\alpha$ is the only pole of $\tilde{g}_2$ and $\tilde{g}_2$ is bounded outside of a certain neighbourhood of this pole . Hence, we conclude that $\tilde{g}_2$ is the rational function of the form :

$$(3) \qquad \tilde{g}_2(z) = \frac{P_k(z)}{(z-i\alpha)^k} , \qquad k \in \mathbb{N}$$

where P_k is the polynomial of the k-th degree at most .

In the same way we can state that if $s = -i\beta$, $\beta > 0$ is the solution of (2) then the function $\tilde{g}_1$ has at the point $z = -i\beta$ its only pole, and moreover, $\tilde{g}_1$ is the rational function of the form :

$$\tilde{g}_1(z) = \frac{Q_l(z)}{(z+i\beta)^l}, \quad l \in \mathbb{N} \tag{4}$$

where Q_l is the polynomial of the l-th degree at most .

Further on , it can be proved that $l = k = 1$ and $\alpha = \beta$, so φ_1 and φ_2 are characteristic functions of the exponential distributions

$$\varphi_1(t) = \frac{\alpha}{\alpha - it}, \quad \varphi_2(t) = \frac{\alpha}{\alpha + it}, \quad \alpha > 0 .$$

In the case, when one of the equations (1)-(2), let us say (1), has no solution we can state that $\tilde{g}_2$ is an entire function, while $\tilde{g}_1$ is of the form (4) . On the other hand, by (D_z)

$$\tilde{g}_2(z) = \frac{Q_l(z)}{2\,Q_l(z) - (z+i\beta)^l},$$

so that

$$Q_l(z) = \frac{1}{2}(z + i\beta)^l + A, \quad A \in \mathbb{R} .$$

Therefore

$$\tilde{g}_2(z) = \frac{1}{2} + \frac{1}{2}\left(1 - \frac{iz}{\beta}\right)^l, \quad \text{as } \tilde{g}_2(0) = 1$$

and $\tilde{g}_2$ is not bounded on $\mathbb{R}$, which leads to contradiction .

In the last case , when neither (1) nor (2) has solution we have

$$\lim_{s\to\infty} \tilde{g}_1(is) = \lim_{s\to\infty} \int_0^\infty e^{-sx} dF_1(x) \geq \frac{1}{2}, \quad \lim_{s\to-\infty} \tilde{g}_2(is) = \lim_{s\to-\infty} \int_{-\infty}^0 e^{-sx} dF_2(x) \geq \frac{1}{2}$$

so F_1 has the saltus $p \geq \frac{1}{2}$ at the origin, and F_2 has the saltus $q \geq \frac{1}{2}$ at the origin. Since in this case $\tilde{g}_1$ and $\tilde{g}_2$ are entire functions, bounded for $p > \frac{1}{2}$, $q > \frac{1}{2}$, so they are constant functions. If $p = \frac{1}{2}$ and $q = \frac{1}{2}$ then we obtain lattice distributions with two discontinuity points, in contradiction to our assumption .

It can be verified that if F_1, F_2 determine lattice distributions with supports on different semi-axes, and if their characteristic functions satisfy (D), then these distributions are defined on the same lattice with the step $h>0$ and with the origin as a lattice point. So we can write their characteristic functions as follows:

$$\varphi_1(t) = \sum_{k=0}^{\infty} p_k e^{itk}, \qquad \varphi_2(t) = \sum_{k=0}^{\infty} q_k e^{-itk}, \qquad (h=1)$$

where $0 \leqslant p_k \leqslant 1$, $0 \leqslant q_k \leqslant 1$, $k=0,1,2,\ldots$ and $\sum_{k=0}^{\infty} p_k = 1$, $\sum_{k=0}^{\infty} q_k = 1$

Now we introduce functions:

$$\tilde{g}_1(z) = \begin{cases} g_1(z), & |z| \leqslant 1 \\ \dfrac{g_2(z)}{2g_2(z)-1}, & |z|>1 \end{cases}; \qquad \tilde{g}_2(z) = \begin{cases} \dfrac{g_1(z)}{2g_1(z)-1}, & |z|<1 \\ g_2(z), & |z| \geqslant 1 \end{cases}$$

where $g_1(z) = \sum_{k=0}^{\infty} p_k z^k$, $g_2(z) = \sum_{k=0}^{\infty} q_k z^{-k}$, $z = re^{it}$, $r,t \in \mathbb{R}$.

Note that $\tilde{g}_1$ is analytic inside the circle $|z|=1$, $\tilde{g}_2$ is analytic outside the circle $|z|=1$, they are both continuous on this circle and

$$\tilde{g}_1(z) = \varphi_1(t), \qquad \tilde{g}_2(z) = \varphi_2(t), \qquad |z| = 1.$$

Thus the above formulas define meromorphic extensions of the characteristic functions φ_1, φ_2 for which (D_z) is satisfied for the whole complex plane. In the way of analytic, detailed discussions of cases:

$$1^\circ \ q_o > \tfrac{1}{2}, \qquad 2^\circ \ q_o = \tfrac{1}{2}, \qquad 3^\circ \ q_o < \tfrac{1}{2} \qquad \left(q_o = \lim_{|z|\to\infty} g_2(z)\right)$$

we can obtain the exact form of colutions of (D_z) on the circle $|z|=1$, i.e. the lattice solutions (i) - (vi) of the Dugué problem.

Finally, it is not difficult to state that there exists only a trivial solution of the Dugué problem for characteristic functions with supports on the same semi-axis.

By a similar reasoning to the argument we sketch above, the solution of equation

$$(D_c) \qquad (1-c)\varphi_{1,c}(t) + c\varphi_{2,c}(t) = \varphi_{1,c}(t)\varphi_{2,c}(t), \qquad 0<c<1$$

for characteristic functions of distributions on semi axes was obtained by Wolińska-Welcz (1985).

However, it is possible to show other solutions of (D). As an example we take the following characteristic functions of distributions with supports on the whole real line

$$\varphi_1(t) = \frac{1}{(1-it)^2\left(1+\frac{it}{2}\right)}, \qquad \varphi_2(t) = \frac{1}{(1+it)^2\left(1-\frac{it}{2}\right)}$$

for which (D) holds.

We give one of methods of construction of such solutions of (D), based on the solutions known before. Other methods and examples were presented earlier in Wolińska (1982).

THEOREM 2. If $(\varphi_{1,c}, \varphi_{2,c})$ are characteristic functions satisfying the condition (D_c) for each $c\in(0, 1/4)$ then functions

$$\psi_{1,c}(t) = \varphi_{1,c/4}(t)\varphi_{2,c/4}(t),$$

$$\psi_{2,c}(t) = \left[\frac{\left(1-\frac{c}{2+\sqrt{c}}\right)\varphi_{1,c/4}(t)}{\varphi_{1,c/4}(t)-\frac{c}{2+\sqrt{c}}}\right]\left[\frac{\left(1-\frac{c}{2-\sqrt{c}}\right)\varphi_{1,c/4}(t)}{\varphi_{1,c/4}(t)-\frac{c}{2-\sqrt{c}}}\right]$$

are also the characteristic functions satisfying this condition for each $c\in(0,1)$.

Moreover, if $\varphi_{2,c}$ is an infinitely divisible characteristic function then also $\psi_{2,c}$ is an infinitely divisible characteristic function.

In a particular case functions:

$$\psi_1(t) = \varphi_1(t)\varphi_2(t), \qquad \psi_2(t) = \left[\frac{\left(1-\frac{1}{4+\sqrt{2}}\right)\varphi_1(t)}{\varphi_1(t)-\frac{1}{4+\sqrt{2}}}\right]\left[\frac{\left(1-\frac{1}{4-\sqrt{2}}\right)\varphi_1(t)}{\varphi_1(t)-\frac{1}{4-\sqrt{2}}}\right]$$

are characteristic functions satisfying (D) if $\frac{7}{8}\varphi_1(t) + \frac{1}{8}\varphi_2(t) = \varphi_1(t)\varphi_2(t)$.

We finally conclude from this theorem, that starting with one no-trivial solution of the Dugué problem on semi-axes we can generate many couples of characteristic functions satisfying this condition of other distributions, such that at least one of them is not concentrated on semi-axes.

REFERENCES

Dugué D. (1957): Arithmétique des lois de probabilités. Mémorial des Sciencis Math., v. 137.

Kubik L. (1969): Sur un probleme de M.D. Dugué. Comment. Math., v. XIII, 1-2.

Puri P., Rubin H. (1970): A Characterization based on the Absolute Difference of two i.i.d. Random Variables. Ann. Math. Statist., v. 41, No 6, 2113 - 2122.

Rossberg H.-J. (1972): Characterization of the Exponential and Pareto Distributions by Means of some Properties of the Distributions which the Differences and Quotients of Order Statistics are subject to. Math. Operationsforsch. Statist. Ser. Statist. 3, No 3, 207 - 216.

Vallander S., Ibragimov I., Lindtrop N. (1969): On Limiting Distributions for Moduli of Sequential Differences of Independent Variables. Teorja Verojatn. i ee Primen., v. XIV, No 4, 693 - 707.

Wolińska A. (1982): On a Problem of Dugué. Lect. Notes Math., v. 982, 244 - 253.

Wolińska-Welcz A. (1985): On a Solution of the Dugué Problem. Bull. Acad. Polon. Sci. Ser. Sci. Math., v. 33, No 7 - 8, 421 - 423.

(1986): On a Solution of the Dugué Problem. Probab. Math. Statistics, v. 7, 169 - 185.

Institute of Mathematics
Maria Curie-Skłodowska
University
20-031 Lublin, Poland.

MIX
Papier aus verantwortungsvollen Quellen
Paper from responsible sources
FSC® C105338

If you have any concerns about our products,
you can contact us on
ProductSafety@springernature.com

In case Publisher is established outside the EU,
the EU authorized representative is:
Springer Nature Customer Service Center GmbH
Europaplatz 3, 69115 Heidelberg, Germany

Printed by Libri Plureos GmbH
in Hamburg, Germany